13th International Conference on Films and Coatings (ICFC13)

Journal of Physics: Conference Series Volume 857

St. Petersburg, Russia
18 – 20 April 2017

ISBN: 978-1-5108-4193-2
ISSN: 1742-6588

Printed from e-media with permission by:

Curran Associates, Inc.
57 Morehouse Lane
Red Hook, NY 12571

Some format issues inherent in the e-media version may also appear in this print version.

Printed by Curran Associates, Inc. (2017)

For permission requests, please contact the Institute of Physics
at the address below.

Institute of Physics
Dirac House, Temple Back
Bristol BS1 6BE UK

Phone: 44 1 17 929 7481
Fax: 44 1 17 920 0979

techtracking@iop.org

Additional copies of this publication are available from:

Curran Associates, Inc.
57 Morehouse Lane
Red Hook, NY 12571 USA
Phone: 845-758-0400
Fax: 845-758-2633
Email: curran@proceedings.com
Web: www.proceedings.com

Table of contents

Volume 857

13th International Conference on Films and Coatings
18–20 April 2017, St. Petersburg, Russian Federation

Accepted papers received: 25 May 2017
Published online: 1 June 2017

Preface

Papers

IOP Conf. Series: Journal of Physics: Conf. Series **857** (2017) 011001 doi:10.1088/1742-6596/857/1/011001

Analytical Review of the Reports Presented at the 13th International Conference on Films and Coatings (ICFC13)

V G Kuznetsov[1] **and D K Kostrin**[2]

[1] Institute of Problems of Mechanical Engineering, Russian Academy of Sciences, 199178, Saint Petersburg, Russia
[2] Saint Petersburg Electrotechnical University "LETI", 197376, Saint Petersburg, Russia

E-mail: kvgipme@gmail.com

Abstract. In this preface main trends in the development of films and coatings technology presented at the 13th International Conference on Films and Coatings (ICFC13) that was held on 18–20 April 2017 in Saint Petersburg, Russia are discussed.

During the Conference (figure 1) were presented the results of theoretical and experimental research in the physics and mechanics of condensed matter, physics of low temperature plasma, formation of films and coatings using plasma and related methods that were obtained over the past two years that have passed since the previous meeting.

Figure 1. Introduction to the 13th International Conference on Films and Coatings (ICFC13).

Special attention was paid to the study of properties of surfaces and coatings, methods of their research, new coating materials, including nanomaterials, new areas of their application, development of modern equipment and technological processes, surface preparation and many other issues.

During the Conference were presented more than 150 reports that were distributed in 8 sections:

- vacuum ion-plasma methods;
- thermal spray coating methods;
- physical processes at all stages of coatings formation;
- equipment for application of films and coatings;
- materials for sputtering and deposition;
- nanomaterials and nanotechnologies;
- properties of coatings and films and methods of their evaluation;
- preparation of surfaces before application and methods of post-processing of coatings and films.

For the first time the Conference had a special section for young scientists, where presentations were made by students, postgraduates, candidates and doctors of science aged up to 35 years (figure 2). Among the participants were representatives of Russia, Germany, Czech Republic, Poland, Belarus, Ukraine, China, Vietnam, Myanmar and Kazakhstan.

Figure 2. Participants of the 13th International Conference on Films and Coatings (ICFC13).

In their reports the authors reflected the issues of thin films growth, their compositions and structures, including nanostructures. Were presented original researches in the field of thermodynamics and kinetics of phase transitions of the first kind in multicomponent systems. Particular attention was paid to the kinetic theory of initial stages of brittle destruction of solids. Great attention was paid to the mechanical properties of films and coatings. Was presented a two-dimensional experimental-theoretical approach to determining the integral stiffness tensile for non-planar film or membrane compositions of complex structure. Was proposed a variant of the asymptotic expansion of the differential equations of plane problem of elasticity theory, describing the joint deformation of isotropic beams and applied on it layer of the elastic coating. Algorithm for calculating the stresses during bending of the connection was created.

In a number of reports were presented the issues of optimization of the growth processes of thin films during vacuum ion-plasma and gas-thermal spraying using computer methods. Further development have acquired the theoretical and experimental researches in the field of physics and mechanics of formation of hollow powders for thermal spray methods of coatings deposition. On the example of thermal spray coating process of a hollow microsphere Al_2O_3 powder was experimentally confirmed a significant advantage of this class of powders to enhance the performance properties of coatings.

Special attention was paid to the reports on the implementation of the arc discharge in vacuum and possibilities of its use for coating deposition and for the solution of special technological problems. Further development has received the direction of research associated with the processing of materials surface with a cathode spot of a vacuum-arc discharge. For the first time this process was used to clean the surface of metals from radionuclides. The coefficient of purification at the level of 20 000 is the maximum compared to all existing methods.

One of the new trends in the development of science and technology, reflected in the reports is the formation of pyrocarbon coatings in plasma of a vacuum arc discharge. For the first time such coating has been researched and applied to the electrodes of powerful generator tubes as antiemission coatings. Theoretically and experimentally were investigated the thermal processes during treatment of the inner surface of the cylindrical cavity by the cathode spots of a vacuum arc discharge.

In a number of reports were reflected the characteristics of magnetron sputtering systems and principles of coatings deposition on their basis. The characteristics, technological aspects of production and results of testing of gradient coatings for aerospace optics were discussed. Promising technology of pulsed magnetron sputtering was noted. Possibilities of application of multilayer composite coatings in the systems of radiation protection of spacecraft were reviewed. Were shown the advantages of composite coatings before traditionally used in space technology aluminum alloys.

At this Conference many reports were devoted to the formation of oxide coatings by different methods and for different fields of application. For example, the results of comparative studies of the original and processed in the plasma flow oxide microcomposites, consisting of TiO_2, SiO_2, Al_2O_3, and also plasma coatings from them – materials with amorphous-crystalline structure and a reinforced ultrafine phases of stishovite.

It was shown that a reliable method of forming a specified surface nanorelief is a direct resistless lithography by a focused ion beam. The use of ions of different masses and energies significantly expands its abilities for nanoconstruction and nanoengineering of thin-layer structures. Examples of the use of different ions for the nano-processing of coatings and surfaces of solids were shown.

In the series of reports were presented the results of studies on the effect of layout schemes of technological installation on the structure parameters and mechanical properties of wear-resistant coatings. The physical aspects of the developed by the authors low temperature process of plasma hardening of the surface layer of geometrically-complex products, characterized by the low-energy plasma particles that interact with surface atoms according to the scheme of heterogeneous recombination were discussed.

The possibility of application of film coatings rich with hydrogen for storage and desorption of hydrogen were analyzed. Was proposed a film hydrogen accumulator for use in the mobile objects on hydrogen fuel cells. Were discussed the method of its production and its application prospects.

Possibilities of acquisition of the graphene materials during the decomposition of acetylene with the plasma torch with DC power up to 35 kW were shown. The influence of the coating material and the geometric shape of the reactor on the morphology of graphene were determined.

Results were presented at the modeling and study of the processes of reactive ion-plasma etching. The results of stochastic modeling of the formation of the layered structure and the characteristics (microstructure, porosity, roughness) of the ceramic coatings from powders of corundum (α-Al_2O_3 phase) under conditions typical for the atmospheric plasma spraying and also detonation and supersonic flame spraying were shown. Fundamentals of physico-chemical model of reactive sputtering of the hot metal target in nitrogen environment were stated. The system of equations describing the model was solved for spraying of titanium in nitrogen environment. The features of the numerical method in modeling of processes of thin films formation were discussed. Was demonstrated the procedure of application of numerical methods to improve the accuracy and efficiency of computational processes.

IOP Conf. Series: Journal of Physics: Conf. Series **857** (2017) 011001 doi:10.1088/1742-6596/857/1/011001

Many of the results and designs of the authors were at the world level. Many developments are protected by patents for inventions. The design and main parameters of vacuum ion-plasma systems for applying functional coatings were presented. Examples of the application of superhard nanostructured coatings on materials and products were shown. Prospects for the use of the developed equipment in the industry were discussed. Also during the Conference were discussed issues of preparation of young specialists and scientific personnel.

Thus, the 13th International Conference on Films and Coatings contributed to the further development of this scientific and applied direction of research and professional development of the participating scientists.

Peer review statement

All papers published in this volume of *Journal of Physics: Conference Series* have been peer reviewed through processes administered by the proceedings Editors. Reviews were conducted by expert referees to the professional and scientific standards expected of a proceedings journal published by IOP Publishing.

IOP Conf. Series: Journal of Physics: Conf. Series **857** (2017) 012001 doi:10.1088/1742-6596/857/1/012001

Fracture toughness of $Al_2O_3/ZrSiO_4$ coatings obtained by multi-chamber gas-dynamic accelerator

M Yu Arseenko, M G Kovaleva, M S Prozorova, O N Vagina and M N Yapryntsev

Belgorod National Research University 308015, Belgorod, Russia

E-mail: kovaleva@bsu.edu.ru

Abstract. In the current work the research of fracture toughness of $Al_2O_3/xZrSiO_4$ composite coatings have been realized using automatic system for microhardness analysis. $Al_2O_3/xZrSiO_4$ (x = 0, 3, 25 wt. %) coatings were produced from a mixture of cheap raw materials (alumina and zircon $ZrSiO_4$) on the surface of stainless steels by a new multi-chamber gas-dynamic accelerator. It has been experimentally established that adding of $ZrSiO_4$ increases fracture toughness of alumina coatings.

Zirconium silicate ($ZrSiO_4$) and aluminum oxide (Al_2O_3, alumina) are some of the most used and cheapest materials for coatings production on the heat-resistant surfaces [1]. Microhardness and fracture toughness of Al_2O_3 coatings can be improved by adding of other components with a goal to compensate decreasing of these properties owing to transformation of Al_2O_3 to metastable condition. For instance, an addition of $ZrSiO_4$ (zircon) increases fracture toughness of Al_2O_3 coatings [2]. In the current work influence of $ZrSiO_4$ addition to Al_2O_3 composite coatings (0, 3, 25, wt. %) on fracture toughness was investigated.

In the present work the mixture of alumina (average size of particles is 34.2 μm) and zircon (average size of particles is 18.3 μm) was used as feedstock materials for coatings deposition on the surfaces of 304L stainless steel. The powders were mixed and grinded during 3 h with ethanol in a ball mill. The alumina powder consists of γ-Al_2O_3 with a cubic lattice, α-Al_2O_3 with a trigonal lattice and TiO_2 (rutile) with a tetragonal lattice. The phase analysis has shown that the main phase in zircona powder is $ZrSiO_4$ with a tetragonal lattice. The phase composition of the powders and the coatings was investigated by X-Ray diffractometer Rigaku Ultima IV. $Al_2O_3/xZrSiO_4$ composite coatings (80–190 μm) were obtained by a multi-chamber gas-dynamic accelerator [3–5]. The porosity of the coatings is less than 1 % (0.30 ± 0.01 %). The investigation of fracture toughness was provided with using of an automatic system for a microhardness analysis on the base of a motorized microhardness tester DM-8 and scanning electron microscope Quanta 200 3D (SEM). Five imprints were made with the load of 300 g in a cross-section in the coating layer. Then, the imprints and cracks were photographed with help of SEM and the length of all cracks and imprints were measured. The magnitude of fracture toughness was calculated with the following formula:

$$K_C = 0.079\, P/a^{3/2} \log 4.5(a/c),$$

where P – the load of the indenter (μN); a – the average value of a half of the imprint length (μm); c – the sum of the average value of a crack length with a half of the imprint length (μm) [6, 7].

The type of cracks formation is determined as I type (longitudinal crack) (figure 1), where the dominating cracks occur parallel to a substrate and spring from one or two horizontal angles of the imprint.

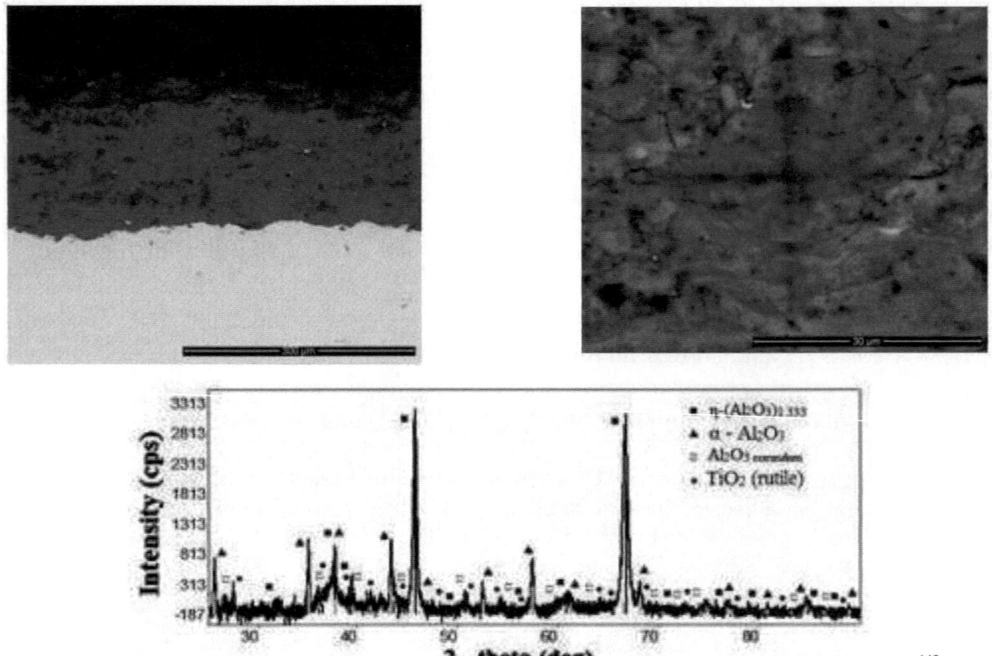

$K_C = 1.68 \pm 0.54\,\text{MPa} \cdot \text{m}^{1/2}$

(a)

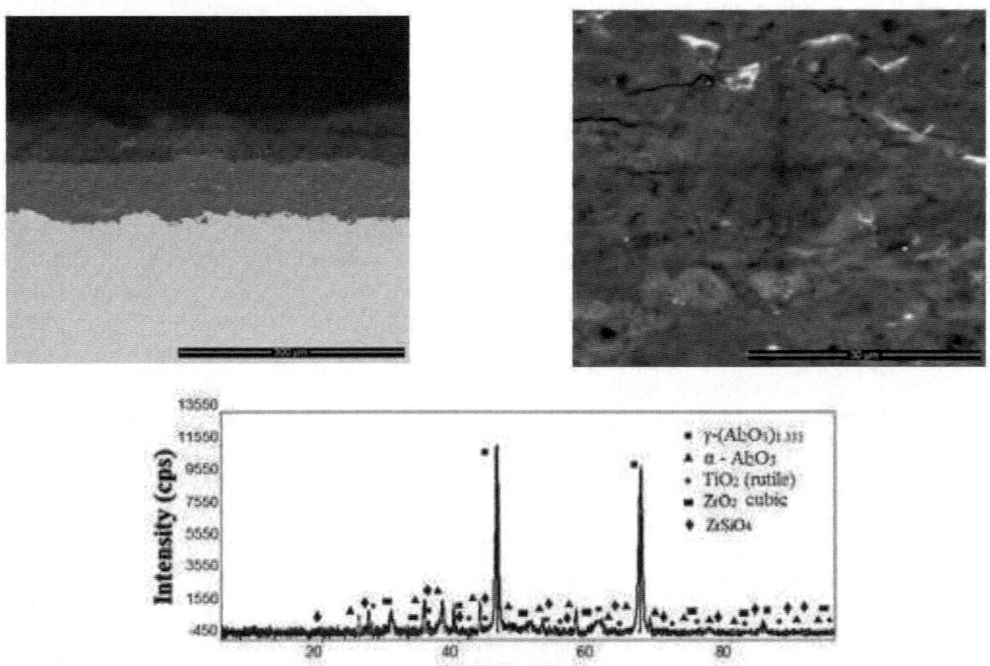

$K_C = 1.87 \pm 0.48\,\text{MPa} \cdot \text{m}^{1/2}$

(b)

Figure 1. SEM microphotographs, X-Ray pattern and typical views of imprints and cracks (cross-section) for: **(a)** – Al_2O_3; **(b)** – $Al_2O_3/3ZrSiO_4$.

The fracture toughness of the coatings on the base of alumina noticeably increases from 1.68 to 2.45 MPa·m$^{1/2}$ with $ZrSiO_4$ addition. A bigger zirconium dioxide grains was found in the structure of $Al_2O_3/25ZrSiO_4$ coating that lead to high content of monoclinic phase (figure 2). This mechanical strength can improve resistance to cracks propagation in zirconium dioxide owing to hardening mechanism as a result of transition of one modification to another [8, 9].

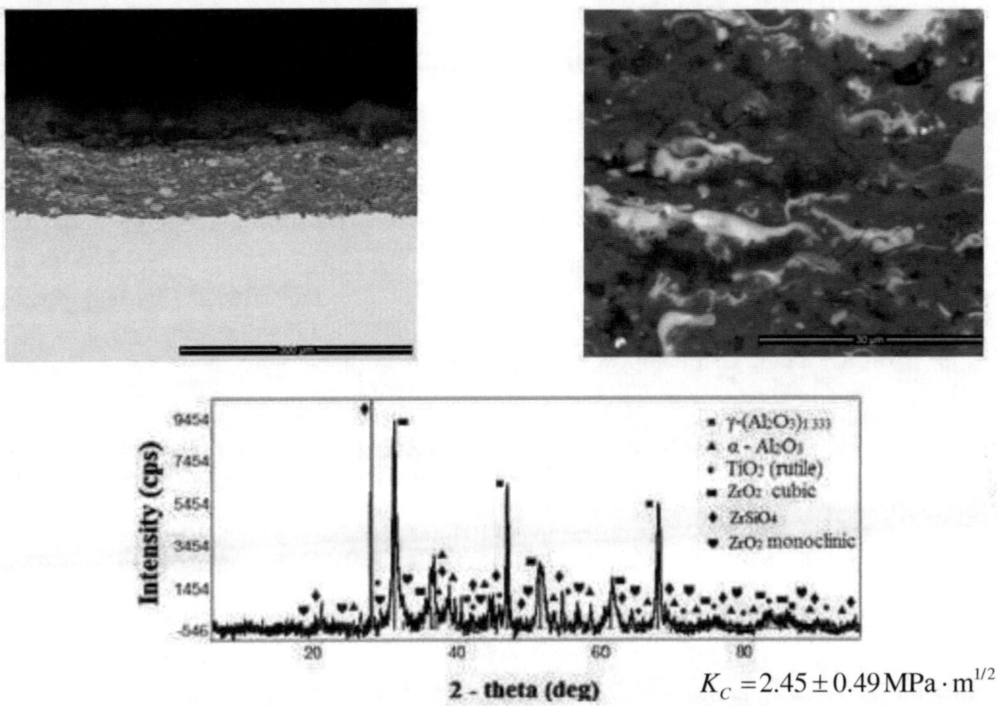

Figure 2. SEM microphotographs, X-Ray pattern and typical views of imprints and cracks (cross-section) for $Al_2O_3/25ZrSiO_4$.

Acknowledgment

The study was financial supported by the Russian Science Foundation, under grant No 15-19-00189. All of studies were carried out on the equipment of the Joint Research Center of Belgorod State National Research University «Technology and Materials».

References

[1] Chráska P, Neufuss K and Herman H 1997 *Journal of Thermal Spray Technology* **6** 445–8
[2] Li Y and Khor K A 2002 *Surface and Coatings Technology* **150** 143–50
[3] Vasilik N, Tyurin N and Kolisnichenko O 2012 Method for gas-dynamic detonating speedup of powders and device for its implementation *RU Patent 2506341*
[4] Kovaleva M G, Prozorova M S, Arseenko M Yu, Yapryntsev M N, Vagina O N 2016 *International Journal of Applied Chemistry* **12** 593–601
[5] Kovaleva M G, Tyurin Yu N, Vasilik N Ya, Kolisnichenko O V, Prozorova M S, Arseenko M Yu and Danshina E P 2013 *Surface and Coatings Technology* **232** 719–25
[6] Ponton C B and Rawlings R D 1989 *Materials Science and Technology* **5** 865–72
[7] Lee C W, Han J H, Yoon J, Shin M C and Kwun S I 2010 *Surface and Coatings Technology* **204** 2223–9
[8] Bocanegra-Bernal M H and Díaz de la Torre S 2002 *Journal of Materials Science* **37** 4947–71
[9] Kern F and Palmero P 2013 *Ceramics International* **39** 673–82

IOP Conf. Series: Journal of Physics: Conf. Series **857** (2017) 012002 doi:10.1088/1742-6596/857/1/012002

Nanocomposite films of cobalt-containing polyacrylonitrile as a basis of gas-sensitive material for resistive type sensors

T A Bednaya[1] and S P Konovalenko[2]

[1] Don State Technical University, Rostov-on-Don, Russia
[2] Rostov State University of Economics, Rostov-on-Don, Russia

E-mail: bednayat@mail.ru, svetlana_s12@mail.ru

Abstract. The structure of the metal-carbon nanocomposite based on cobalt-containing polyacrylonitrile (PAN) is studied. The morphology of a surface with the theory of self-organization was analysed. The elemental composition, chemical and electronic states of the elements composing the material films are determined by the X-ray photoelectron spectroscopy (XPS) method. The X-ray diffraction (XRD) method shows that the obtained materials contain crystalline inclusions of CoO, Co_3O_4 and CoO (OH) in the organic matrix of PAN. Gas-sensitive characteristics of the obtained films.

1. Introduction

In advanced electronics we use materials that are nanoscale dispersions of inorganic substances (with particle sizes of approx. 1–100 nm) in the carbon matrix and are metal-carbon nanocomposites. Metal-carbon nanocomposites combine the beneficial properties of organic and inorganic substances, useful from the point of view of practical application. Organic materials are used for design new instruments, including gas sensors and detection of contaminants in the air. Heat treated polyacrylonitrile (PAN) is one of such materials. Is possesses has semiconducting properties, and it is electrical conductivity can be varied from dielectric to metal. To modify the physical and chemical properties of PAN and to obtain its nano-formations, the mechanism of self-organization of the structure as a result of the action of IR-radiation on the polymer was used. PAN is able to adsorb gases efficiently due to the developed morphology of the surface.

2. Methodology

The XPS method was used to determine chemical bonds and compounds in the structure of the organic matrix of cobalt-containing PAN films, the concentration and chemical state of cobalt in the near-surface region of the films. The XPS spectra were registered using a K-Alpha X-ray photoelectron spectrometer by Thermo Fisher Scientific. The source of the X-ray radiation was $K\alpha 12$-line Al ($h\nu = 1486.6$ eV). The vacuum in the analyzer chamber was $2.5 \cdot 10^{-9}$ mbar. The accuracy in determining the binding energy was ±0.2 eV, and that of the element composition was ±0.5 at. %. The XRD was performed using a DRON-6 device. The wavelength of the X-ray radiation of Cu$K\alpha$ was 1.54051 Å.

To study the gas-sensing properties of a material of the received cobalt-containing PAN films measured their resistance on teraommetre E6-13A. Sensitivity of the films estimated by means of

IOP Conf. Series: Journal of Physics: Conf. Series **857** (2017) 012002 doi:10.1088/1742-6596/857/1/012002

factor of gas-sensitivity S, which is calculated as: $S = (R_o - R_g)/R_o$, when $R_o > R_g$, where R_o – value of resistance of a film on air; R_g – value of resistance of a film in the atmosphere of detected gas.

3. Results

Nanoscale films of cobalt-containing PAN were obtained from a film-forming solution under the influence of incoherent infrared radiation at low vacuum ((5–10)·10^{-2} mm Hg) in two stages [1]. The intensity and duration of exposure to IR-radiation allow us to control the properties of the film material by changing the molecular structure of the polymer, which made it possible to obtain thermo-structured PAN films with different electrical conductivity [2–4].

In the process of IR heating a number of chemical transformations occur in PAN and nanocomposites. In composites, the cyclization of nitrile groups and the formation of the C=C and N=N poly-conjugate system occur at low temperatures. Cobalt introduced into the polymer in the form of chloride primarily forms complexes with the –C=N– poly-conjugation system [5–6].

The study of the surface morphology of the film samples of Co-containing PAN was performed using the atomic force microscopy method (AFM) using a Solver P47 Pro (NT-MDT) microscope in a semi-contact mode across areas of 5×5 μm^2 (figure 1). The statistical parameters of the surface morphology were determined using the Image Analysis NT-MDT program. The analysis of morphology of a surface with the theory of self-organization with use of a method of an investment of Takens and algorithm of Grassberger-Procaccia is carried out. It is established that modification of polymer film by cobalt metal materials with different percentage concentration and using of various time-temperature modes of her formation leads to change of morphology of a surface of films. The increase in the percentage content of the modifying additive in the film leads to disappearance of protrusions of the micro-relief, the microstructure acquires a homogeneous character.

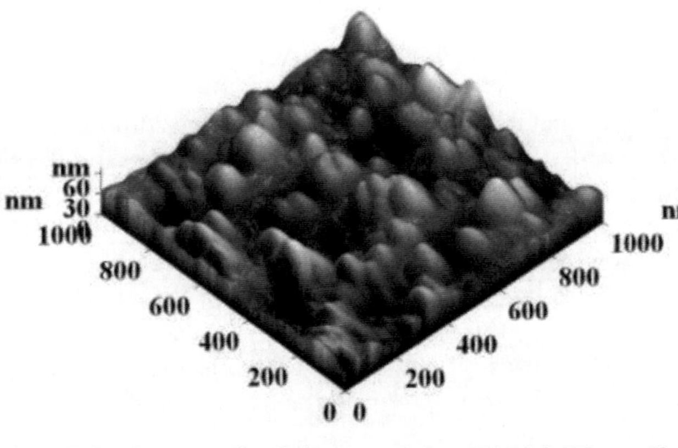

Figure 1. AFM micrograph of Co-containing PAN 0.75 wt. % of cobalt.

The values of the fractal dimension D_f self-affine surface of Co-containing PAN range from 2.41 to 2.50, so the surface of the investigated films are close to two-dimensionality (D_f = 2), but also have some of the properties of three-dimensionality [7].

The study estimated the effect of temperature of material production film samples and the concentration of modifying component formation to fractal structures and value of the coefficient of the gas sensitive films [8].

The XPS analysis of cobalt-containing PAN films showed that the main peaks related to C 1*s* (~ 285.7 eV), N 1*s* (~ 399 eV), O 1*s* (~ 531 eV) and Co 2*p* (~ 781 eV), can be expanded into several components.

In this study of the films, we took pictures of the spectra of high-resolution of the basic elements of the components that were obtained from the overview spectrum: C, O, N and Co. X-ray phase analysis made it possible to confirm the XPS data on the state of cobalt in the composite film materials. The X-ray phase analysis showed that the samples studied contain crystalline inclusions of CoO, Co_3O_4, and CoO(OH), whose average size was calculated using the Scherrer formula. Thus, the films are nanocomposites in which crystals of cobalt compounds are distributed in an amorphous matrix. The manufactured composite film materials demonstrate semiconducting properties, which is supported by their temperature dependences (figure 2(a)).

The fabricated film materials are generally defined as those having long term properties of metal-containing organic conjugated semiconductor polymer that depend from the IR-pyrolysis temperatures and the weight concentration of the modifying additive in film-forming solutions (figure 2(b)).

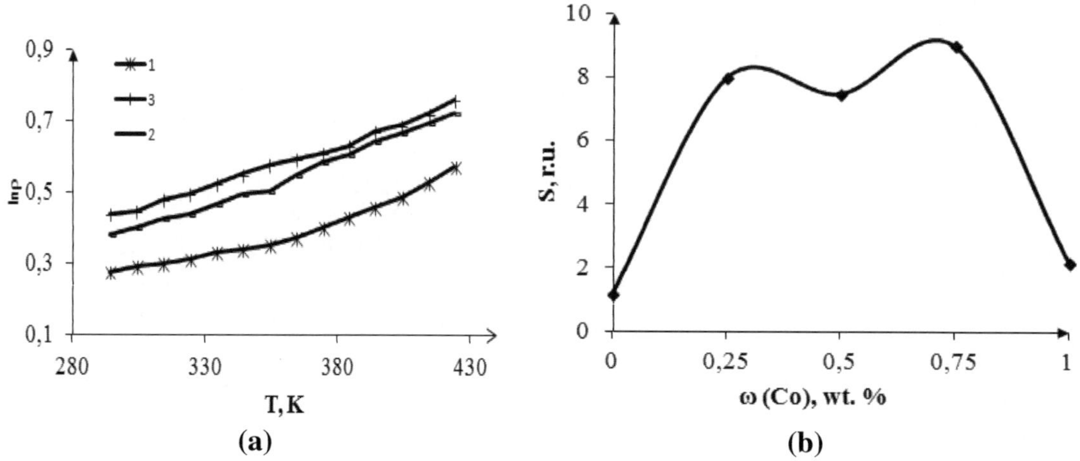

Figure 2. The physico-chemical properties of Co-containing PAN: Electrical Conductivity a function of temperature **(a)**: 1 – 0.5 wt. %; 2 – 0.75 wt. %; 3 – 1 wt. %; Gas-sensitivity as a function of metal concentration **(b)**.

4. Conclusion

The structure and morphology of metal–carbon nanocomposite based on PAN with intercalated cobalt have been studied. It is shown that IR-pyrolized PAN films can be applied as a sensitive layer of power-effective gas sensors of resistive type.

The films, whose self-organizing processes are most apparently expressed, show the highest gas-sensitive properties. Processes of self-organizing confirm presence of structures of several correlation dimensions. The parameters of self-organization are dopant concentration and the technological parameters of forming Co-containing PAN material. Studying of influence of technological parameters on properties of materials will allow to operate process of growth of disordered materials effectively, and also to program the synthesis of materials for micro- and nano-electronics with new unique properties.

Acknowledgments

The work was carried out with partial financial support of the Grant LV Krasnichenko on the topic "Modeling and optimization of technological processes for obtaining thin films of cobalt-containing polyacrylonitrile for the creation of nitrogen dioxide and chlorine sensors".

References

[1] Konovalenko S P and Bednaya T A 2017 *Perspektivnye materialy* **3** 33–40
[2] Konovalenko S P and Semenistaya T V 2013 *Izvestiya SFedU. Engineering Sciences* **1** 178–83
[3] Semenistaya T V, Petrov V V, Bednaya T A and Zaruba O A 2015 *Mat. Tod.: Proceedings* **2** 77–84
[4] Bednaya T A, Konovalenko S P, Semenistaya T V, Perov V V and Korolev A N 2012 *Izv.Vyssh. Uchebn. Zaved. Electronics* **4** 66–71
[5] Kozlov V V, Karpacheva G P, Petrov V S and Lazovskaja E V 2001 *High-molecular connections. Series. A* **43**(1) 1–7

[6] Zaporotskova I A, Kozhitov L V, Anikeeva N A, Davletova O A, Popkova A V, Muratov D G and Yakushko E V 2015 *Modern Electronic Materials* **1** 43–9
[7] Torkhov N A, Bozhkova V G, Ivonin I V and Novikov V A 2003 *Semiconductors* **43**(1) 33–41
[8] Konovalenko S P and Bednaya T A 2016 *J. Nano- Electron. Phys.* **4**(1) 04024

Influence of high power density plasma irradiation on the boron carbide coating on tungsten

L B Begrambekov, A V Grunin, N A Puntakov, Ya A Sadovskiy and N S Utkov

National Research Nuclear University (MEPhI), 115409, Moscow, Russia

E-mail: lbb@plasma.mephi.ru

Abstract. The paper considers an influence of T-10 tokamak plasma disruption on boron carbide (B_4C) coating on tungsten. The power density of coating irradiation reached 100 MW/m^2. The relief and composition of the boron carbide coating sample areas heated up to different temperature due to influence of disruption is determined. Conclusion is made that B_4C does not change integrity, when heated to temperatures of up to 2000 K. Local melting was observed in areas heated up to 2500 K. In the range of 2500–3600 K most of the coating was melted and collected into droplets. Composition rate is reduced to B:C = 1:1. In all temperature ranges the coating remained continuous and provided protection of tungsten from direct plasma irradiation.

1. Introduction

Tungsten is chosen as a plasma-facing material for the ITER tokamak divertor. Recent investigations have shown that tungsten undergoes cracking, intensive erosion, macroscopic particle emission and melting at relatively small temperatures under high power density plasma radiation, which ultimately lead to accelerated destruction of ITER tungsten divertor tiles [1–5]. It was shown in paper [6] that plasma-facing materials can be protected from erosion and destruction by using in-situ renewable B_4C coating. Deposition of protective film can be done via a regular discharge using the vapors of non-toxic explosion-proof carborane ($B_{10}C_2H_{12}$) [6]. B_4C has a high melting point (2673 °C), low sputtering coefficient at temperatures below 1400 °C, with no chemical sputtering in particular [7], low hydrogen trapping capability [8]. A periodically renewed film will prevent tritium trapping in the tungsten substrate. As such, it is possible to prevent accumulation of large quantities of tritium in the tokamak chamber.

At the same time, the behavior of B_4C films on tungsten during high power density discharges in thermonuclear facilities has not been studied. In this paper, the effect of plasma disruption on the B_4C coating on tungsten substrates in the T-10 tokamak ("Kurchatov Institute" Research Centre) has been investigated.

2. B4C coating deposition on tungsten and its irradiation in the T-10 tokamak

The method of B_4C coating deposition on tungsten in the conditions imitating tokamak first walls and divertors of existing thermonuclear facilities and the upcoming ITER tokamak has been developed in our previous work [9]. The film was deposited on the Coating Deposition and Material Testing (CODMATT) stand (NRNU MEPhI). The coating is formed by deposition of boron and carbon atoms sputtered by plasma ions from their respective targets, 5 μm thick boron carbide coating was deposited on 15×15×1 mm^3 tungsten substrates for testing in a T-10 tokamak. The temperature of the tungsten substrates during deposition was equal to 1000 K. The films had smooth surface with no cracks, exfoliations and other signs of poor adhesion properties. The composition of the coating (B 76 %, C 22 %, W 2 %) obtained via energy dispersion X-ray spectroscopy (EDX) on the Tescan Vega scanning electron microscope (SEM) with the INCA X-act console was practically identical to the stoichiometric composition of boron carbide. Tungsten atoms signal in the data of EDX analysis were from the tungsten substrate.

Tungsten samples with the coating were mounted on a T-10 tokamak diagnostic probe. The following discharge parameters are obtained during the T-10 tokamak regular discharge: magnetic field 2.4 T, plasma current 220 kA, electron density (1–2)·10^{19} m^{-3}, discharge duration 0.1 s. The expected value of heat flux density on the sample during the discharge was 1–2 MW/m^2.

Figure 1 shows one of the samples after testing in the T-10 tokamak. It is possible to note the melting zone of the sample, the coating zone subjected to irradiation by plasma streams and the sample zone not exposed to plasma flows (the place of attachment of the sample in the holder).

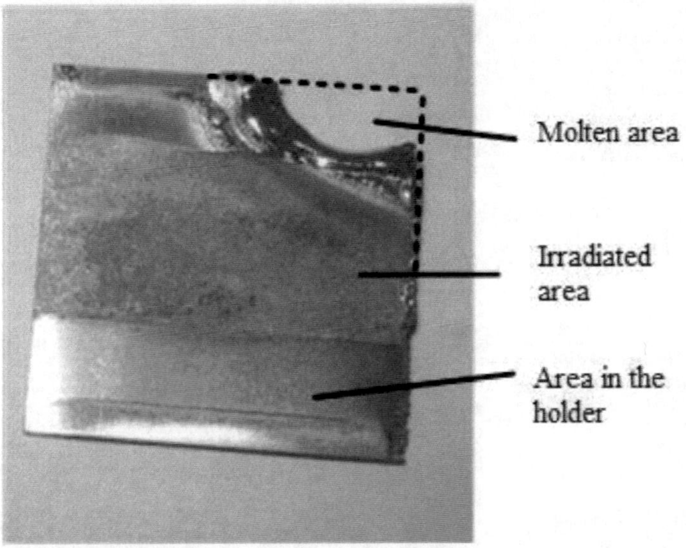

Figure 1. Tungsten sample with the B_4C coating after testing.

The power of plasma irradiation during a regular discharge is far from sufficient to melt the sample. During the analysis of samples after the tests power density of irradiation was calculated, as well as temperature distribution. The calculations showed that the power of plasma irradiation during a regular discharge is far from sufficient to melt and remove of part of the sample and heat of the adjacent section to a temperature $T \geq 2800$ K, at which the coating melts (see below). Conclusion was made that it could not occur under regular discharge, but was the result of plasma disruption. The duration of the disruption according to the change in plasma current characteristics during the discharge was approximately 70 ms. During that time, the heat load in the molten area of the sample reached 100 MW/m^2.

3. The behavior of B4C coating on tungsten under the T-10 tokamak plasma disruption

The change in the morphology of coating was studied using SEM, and the change in the composition was studies via EDX. Comparison of the results of surface composition and morphology analyses with temperature distribution on the sample surface during the disruption allowed for having a conclusion on the sample modification during plasma irradiation for varying temperatures.

- the coating completely evaporated from the surface of tungsten, which melted but remained on the sample. The figure 2 presents this regions. One can see micro grains formed during tungsten solidification. The pores and cracks are created on their boundaries;
- in the area with the average temperature at the heat maximum of 2800 K (coating melting area) the layers of coating closest to the surface melted and accumulated into droplets (figure 3(a)). There

IOP Conf. Series: Journal of Physics: Conf. Series **857** (2017) 012003 doi:10.1088/1742-6596/857/1/012003

was no exfoliation and cracks on the coating remaining on tungsten. The remaining film and droplets have lost most of boron, with the composition being B:C ≈ 1:1. However, it is known that the structure of boron carbide remains intact at this concentration [10];

- in the area with the maximum temperature between 2000 and 2500 K only partial melting on the coating film could be observed (figure 3(b)). The molten zones covered most of the area on the part with the higher temperature, while there was almost none such instances on the opposite side. Boron evaporation on those areas is much smaller, with the composition being approximately B:C ≈ 3:1 on average;
- the area of the sample on the "hot" borderline with the temperature of about 2000 K retained the coating completely, with its composition identical to the initial one. Signs of ion bombardment could be observed on the surface (figure 3(b)). There was no observable signs of exfoliation or melting on that part of coating.

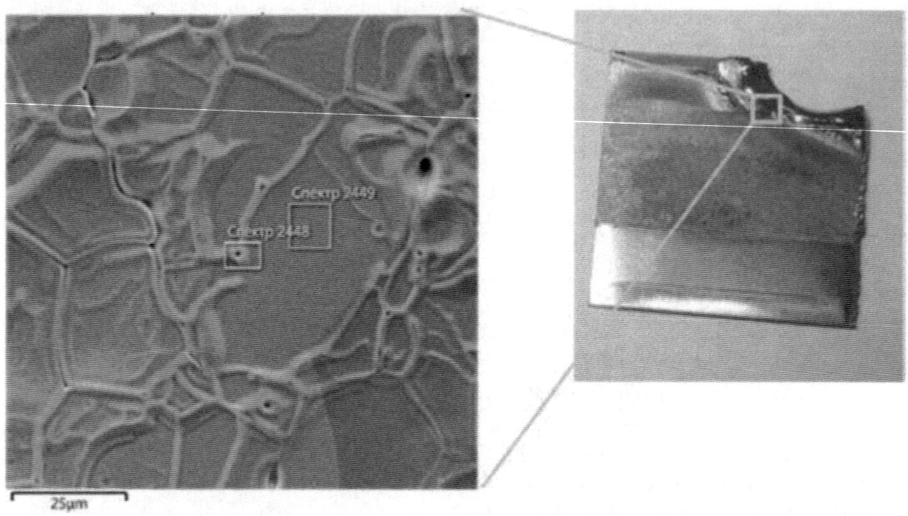

Figure 2. The surface of melted and solidified tungsten.

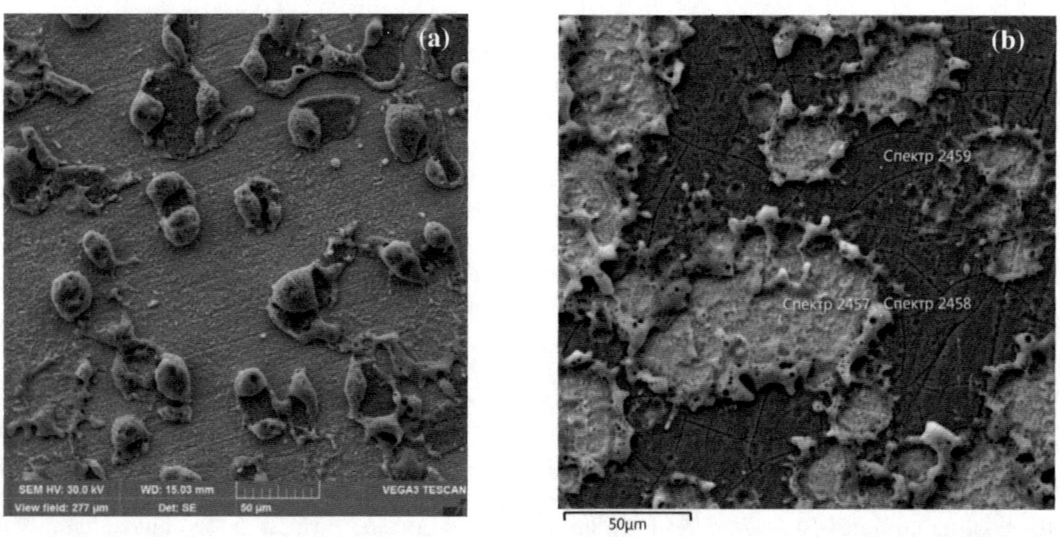

Figure 3. The sample surface with molten coating **(a)** and the sample surface area with partially melted coating **(b)**.

4. Conclusions

A study on the effect of T-10 tokamak plasma disruption on B4C coating on tungsten was conducted. The power density of coating irradiation reached 100 MW/m^2.

Surface composition and morphology analysis has shown that boron carbide coating retains its composition, adhesive properties and integrity, when heated to temperatures of up to 2000 K. Local melting can be observed in areas heated within the range of 2000–2500 K, with the coating composition remaining close to stoichiometric. In the range of 2500–3600 K, most of the coating is melted and collected into droplets. The structure of droplets and the remaining film retains that of boron carbide, but the composition rate is reduced to B:C = 1:1.

In all temperature ranges, boron carbide coating does not crack, retains adhesion to tungsten and provides protection from direct plasma irradiation for temperatures up to melting temperature of tungsten.

The retention of boron carbide surface layer contacting with tungsten for temperatures up to melting point allow for in situ renewing the coating on its surface even after interactions with high-energy plasma using the vapors of non-toxic explosion-proof carborane ($B_{10}C_2H_{12}$).

Acknowledgment

This work was supported by Competitiveness Growth Program of the Federal Autonomous Educational Institution of Higher Education National Research Nuclear University MEPhI.

References

[1] Krieger K, Geier A, Gong X, Maier H, Neu R, Rohde V and ASDEX upgrade team 2003 *J. Nucl. Mater* **313–316** 327–32

[2] Buzi L, de Temmerman G, Unterberg B, Reinhart M, Litnovsky A, Philipps V, van Oost G and Möller S 2014 *J. Nucl. Mater.* **455** 316–9

[3] Krasheninnikov S I, Smirnov R D and Rudakov D L 2011 *Control. Fusion* **53**(8) 083001

[4] Shu W M , Nakamichi M, Alimov V Kh, Luo G-N, Isobe K and Yamanishi T 2014 *J. Nucl. Mater.* **390–391** 1017–21

[5] Maier H, Greuner H, Balden M, Böswirth B, Lindig S and Linsmeier C 2013 *J. Nucl. Mater.* **438** 921–4

[6] Buzhinskij O I, Otroschenko V G, Whyte D G et al. 2003 *J. Nucl. Mater.* **313–316** 214

[7] Roth J, Tsitrone E, Loarer T, Philipps V, Brezinsek S, Loarte A, Counsell G F, Doerner R P, Schmid K, Ogorodnikova O V and Causey R 2008 *Plasma Phys. Control. Fusion* **50**(10) 103001

[8] Begrambekov L B, Buzhinsky O, Gordeev A, Miljaeva E, Leikin P and Shigin P 2004 *Physica Scripta* **108** 72–6

[9] Sadovskiy Y, Begrambekov L, Ayrapetov A, Gretskaya I, Grunin A, Dyachenko M and Puntakov N 2016 *Journal of Physics: Conference Series* **748**(1) 012003

[10] Sezer A O and Brand J I 2001 *Materials Science and Engineering: B* **79**(3) 191–202

About properties of ZrO_2 thermal protective coatings obtained from spherical powder mixtures

O B Berdnik, I N Tsareva and Yu P Tarasenko

Institute for Problems in Mechanical Engineering of Russian Academy of Sciences (IPME RAS), 603024, Nizhny Novgorod, Russia

E-mail: tribonika@yandex.ru

Abstract. It is developed the technology of high-energy plasma spraying of the zirconium dioxide (ZrO_2) thermal protective coating on the basis of ZrO_2 tetragonal and cubic phases with the spheroidal grain shape and the columnar substructure, with the total porosity $P = 4$ %, the hardness HV = 12 GPa, the roughness parameter $R_a \sim 6$ µm, the thickness 0.3–3 mm. As a sublayer it is used the heat-resistant coating of "Ni-Co-Cr-Al-Y" system with an intermetallic phase composition and the layered microstructure of the grains.

1. Introduction

In the course of technology development the up-to-date method of high-energy plasma powder spraying has been used. To create the zirconium dioxide thermal protective coating with improved performance properties the IPME RAS and the NPC "Tribonika" have updated the technology of plasma powder spraying through the use of plasmatron (capacity 50–70 kW), the design of which is improved via the incorporation of the sectionalized insert and the loop scheme of powder mixture introduction. The plasmatron modernization has allowed to stabilize the arc current and to increase the plasma flow energy characteristics (mass-average velocity of plasma jet V = 1200–2400 m/s, temperature of plasma flow T = 6500–7000 K) [1]. So, the advantage of high-energy plasma spraying is the heating effectiveness increase of powder mixture when passing through plasma flow and this increase contributes to the fusion of powder particles edges and to their spheroidizing.

2. Researches technique

The targets of researches were the ZrO_2 thermal protectives coatings on the intermetallic sublayer of "Ni-Co-Cr-Al-Y" system [2, 3], which are formed in spherical ZrO_2 powder mixtures with different content of yttrium oxide stabilizing additive: 1 – powder mixture (ZrO_2 + 7 % Y_2O_3) in solid spherical particles; 2 – powder mixture (ZrO_2 + 5.5 % Y_2O_3) in hollow microspheres; 3 – powder mixture (ZrO_2 + 16 % Y_2O_3) in hollow microspheres.

The phase composition of coatings has been examined by X-ray structural analysis using the diffractometer "Drone-3M". The metallographic examinations were carried out on the scanning electron microscope "VEGA TESCAN II". The Vickers hardness (HV) measurements have been performed with the use of the ultrasonic hardness tester with the load on indenter 1 N in accordance with GOST 9450-76 Standard. The coating surface roughness (R_a parameter) has been determined using the profile meter TR 200 in accordance with GOST 2789-73 Standard. The strength of coatings under formation has been evaluated by scratch hardness testing with diamond indenter (Rockwell) on the automated installation Revetest RST with the indenter displacement speed 1mm/min and with the continuous load increase on it from 0 to 200 N. The coatings porosity has been determined by the use of mathematical processing methods (software «Good Grains») and also of electronic and microscopic pictures of coatings cross sections.

3. Results of researches

When developing the heat-resistant coating the powder mixture of NiCr20Co20Al13 grade with particles dispersion 40/100 μm has been selected. By X-ray structural analysis it was established that the main phase of the powder mixture is an intermetallic compound β-(Ni, Me)Al. In case of high-energy plasma spraying the coating is formed with two-phase intermetallic composition: β-(Ni, Me)Al + γ'-(Ni, $Me)_3Al$. The γ'-phase formation in the coating is a result of the phase transformation β→γ' running due to the high-temperature effect of plasma flow. The obtained intermetallic sub-layer has the thickness of ~150 μm with a size of internal pores of 5–10 μm, the hardness HV = 7.5 GPa and the surface roughness R_a = 4.6 μm.

The powder mixture obtained by plasma technology and consisting of solid spherical particles (figure 1(a)) and of hollow microspheres (figures 1(b), 1(c)), has a greater spread of particle size distribution. The main defects are particles of irregular shape with a cut, ellipsoid granules and particles from the chipped surface of the shell (figure 1(a)). On the particles surface of coarse fraction (~90–100 μm) there are the accumulations of micro-particles. Large particles have a microstructure of grains in the form of polyhedra with well-developed subgrain structure inside (figures 1(b), 1(c)).

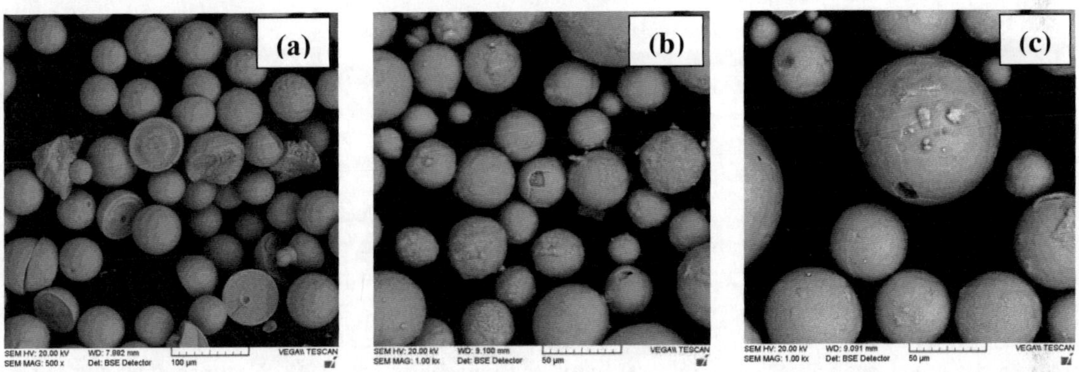

Figure 1. Microstructure: (a) – of powder mixture particles (ZrO_2 + 7 % Y_2O_3) in solid spherical particles (x500); (b) – of powder mixture particles (ZrO_2 + 5.5 % Y_2O_3) in hollow microspheres (x1000); (c) – of powder mixture microspheres (ZrO_2 + 16 % Y_2O_3) in hollow microspheres (x1000).

When forming an external ceramic layer the spherical powder mixtures have been used with maximum particles size ~100 μm. By X-ray structural analysis it is determined that the powder had a phase structure on the basis of tetragonal phase (T)-ZrO_2. In case of high-energy plasma spraying the two-phase coating is formed from the tetragonal (T)-ZrO_2 phase and the cubic (K)-ZrO_2 phase (~10 %). Figure 2 shows the ceramic coating microstructures with columnar grains ZrO_2 size in cross section ~100–200 nm. In coating cross section are visible the layers (thickness ~2.5 μm) with a columnar structure of grains. The cross dimension of columns is in the interval from 0.1 to 1μm. The physical-mechanical characteristics of coatings, obtained in different powder mixtures are given in table 1.

At present time high demands are required to the life-time of thermal protective coating with a view to the possibility of its premature destruction under high operating temperatures (up to 1100 °C). To increase the coating reliability the great importance is to solve the problem to enhance its adhesion strength with a basic material, since the coating performs its protective function while the coating maintains its integrity. In terms of failure mechanism more preferable is the cohesive nature of failure, accompanied by partial or fragmentary detachment and which promotes the extension of coating life cycle.

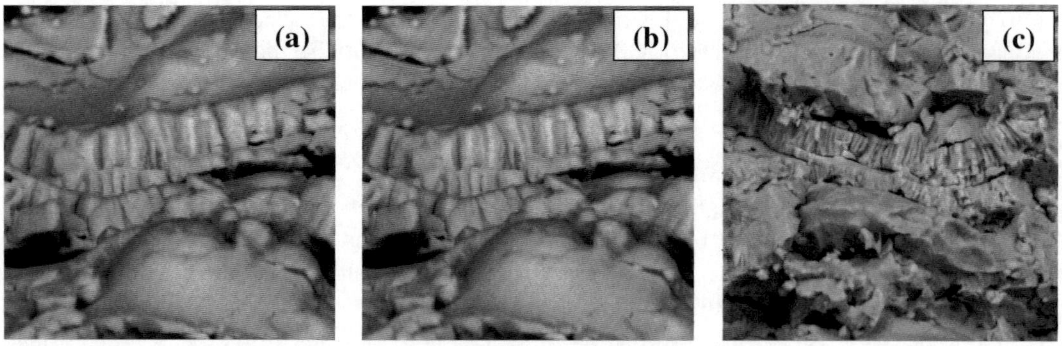

Figure 2. Cross section of zirconium dioxide thermal protective coating, formed from the powder mixture: **(a)** – (ZrO_2 + 7 % Y_2O_3); **(b)** – (ZrO_2 + 5.5 % Y_2O_3); **(c)** – (ZrO_2 + 16 % Y_2O_3) on intermetallic sublayer of «Ni-Co-Cr-Al-Y» system (x20 000).

Table 1. Physical-mechanical characteristics of thermal protective coatings formed in different powder mixtures

Coating material	Porosity, %	Surface roughness R_a, μm	Hardness HV, GPa
Coating in spherical solid powder (ZrO_2 + 7 % Y_2O_3)	4.1	6.3	12
Coating in powder of hollow microspheres (ZrO_2+ 5.5 % Y_2O_3)	4.7	7.5	9.65
Coating in powder of hollow microspheres (ZrO_2+16 % Y_2O_3)	2.1	7.6	9.45

In case of adhesive bond violation with the substrate under a great detachment (from the substrate), the access for aggressive medium opens and the destruction of a basic material begins. For the coating, obtained from the solid spherical powder, in case of scratch hardness testing, the low-amplitude acoustic signal, caused by the process of fragmentary micro-shearing is registered. With the growth of indenter depth penetration under increased load the signal amplitude practically does not change and this fact shows the cohesive nature of failure. Even at a maximum load (200 N) there is no violation of the adhesive bond and there is no coating detachment from the substrate.

4. Conclusions

As a result of researches performed is developed the zirconium dioxide thermal protective coating of enhanced thickness (0.3–3 mm), which is formed on an intermetallic sublayer of "Ni-Co-Cr-Al-Y" system in spherical powder mixtures with a different content of yttrium oxide. This coating and intended for parts protection working at high temperatures (up to 1000 °C).

The developed thermal protective coating has been put to an evaluation test at the NPC "Tribonika" (Nizhny Novgorod city, Russia) as part of repair and recovery technology to extend the service life of working and guide blades of SGT 800 Siemens gas turbine (for "Cityenergo" LLC, Moscow).

References

[1] Tarasenko Y P, Tsareva I N, Berdnik O B, Fel J A and Krivina L A 2014 *Development of heat-resistant intermetallic bond coat for a thermal barrier coating of turbine blades of gas turbine engines* 3
[2] Tarasenko Y P, Tsareva I N, Berdnik O B and Fel J A 2014 *N. E. P. heat-resistant intermetallic coatings obtained by high-energy plasma spraying for turbine blades of gas turbine engines* **4**
[3] Kuzmin V I, Card E V, Sergachev D V, Kornienko E E and Lapushkina E J 2014 *Air plasma spraying of powder coatings with gas-dynamic focusing heterogeneous flow*

IOP Conf. Series: Journal of Physics: Conf. Series **857** (2017) 012005 doi:10.1088/1742-6596/857/1/012005

Use of coatings for protection of welded joints of steels, their structure and properties

V P Bezborodov[1,2] **and Yu N Saraev**[1]

[1] Institute of Strength Physics and Materials Science, SB RAS, Tomsk, Russia
[2] National Research Tomsk Polytechnic University, Tomsk, Russia

Abstract. The paper studies the structure and demonstrates the efficiency of application of eutectic nickel coatings for protection of welded joints of the 10G2S-type steels from corrosive action. It increases simultaneously with the increase of eutectic content and chemical compounds along the grain boundaries of the γ-solid solution based on nickel. The refinement of the coatings' structure and the reduction of their heterogeneity allow enhancing the protective properties and resistance of welded joints to corrosion.

1. Introduction

The operational life of welded joints made from 10G2S-type steels is defined by both their material strength and surface properties [1–4]. The practice of technical maintenance of the equipment shows that the majority of products function under conditions of corrosive environment and varying loadings. Therefore, the reliability and durability of the operation of such products are determined by corrosion resistance of their working surfaces, by strength and endurance of welded joints.

The increase of the operation life of the products made from the 10G2S steel with welded joints by means of protecting their surfaces from corrosion is an urgent scientific-technical problem. One of the efficient methods of the operating life increase is the use of protective coatings. The use of the coatings made from nickel-based alloys for protection of welded joints from corrosion is limited by their susceptibility to cracking and pore formation.

The objective of the paper is the research of reasons of defect precipitation in the form of cracks and pores reducing the protective properties of coatings, as well as the assessment of possibilities of using coatings made from eutectic nickel alloys for the purpose of protection of welded joints' surfaces from corrosion.

Welding of pipe fragments made from 10G2S steels was performed using inverter-type power supply MP 2400 by "KEMPPI" Company. Powder materials made from Ni-Cr-B-Si-alloys were used to apply the coatings. Low-carbon steels were used as base materials. The coatings were applied by means of facing or argon-nitric plasma with subsequent flowing. A qualitative assessment of the influence of the flowing on the coatings' structure was made.

In the production, high activity of process media exacerbates the combined value of wear of parts and equipment. The wear intensity of parts depends on the velocity of the flow, the chemical activity of the medium, duration of the running period, the presence of abrasive inclusions, and other factors. Under conditions of dynamic impact of corrosive and corrosive-abrasive media, the operation life of products reduces drastically.

Many products, operating under conditions of corrosive media, are simultaneously subjected to pitting or dangerous intercrystalline corrosion and intensive abrasive wear, as a result of which the wear-and-tear of the working surface, accompanied by failure, occurs.

The paper suggests application of the technology of combined protection of surfaces of large-sized products by combining the operations of welding, facing, and flowing of coatings made from nickel alloys [5–7]. Such method allows extending the operation life of the products. The problem can be solved by applying protective coatings on the area of the welded joint with their subsequent flowing (figure 1).

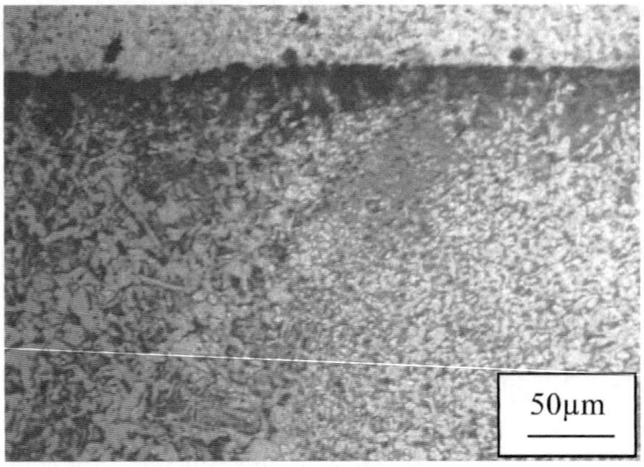

Figure 1. The structure of the composition "flowing coating PG-12N-02 – the welded joint of the 10G2S steel".

In steels, when erosion occurs by means of craters' formation and ductile fracture of their edges, grains' sizes do not have much influence if they are greater than the crater diameter. However, if these sizes become less than the craters' diameter, erosion resistance increases drastically. Based on the qualitative index of corrosion resistance (changing of the surface appearance), the coatings were assessed as: "stable" – "unstable"; "acceptable" – "unacceptable" (table 1). The corrosion stability was assessed in numbers of conventional scales (according to GOST 9.076-77). The working capacity of the coating is defined by its structural peculiarities. After flowing, the grains' sizes are 30–40 μm. The grains' sizes of the γ-solid solution strongly depend on the composition of the alloy and the flowing temperature, as well as on the coating thickness: the thicker it is, the coarser the coating structure is, which is related to the increase of the coating crystallisation period.

Table 1. Chemical resistance of steels and coatings

Corrosion medium	Medium temperature, °C	Material				
		Carbon steels	PG-10N-01	PG-10N-03	PT-19N-01	PS-12N-VK
H_2SO_4	60	D	B	D	B	D
HNO_3	50	D	C	D	C	C
HCl	30–70	D	D	D	D	D
H_2CO_3	120	C	B	B	B	B
NaOH	30–40	B	D	D	B	D
NaCl	30–50	C	B	D	B	B
H_2O	20–140	C	A	B	B	A
NH_3	40–120	C	A	D	B	C

Note: A, B, C, and D are the groups of materials' resistance to corrosion, which correspond to corrosion rates (mm/year): 0.00–0.05; 0.05–0.1; 0.1–0.5; > 0.5.

IOP Conf. Series: Journal of Physics: Conf. Series **857** (2017) 012005 doi:10.1088/1742-6596/857/1/012005

Heat removal in the coating area, adjoining the basis, occurring at a high rate causes the growth of column dendrites oriented normally relatively the basis by the axes of the first order. As the content of alloying elements in the alloys increases, the formation of a more finely-dendrite structure occurs. As a rule, the coating after flowing represents a material with features of both hypoeutectic and hypereutectic alloys. Formation of either structure depends predominantly on the initial composition of the powder material and conditions of coating obtainment. More often, in the coating, all types of structures are simultaneously fixed on different areas. When increasing the content of alloying elements in the coating by application of the more alloyed powder material, the content of carbide and boride phases increases and that of the γ-solid solution decreases. Dilute alloys are mainly characterised by the hypoeutectic structure; high alloys – by the hypereutectic structure with the high content of carbides and borides. Formation of coatings with the hypoeutectic structure leads not only to an increase of the plasticity of the material, but also to a certain reduction of hardness and strength. This allows using the material with such coating as a basis when creating compositions withstanding substantial mechanical loads. The structure of hypoeutectic nickel alloys consists mainly of the matrix represented by grains of the γ-solid solution and γ-Ni + Ni_3B eutectic with the volume content of up to 20 % filling the spaces between nickel grains (figure 2).

The coatings cooled at different rates contain different amount of eutectic. An increase of the cooling rate owing to an increase of the thickness or volume of the basis metal results in a reduction of sizes of eutectic colonies.

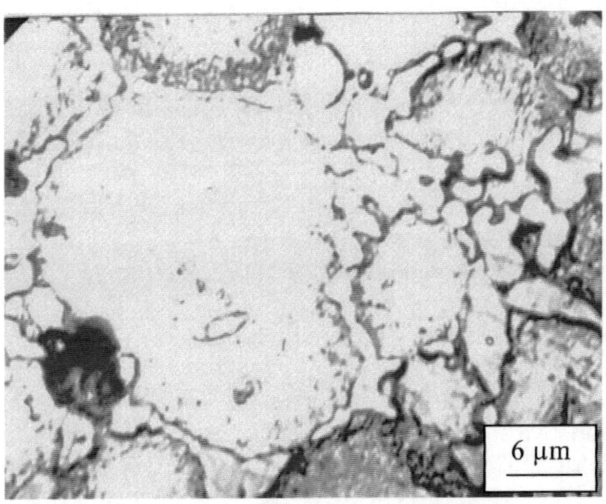

Figure 2. The structure of flowed coating PG-10N-04 with γ-Ni+Ni_3B eutectic along γ-Ni grains' boundaries after cooling.

Large heterogeneity of the coating by dispersity and phase distribution predetermines nonuniformity of the corrosion processes' behaviour and leads to formation of cracks in the defect areas with subsequent pitting of fragments. When forming the coating with the refined structure and more uniform distribution of carbide phases throughout the volume, the reduction of fragments' sizes and enhancement of service properties occur. High resistance of the coating is determined by its structure and high strength.

With the increase of the coating thickness, the intensity of its heating and cooling during crystallisation, the probability of crack formation increases. Prevention of the crack formation process is ensured by selection of heating and cooling modes of the coating. The application of volumetric methods of the heating and deep heating-through of the basis with subsequent relatively slow cooling allows avoiding cracks formation. This is also facilitated by isothermal soaking at a temperature of intensive austenite transformation and refinement of the coating structure accompanied by the reduction of the degree of chemical and structural heterogeneity. Stereometric analysis of the coating showed that its porosity including the pore sizes, quantity, shape and distribution throughout the volume strongly depend on the flowing modes. The coating flowed at temperatures of ≤1050 °C and short soaking times (1 min) has significant porosity (10–12 %, vol.). The pores are of spherical shape, up to 20 μm in sizes, and are uniformly distributed throughout the coating volume. At higher temperatures (1050–1150 °C) and soaking times of 1–5 min, low-porosity coatings (4–6 %, vol.) form.

2. Conclusions

1. The probability of crack formation in the coatings used for protection of welded joints from corrosion increases with the thickness growth of these coatings, heating and cooling intensity during crystallisation, as well as during increasing the content of chemical compounds.

2. Crack formation in the coating can be prevented only by selecting the heating and cooling modes of the composition, or by soaking at a temperature of intensive austenite transformation.

3. Modification of the structure of these coatings by refining and reducing heterogeneity allows increasing resistance of the coatings to crack formation and enhancing protective properties of welded joints from corrosion.

4. The increase of the content of alloying elements leads to the growth of the volume content of carbide phases, to the increase of the strength and corrosion resistance of protective coatings, made from Ni-Cr-B-Si-alloys, and welded joints of steels on the whole.

Acknowledgement

The work has been performed on the project of the Russian Scientific Fund, No. 16-19-10010.

References

[1] Vinokurov V A, Kurkin S A and Nikolaev G A 1996 *Welded Constructions. Fracture Mechanics and Standards of Working Capacity* ed. Paton B E (Moscow: Mechanical engineering) 576
[2] Saraev Yu N, Sleptsov O I, Bezborodov V P, Nikonova I V and Tyutev A V 2005 *Installation Work and Special Services in Construction* **8** 18
[3] Saraev Yu N, Bezborodov V P, Selivanov Yu V and Nikonova I V 2007 *Metal Processing* **2** 33–6
[4] Larionov V P, Sleptsov O E, Saraev Yu N and Bezborodov V P 2008 *Herald of the Academy of Military Sciences* **3** 67
[5] Sbrizher A G 1988 *Physical Metallurgy and Thermal Treatment of Metals* **4** 42–4
[6] Doroshenko L K, Borisova A L, Grigorenko G M et al. 1990 *Automatic Welding* **10** 22–3
[7] Antonova Ye A, Andrushenko N S and Sinai L M 1971 *Metal Protection* **7**(2) 137–42

The method of measuring the thermoelectric power in the thin films of the semimetals and narrow-gap semiconductors formed on the thin substrates

E V Demidov, V M Grabov, V A Komarov, A V Suslov and M V Suslov

Herzen University, 191186, Saint Petersburg, Russia

E-mail: demidov_evg@mail.ru

Abstract. In our paper we propose the method for measuring the thermoelectric power of thin semimetals films and narrow-gap semiconductors on the thin substrates. This method eliminates the occurrence of mechanical stresses due to different thermal expansion coefficients of the substrate material and the gradient plates. We compared the thermopowers' temperature dependences obtained by the new method of measuring and by the standard method for thin films $Bi_{1-x}Sb_x$ on a substrate of mica. With the help of our method it is possible to get more reliable results for thin films of semimetals and narrow-gap semiconductors formed on thin substrates.

1. Introduction

Currently, studies are being conducted actively to increase the efficiency of thermoelectric materials suitable for the creation the thermoelectric converters. One way to improve the efficiency of such converters is the use of thermoelectric thin films [1–2]. When conducting research in this area there are a number of difficulties, especially in the measuring the thermopower.

Studying low-dimensional systems' thermoelectric power is a difficult task, since it is necessary to create and accurately measure the temperature difference between "cold" and "hot" contacts for miniature objects. The complexity of these investigations increases when study the thin films on the thin substrates (5–40 μm) in a wide temperature range. The classical method of creating the controlled temperature gradient along the film is placing it on a sufficiently thick metal plate, along which there is measured temperature gradient. To improve the thermal contact of the substrate with film and plates liquid lubricant are used. However, this method gives satisfactory results only in the temperature range in which the plasticity of thermally conductive grease is stored. When measuring the thermopower at temperatures less than 200 K, especially for narrow-gap semiconductors and semimetals films, can be obtained substantially distorted results. It is known that the properties of these materials are heavily influenced by mechanical deformation [3–7]. The solidified lubricant is rigidly connected the substrate with film and the "gradient" plate that, due to different coefficients of thermal expansion (CTE) of the used materials give a rise to the deformation in the system, and thus in the film. So when using the gradient plates, we study not the film on the substrate, but the film on the substrate fixed on a gradient plate.

2. The basic idea of the new method of measuring the thermopower

The main idea of our method is to create an additional film structure on the back side of the substrate with respect to the film. This structure contains a copper resistance thermometer with four probe resistance measurement method and film heater (figure 1). The resistance thermometer located just below the hot contacts to the film. The film heater is located close to the resistance thermometer. Take into account the small thickness of the substrate, a resistance thermometer's temperature and the temperature of the film in the contact area equal.

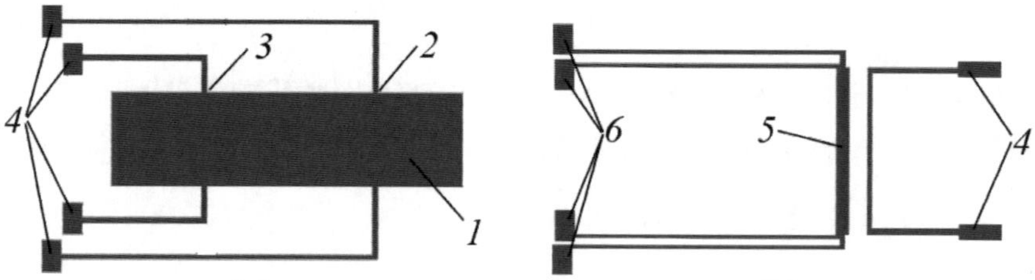

Figure 1. The topology of the object. Front side **(a)**: *1* – studied thin film; *2* – "hot" contact; *3* – "cold" contact; *4* – contacts for measuring the thermopower. Back side **(b)**: *4* – film heater; *5* – thermistor; *6* – lead contacts to the thermistor.

As the material for the film resistance thermometer is selected a copper as it has close to a linear temperature dependence of the resistance in a wide temperature range. The contacts and film heater are made of manganin, which has a low value of the thermoelectric power combined with copper and a small temperature coefficient of resistance. Additional thin film structure is produced by thermal evaporation in a vacuum.

The holder's flat surface on which is mounted the film has a ledge with the height of 0.5–1 mm. Film is installed on this surface so that over the ledge is only "hot" contact of the film, resistance thermometer and heater. Installing the substrate with the film to the holder is either weak mechanical pressure in the region of the contact pads, or a small drop of glue also in the contact area. The process of measuring the thermopower's temperature dependence with the help of our method is carried out according to the algorithm presented below.

1. The film heater is turned off; 2. The holder's temperature is stabilized in accordance with the studying temperature point; 3. The resistance of the resistance thermometer is measured; 4. The studied film's value of the thermopower is measured at the heater off; 5. Turns film heater, pausing to reach steady state (~ 5 s); 6. The current and the voltage drop are measured by the resistance thermometer; 7. The studied film's value of the thermopower is measured at the heater on; 8. The film heater is turned off; 9. The holder's temperature is stabilized in accordance with the next studying temperature point. Next, the algorithm is repeated.

If we study the influence of the magnetic field on the thermopower, points 6 and 7 are repeated with each value of the magnetic field. At the end we process the obtained results.

We calculate the resistance value of the resistance thermometer (RT). Using the values of the holder's temperature at the temperature points and resistance values of RT at these points, we construct the dependence of resistance RT via temperature (figure 2). We find the temperature coefficient of resistance RT by approximating the dependence into the line.

At each temperature point we find the difference of the resistance RT when the heater is on and off and then using the temperature coefficient RT, we calculate the temperature difference formed when the heater is turning on (figure 2). After that we calculate the difference in the thermopower of the film sample with and without a heater and using the temperature difference, we determine the value of the differential thermopower.

The measurement error of the thermoelectric power of the polyimide film on mica substrates with thickness 5–40 μm when the temperature difference across the sample 5–8 K is ~ 5%.

We used this method while studying the temperature dependence of the thermopower of the bismuth and bismuth-antimony films on substrates of mica (muscovite) and polyimide films.

3. Results and discussion

For example, figure 3 shows the dependence of the thermoelectric power via temperature for films $Bi_{92}Sb_8$ $Bi_{95}Sb_5$ on mica. There is the curve obtained with the help of the gradient plate and vacuum oil as a heat contact and the curve obtained when using the proposed method of measurement. For comparison there are results at the increasing and at the decreasing temperature. As it can be seen from figure 3, the thermoelectric power's temperature dependences obtained by the forward and reverse temperature during are almost the same. But when we use the gradient plate the direct and inverse temperature dependence vary considerably. You can see that there is an abrupt increase in absolute value of the thermoelectric power at the area of 150 K on the graph obtained at the decreasing of the temperature. It is connected with exfoliation of the substrate with film from the gradient plate because of the high mechanical stresses caused by the differences in CTE of the substrate material and the gradient plate.

On the graph obtained in the reverse temperature during also there is the sharp changing of the thermoelectric power's value, but at the temperature of 230–250 K. This temperature corresponds to the melting temperature of the vacuum oil and to the recovering of the heat contact between the substrate and the gradient plate.

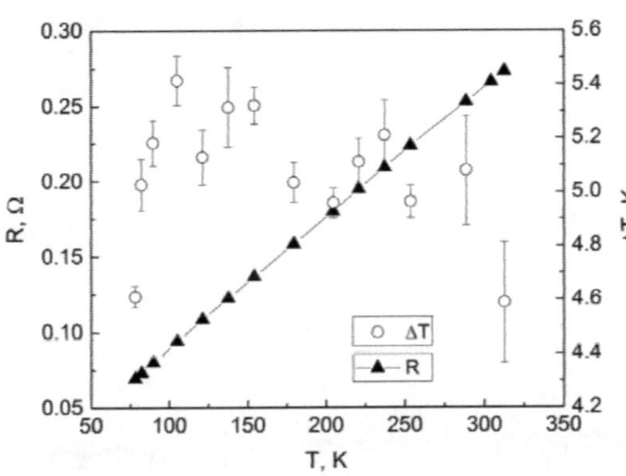

Figure 2. The temperature dependence of the film copper TR's resistance (R) and the temperature difference (ΔT) along the sample at the constant current of the film heater at the studying the thermopower's temperature dependence.

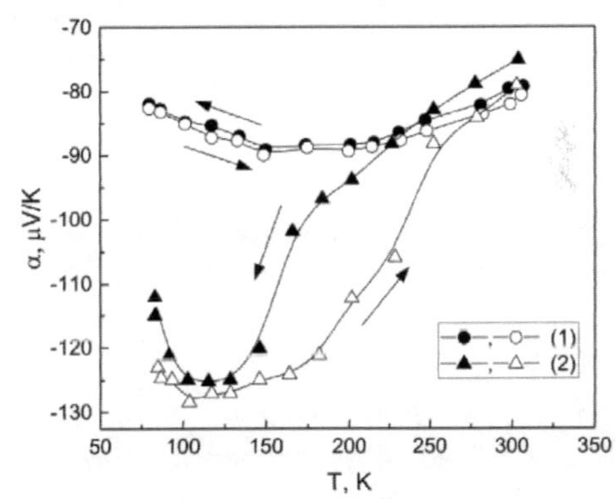

Figure 3: The thermopower's temperature dependence of the thin film $Bi_{92}Sb_8$ with thickness 0.75 μm on mica. This dependence obtained by the proposed method (1). The thermopower's temperature dependence of the thin film $Bi_{95}Sb_5$ with thickness 0.75 μm on mica (2). This dependence obtained with the help of the gradient plate.

4. Conclusions

On the basis of our research we can conclude that the proposed method of measuring the thermoelectric power's temperature dependences of the thin films on thin substrates provides reliable, repeatable results without additional mechanical strains in the sample.

Acknowledgment

This work is supported by Russian Foundation for Basic Research (grant 16-32-00313) and the Ministry Education and Science of the Russian Federation (within the state program, project № 3.4856.2017/BP).

References

[1] Dresselhaus M S, Dresselhaus G, Sun X, Zhang Z, Cronin S B and Koga T 1999 *Phys. Solid State* **41** 679–82
[2] Hicks L D and Dresselhaus M S 1993 *Physical review* B **47** 16631

[3] Kusagaya K, Hagino H, Tanaka S, Miyazaki K and Takashiri M 2015 *Journal of Electronic Materials* **44** 1632–6
[4] Kusagaya K and Takashiri M 2015 *Journal of Alloys and Compounds* **653** 480–5
[5] Komarov V A 2002 *Proc. Int. Conf. Thermoelectrics and their Applications (St. Petersburg)* (Saint Petersburg: Ioffe Institute) p 237
[6] Kablukova N S, Komarov V A, Demidov E V and Khristich E E 2012 *Izvestia: Herzen University Journal of Humanities & Sciences* **153** 13–9
[7] Grabov V M, Komarov V A and Kablukova N S 2016 *Phys. Solid State* **58** 622–8

Structure and magnetic properties of titanium nickelide nanoparticles synthesized by pulsed laser ablation method

A K Drozdova[1], V N Cherepanov[1], V A Svetlichny[1], L I Kveglis[2], M N Volochaev[3] and D A Velikanov[3]

[1] National Research Tomsk State University, 634050, Tomsk, Russia
[2] Siberian Federal University, 660041, Krasnoyarsk, Russia
[3] Institute of Physics n. a. L. V. Kirensky FRC KSC Siberian Branch of the Russian Academy of Sciences, 660036, Krasnoyarsk, Russia

E-mail: anna_drozdova709@mail.ru

Abstract. We have successfully synthesized magnetic titanium nickelide nanoparticles with equiatomic composition by pulsed laser ablation method. Using transmission electron microscopy and microanalysis was revealed that particles represent the homogeneous conglomerates containing identical amount of Ni and Ti atoms. The magnetization value of nanoparticles was estimated.

1. Introduction
Recently multipurpose materials, including alloys with the shape memory effect (SME) and a superelasticity (SE) applied in medicine. The leader among materials with SME is Ni-Ti-based alloy [1]. The combination of well biocompatibility and magnetic properties of material [2] can expand considerably the application area Ni-Ti alloys in medicine. For example, this alloy can be used for creation of magnetic nanoparticles. Furthermore, magnetic nanoparticles are actively applied in medical and biological researches by unique optical, electric and magnetic properties [3]. The particles operated by a magnetic field are used both for address delivery of medicaments and for stress rupture of cages-targets. For this reason structure and magnetic properties characterization of Ni-Ti nanoparticles is perspective.

2. Experiments
Ni-Ti nanoparticles was synthesized from a $Ni_{51}Ti_{49}$ bulk monocrystal by pulsed laser ablation method in water as colloidal solution with further drying up on air at ~50 °C using technique [4]. Structure of Ni-Ti nanoparticles was investigated by transmission electron microscopy (TEM) on HT-7700 (Hitachi, Japan) equipped with energy-dispersive spectrometer XFlash60/6T (Bruker, Germany) at 110 kV accelerating voltage. Magnetic properties were investigated by vibrating magnetometer with Puzey electromagnet [5].

3. Results and discussion
Figure 1 shows that synthesized particles can be divided into two types: spherical with 50–100 nm diameter and flake-like with 5–20 nm size.

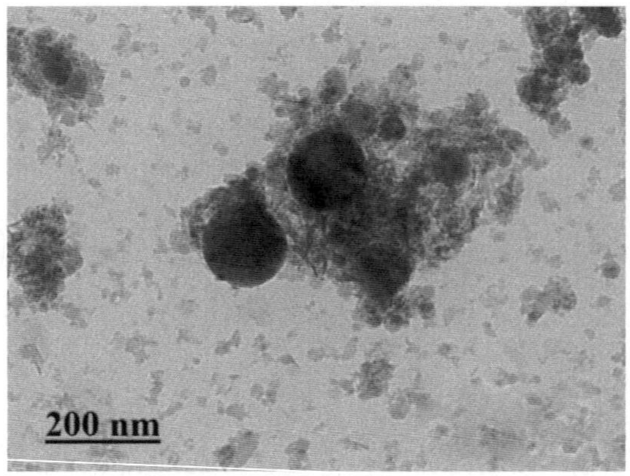

Figure 1. TEM image of Ni-Ti nanoparticles.

Element mapping (figure 2) was carried out for the purpose of chemical element identification and the analysis of their distribution in Ni-Ti nanoparticles. The results of element mapping indicate that the nanoparticles contain Ni, Ti, and also oxygen which are formed on their surface at oxidation. Ni and Ti are distributed in nanoparticles evenly, and their ratio is close to structure of the initial exemplar.

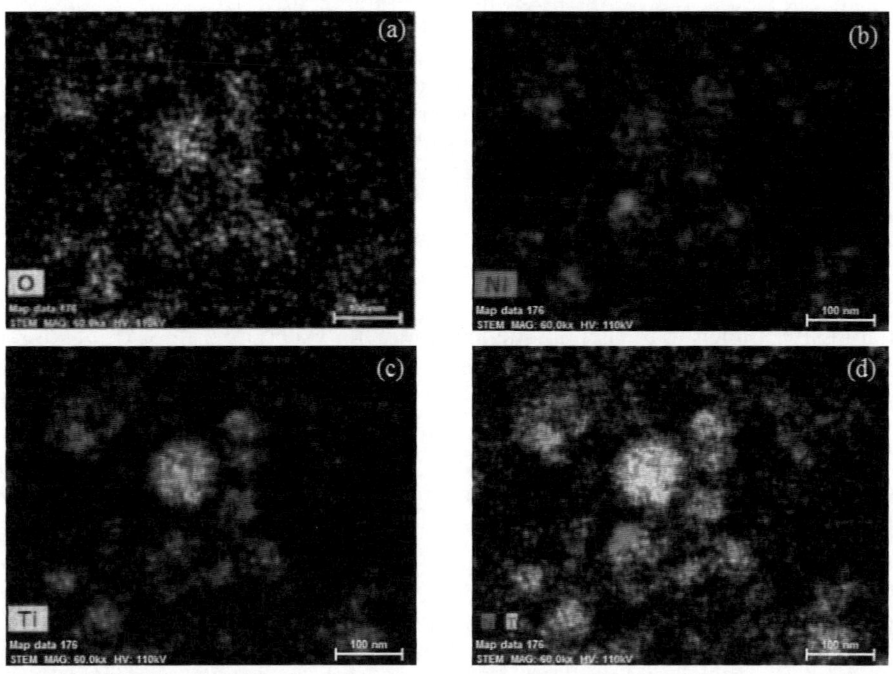

Figure 2. Element mapping received from Ni-Ti nanoparticles: **(a)** – in X-rays of oxygen K_{α}; **(b)** – in X-rays of nickel K_{α}; **(c)** – in X-rays of titanium K_{α}; **(d)** – the combined image in X-rays of titanium and nickel.

Figure 3 represents diffraction patterns from the titanium nickelide nanoparticles synthesized by pulsed laser ablation method (figure 3(a)) and from Ni-Ti particle wich appeared at the junction of three grains in the bulk $Ni_{51}Ti_{49}$ sample subjected to plastic deformation [2]. Also in [2], we found magnetic lenticular crystals of Ni_4Ti_3 phase which appeared after a plastic strain. Interpretation of diffraction patterns showed that lenticular crystals contain icosahedral phases (Frank-Kasper structures). Figure 3(c) shows the combined diffraction pattern (figures 2(a) + 2(b)) which indicates the complete coincidence of diffraction patterns represented on figures 2(a), 2(b). It allows to assume that icosahedral clusters with Frank-Kasper structure present in Ni-Ti nanoparticles synthesized by pulsed laser ablation method.

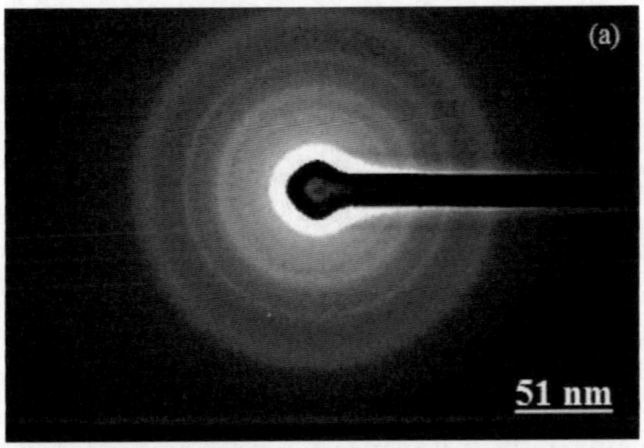

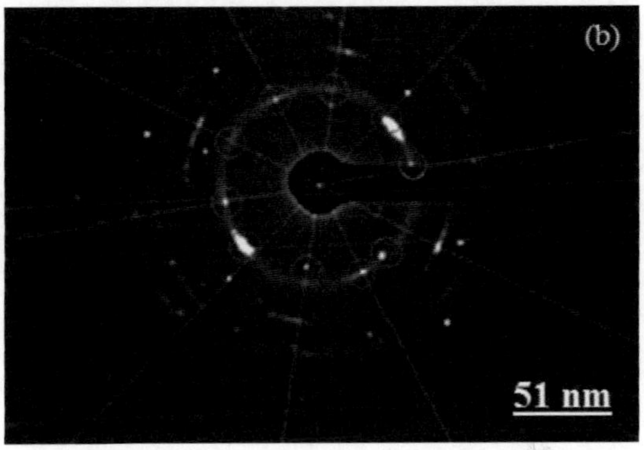

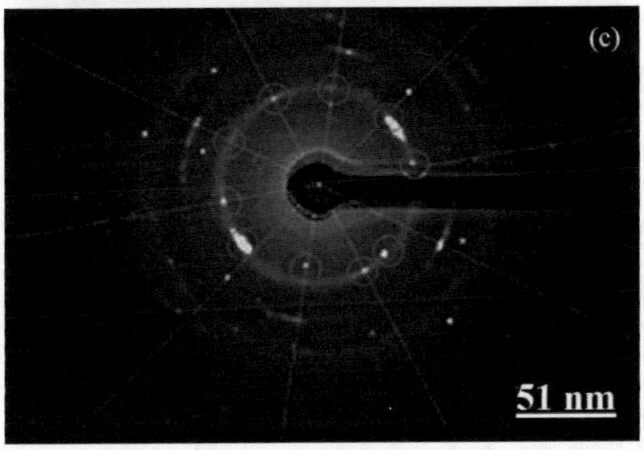

Figure 3. The comparison of diffraction pictures: **(a)** – from the titanium nickelide nanoparticles, received by method of pulse laser ablation; **(b)** – from plastically deformed $Ni_{51}Ti_{49}$ alloy exemplar; **(c)** – the combined diffraction pattern of electrons (figures 2(a) +2(b)).

Icosahedral clusters appear due to the atom displacements under the influence of temperature and pressure gradients that lead to ferromagnetic properties appearance in material. Ferromagnetism in Ni-Ti system disappears when the titanium content is higher than 10 % (at.) according to the phase diagram [6]. However, in our experiments the ferromagnetic phase appears in samples of equiatomic structure both after a plastic deformation and after a laser ablation because the condition of material considerably is removed from equilibrium in fields of tension and temperatures.

Figure 4 shows M-H curves of Ni-Ti nanoparticles (figure 4(a)) at temperatures T = 296 K (1) and 77 K (2) and also the temperature dependence of the magnetic moments at 3 kOe (figure 4(b)).

Magnetic measurements indicated that particles behave as superparamagnetic [7]. Furthermore, superparamagnetism was demonstrated in bulk samples contained the lenticular crystals [8]. The calculation by scattered waves (SW) method [8] showed that not-deformed $Ni_{10}Ti_6$ cluster with Frank-Kasper structure (FK-16) has a small magnetic moment (the average magnetic moment per atom is ~$0.3\mu_B$) and deformed (elongated by 8 % along the selected axis) $Ni_{10}Ti_6$ cluster has the average magnetic moment per atom is ~$1.2\mu_B$ while magnetization of clear nickel is only $0.6\mu_B$. Although, the average magnetic moment of $Ni_{51}Ti_{49}$ alloy samples is equal to zero due to chaotic distribution of clusters in alloy. Increase the magnetization of Ni-Ti alloy is caused by emergence of a resultant magnetic moment due to external influence (mechanical or temperature). Cooperative deformation of clusters with Frank-Kasper structure results in regularity of magnetic moment in system of these clusters [9]. Experimental and theoretical studies carried out this work have shown that the method of pulsed laser ablation allows obtaining the titanium nickelide nanoparticles that have magnetic properties which can be used for production of promising materials to be applied to new nanotechnologies.

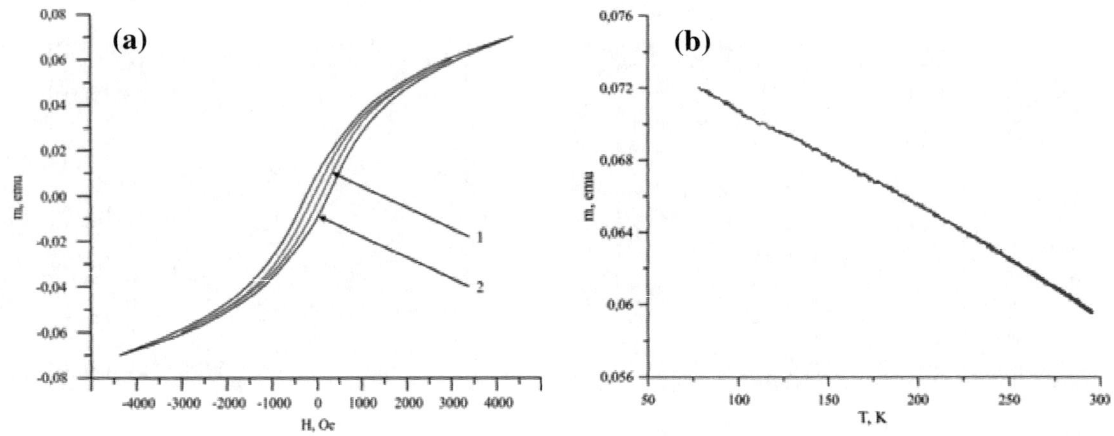

Figure 4. M-H curves for Ni-Ti nanoparticles (**a**): 1 – T = 296 K; 2 – T = 77 K; temperature dependence of magnetic moments (**b**).

4. Conclusions

In summary, we have firstly demonstrated synthesis of magnetic titanium nickelide nanoparticles with equiatomic composition by pulsed laser ablation method. TEM measurements have shown that synthesized Ni-Ti particles with equiatomic composition have a spherical form with 50–100 nm diameter. Magnetic measurements indicated that particles behave as superparamagnetic.

References

[1] Gunter V E, Yasenchuk Y F, Klopotov A A and Khodorenko V N 2000 *Tech. Phys. Lett.* **26** (1) 71–6
[2] Dzhess A V, Noskov F M, Kveglis L I et. al. 2016 *Fund. problems of mat. science* **13** (1) 96–104
[3] Giannaccini M, Giannini M, Calatayud M P et. al. 2014 *Intern. J. Mol.Sci.* **15** 1590–605
[4] Svetlichny V A, Shabalin A V and Lapin I N 2016 *Russian Phys. J.* **59** (12) 30–4
[5] Edelman I S, Patrin G S, Velikanov D A, et. al. 2008 *JETP Lett.* **87** (5) 310–3
[6] Lyakishev N P 2001 *Phase diagrams of double metal systems: Reference book* **3** 655–7
[7] Heard K M 1984 *Physics-Uspekhi* **142** (2) 331–55
[8] Noskov F M, Nyavro A V, Cherepanov et. al. 2017 *Vestnik SibSAU* **18** (1) *in press*
[9] Khomskii D I, Kugel K I, Sboychakov A O and Streltsov S V 2016 *JETP* **122** (3) 484–98

IOP Conf. Series: Journal of Physics: Conf. Series **857** (2017) 012008 doi:10.1088/1742-6596/857/1/012008

Corrosion resistance of the substrates for the cryogenic gyroscope and electrodeposition of the superconductive niobium coatings

A R Dubrovskiy, M A Okunev, O V Makarova and S A Kuznetsov

Institute of Chemistry, Kola Science Centre of the Russian Academy of Sciences, Russia

E-mail: a.dubrovskiy@chemy.kolasc.net.ru

Abstract. The interaction of different materials with the niobium containing melt was investigated. As substrate materials the ceramics, beryllium and carbopyroceram were chosen. Several spherical ceramic and beryllium samples were coated with protective molybdenum and niobium films by magnetron sputtering and PVD, respectively. After the experiment (exposition time 10 min) the exfoliation of molybdenum film from ceramic samples was observed due to interaction of the substrate with the melt. The niobium protective coatings reacted with the melt with niobium oxide formation. The beryllium samples regardless of the shape and the presence of the protective films were dissolved in the niobium containing melt due to more negative electrode potential comparing with niobium one. The carbopyroceram samples were exposed in the melt during 3 and 12 h. It was found that the carbopyroceram not corrodes in the niobium containing melt. The optimal regimes for electrodeposition of smooth uniform niobium coatings with the thickness up to 50 μm on carbopyroceram spheres were found.

1. Introduction

Niobium in the form of thin films and coatings is used in high-tech cryogenic devices using the phenomenon of superconductivity [1, 2]. For their preparation, the niobium ability can be used to form electrolytic structurally dense cathodic films in molten salts [3]. Electrolytic method does not require complex equipment and allows forming uniform coatings with the desired thickness on the inner and outer surfaces of articles with complex configuration. In this context, it is interesting to study the electrolytic niobium coatings on the different materials substrates, for their utilization in superconducting devices for various applications.

The ceramics, beryllium and carbopyroceram (CPC) are potential materials for the substrates.

Ceramics has the following advantages: retention of mechanical properties at high temperatures, high wear-resistance. The drawbacks included the following: a relatively high specific gravity, lack of conductivity and ductility, high hardness, susceptibility to cracking.

Beryllium has a low specific gravity and its disadvantages are a large anisotropy and thermal expansion coefficient, which is almost three times higher than that of niobium. In addition, beryllium is highly toxic and requiring special conditions for ensuring its handling.

Carbopyroceram is isotropic, gas-tight, solid and durable corrosion-resistant material that combines the properties of graphite and metal, able to withstanding repeated rapid heating followed by rapid cooling [4, 5].

There are many methods of niobium coatings deposition. One of the promising methods for deposition of superconductive (SC) niobium coatings is the electrolysis in molten salts, allows obtaining uniform coatings on articles with complex configuration. The electrolyte for electrodeposition of niobium is an aggressive media, so it is necessary to investigate the behavior of the substrate material in contact with the melt.

2. Experimental

In this work, we used chloride-fluoride electrolyte with the composition NaCl-KCl-NaF(10 wt. %)-K_2NbF_7 (8 wt. %) [6].

As substrate materials the ceramic, beryllium and carbopyroceram of different shape (spheres, cylinders, parallelepipeds) for corrosion resistance studies were used. The time of the experiments was varied from 10 min to 12 h.

The current density and time of electrolysis for electrodeposition of smooth uniform niobium coatings were varied from $1{\cdot}10^{-1}$ to $2{\cdot}10^{-1}$ $A{\cdot}cm^{-2}$ and from 8 to 12 h, respectively. All experiments were conducted in an inert atmosphere of argon and the temperature was fixed at 750 °C.

3. Corrosion resistance studies

Since ceramics (Al_2O_3) is not conductive material prior to the experiment, the samples were coated by molybdenum and niobium using magnetron sputtering and PVD, respectively.

Examination of the sample (sphere with diameter 10 mm) showed that the poreless coatings cannot be obtained by magnetron sputtering and pore area (substrate surface non-coated by molybdenum) is on average 200 μm^2 (figure 1).

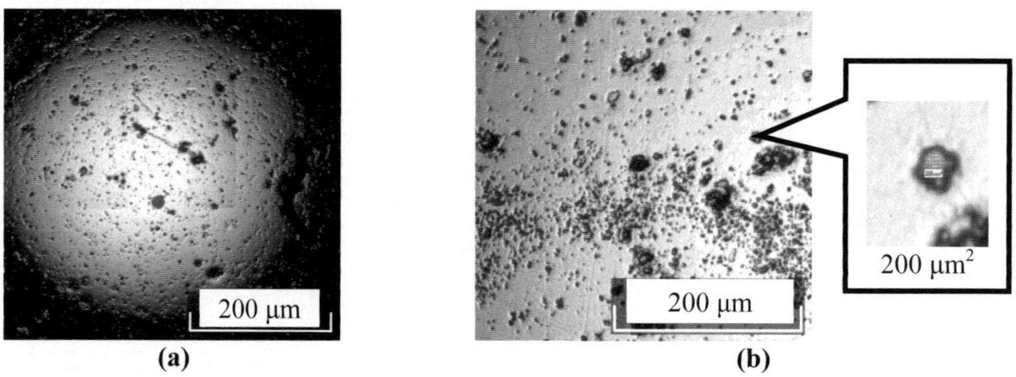

Figure 1. Macro- **(a)** and microstructure **(b)** of ceramic sample with molybdenum protective coating before the experiment. Film thickness is approx. 1 μm.

Investigation of the sample surface after the experiment revealed that an average pore area increased by 27 times (figure 2) and XRD captures the formation of the niobium oxides of various compositions (NbO, Nb_4O_5, NbO_2). To explain this result, the molybdenum corrosion behavior was studied in the same melt. Molybdenum plate (Aldrich 99.9+ %) kept standing for eight hours in the melt then was washed, weighed and analyzed by XRD. Sample weight and the diffraction pattern changes were not observed, indicating the high corrosion resistance of molybdenum in the niobium containing melt.

Niobium oxides are formed between the Al_2O_3 substrate and molybdenum coating during growth provide the mechanical impact on the coating, resulting in its exfoliation [7].

Spherical-shaped ceramics samples coated with niobium by PVD were tested under the same conditions. According to optical microscopy, niobium protective coatings were solid, smooth, mat-gray in color and without obvious defects prior to the experiment. Figure 3(a) shows the diffractogram of the sample prior to corrosion testing, which contains poorly formed peaks. Such distortions in the form of broadened lines can be a consequence of either very small crystal sizes or a change in the unit cell parameters due to interstitial impurities. The identification of all reflections is not possible, but with a high probability it can be stated the following compounds: (▼) Nb_6O and (★) γ-Al_2O_3. Some of the peaks (+) can be related to compounds of niobium with light elements (H_2, N_2, O_2), and some peaks remain unidentified. After the corrosion tests, the increasing of the γ-Al_2O_3 peaks intensity is recorded, indicating a thinning of the protective coating and its interaction with the melt leads to the formation of the NbO_2 compound (figure 3(b)).

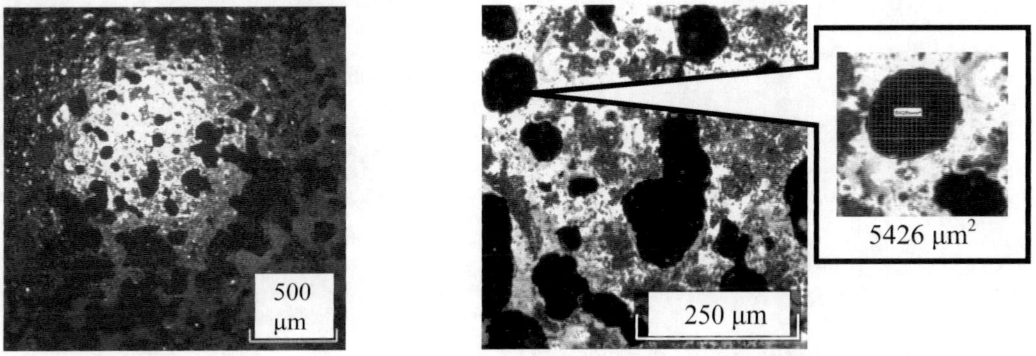

(a) (b)

Figure 2. Macro- **(a)** and microstructure **(b)** of ceramic sample with molybdenum protective coating after the experiment. Conditions: NaCl-KCl-NaF(10 wt. %)-K_2NbF_7(8 wt. %) melt in contact with Nb, $\tau = 10$ min, $T = 750$ °C.

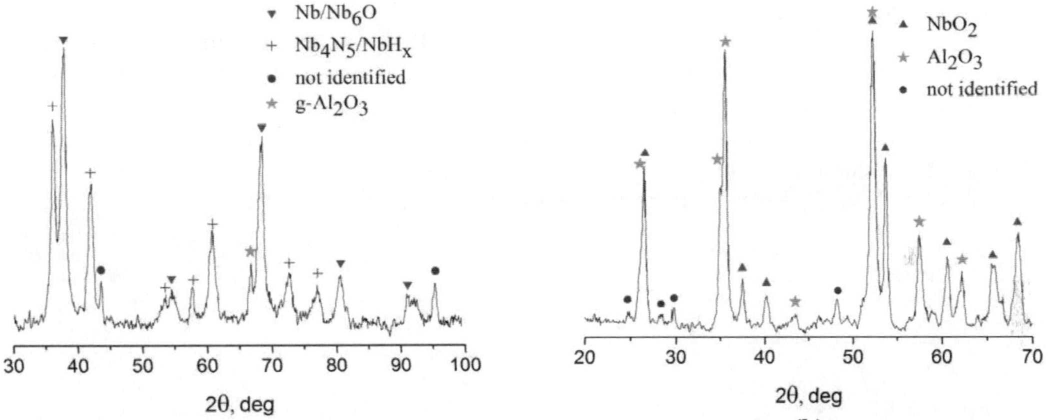

(a) (b)

Figure 3. The XRD data of the sample with niobium protective coating of before **(a)** and after **(b)** corrosion resistance tests.

Spherical and cylindrical beryllium samples with molybdenum coating and without it were tested at the same conditions as samples from ceramics. It was found that dissolution of the substrate occurs at a high speed because its electrode potential is more negative than niobium. Figure 4 shows the image of beryllium cylindrical sample surface. Evident that at the sample surface before the testing clearly expressed by machining grooves (figure 4(a)). After the test, the surface becomes a characteristic of the material subjected to dissolution (figure 4(b)).

Carbopyroceram samples were exposure in niobium containing melt for 3 and 12 h at a temperature of 750 °C. After a three-hour exposure in the melt XRD pattern of the sample remains the same, but after 12 h exposure XRD pattern shows the formation of niobium carbide layer (NbC) on the carbopyroceram surface.

Thus, magnetron sputtering and PVD methods could not provide the efficient metal protection coatings to apply it in molten salt media. Based on the corrosion tests, the carbopyroceram as a substrate for the electrodeposition of niobium coating was selected.

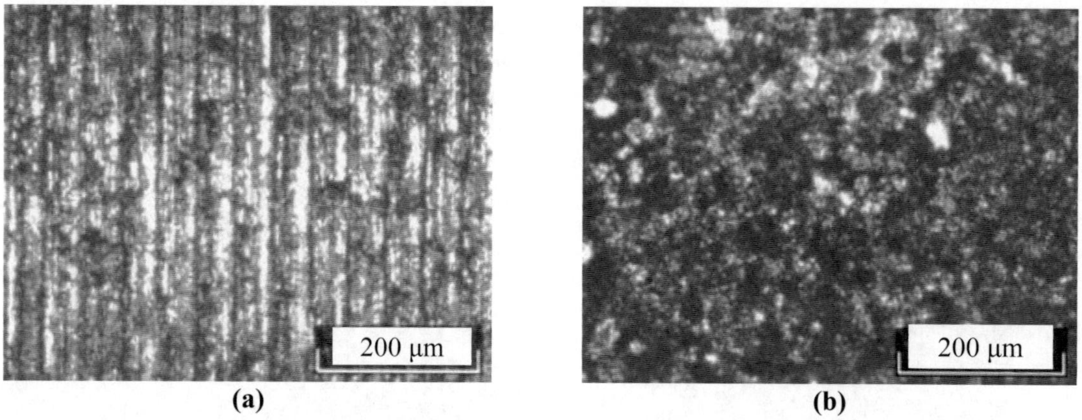

Figure 4. The microstructure of the beryllium cylinder side surface before **(a)** and after **(b)** the experiment. Conditions: NaCl-KCl-NaF(10 wt. %)-K_2NbF_7(8 wt. %) melt, $\tau = 10$ min, $T = 750$ °C.

4. Electrodeposition of niobium coatings

Electrodeposition of the niobium coatings on spheres 10 mm in diameter, manufactured of carbopyroceram, carried out at a temperature of 750 °C. The anodic current density was less than 10^{-3} A cm^{-2} and the cathodic current density during electrolysis was chosen $2\cdot10^{-1}$ A cm^{-2}. The NaCl-KCl-NaF(10 wt. %)-K_2NbF_7(8 wt. %) electrolyte was used for niobium coatings electrodeposition on carbopyroceram spheres. Due a spherical shape of substrates, a special form of the cathode should be used for plating of samples by niobium in molten salts. In the present study the construction of the cathode was developed. It consists of two parts: top part is fixed; bottom part is rotating by digital controlled stirrer. Construction of the cathode provides the electrical contact in a random point of the sample. Due to such contact a uniform coatings can be obtained.The stirrer rotation speed in all experiments was not changed and was chosen to be 30 rpm. Electrolysis time was chosen 8 h to obtain smooth uniform niobium coatings with the thickness 50 µm (figure 5).

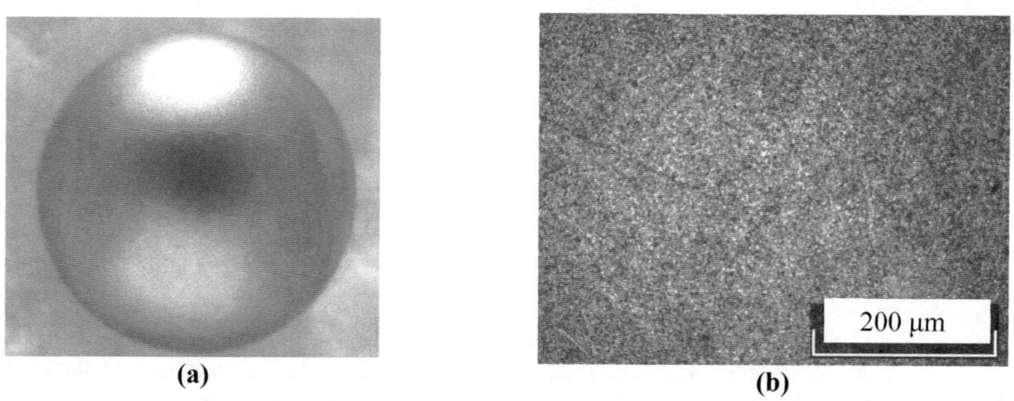

Figure 5. External view **(a)** and specimen surface morphology **(b)** of the carbopyroceram sample coated by niobium. Conditions: NaCl-KCl-NaF(10 wt. %)-K_2NbF_7(8 wt. %) melt, $\tau = 8$ h, $T = 750$ °C.

Spectral quantitative analysis of niobium coatings detected the following contents of impurities (wt. %): Mn < $2\cdot10^{-4}$; Mg < $3\cdot10^{-4}$; Si < 10^{-3}; Fe = $2.3\cdot10^{-3}$; Ni < $5\cdot10^{-4}$; Pb < $5\cdot10^{-4}$; Sn < $5\cdot10^{-4}$; Ti < 10^{-3}; Al < $5\cdot10^{-4}$; Co < 10^{-3}; Mo < 10^{-3}; Ca < 10^{-3}; Zr < $2\cdot10^{-3}$; V< $3\cdot10^{-4}$; Cu ≤ 10^{-3}; Cr < $5\cdot10^{-4}$.

Testing of the samples on the superconducting properties showed the same results as the samples of whole niobium of high purity.

References

[1] Lam S K H, Clem J R and Yang W 2011 *Nanotechnology* **22** 45

[2] Kolosov V N and Sheverev A A 2012 *Inorganic Materials* **48** 2
[3] Sheverev A A and Kolosov V N 2011 *Inorganic Materials* **47** 1
[4] Tatarinov V F and Zolkin P I 1999 *Refractories and Industrial Ceramics* **40** 1
[5] Kubo T, Ohashi Y F and Kinoshita T 2004 *Physica C: Superconductivity* **417** 1–2
[6] Kuznetsov S A, Marenkova E A and Kalinnikov V T 2015 *Doklady Chemistry* **463** 1
[7] Kuznetsov S A and Kalinnikov V T 2008 *Glass Physics and Chemistry* **34** 5

IOP Conf. Series: Journal of Physics: Conf. Series **857** (2017) 012009 doi:10.1088/1742-6596/857/1/012009

Protective coatings preventing hydrogen desorption from titanium during ion irradiation

A E Evsin, L B Begrambekov, S S Dovganyuk, A S Kaplevsky and M I Shutikova

National Research Nuclear University MEPhI (Moscow Engineering Physics Institute), 31 Kashirskoe shosse, Moscow, 115409, Russia

E-mail: evsin@plasma.mephi.ru

Abstract. Effect of yttria and titanium nitride coatings on features of deuterium desorption from titanium layer is investigated. It is shown that both coatings significantly raise the temperature of maximum of deuterium thermal desorption from titanium under linear heating and prevent desorption under prolonged keeping at the operating temperature of a neutron tube target. However, under irradiation with ions of $H_2 + O_2$ plasma the barrier properties of titanium nitride appear to degrade.

1. Introduction

Hydride-forming metals (Ti, Zr, Y, Sc, V, La, etc.) can accumulate significant concentrations of hydrogen isotopes. Thereby, some of these metals are used as hydrogen sorbents in various devices, e.g. in the targets of neutron tubes. The target of the neutron tube consists of a sorbent layer (generally Ti, sometimes Zr or Sc) intended for the accumulation of hydrogen isotopes and a molybdenum substrate. The decrease of hydrogen concentration in the sorbent layer during the operation of the neutron tube leads to the decline in the efficiency of this device and to the reduction of its service lifetime. One of the possible ways for mitigating hydrogen losses from the neutron tube target is the deposition of protective coating on the sorbent layer's surface [1, 2]. Choosing the material for this coating, one should take into account that the coating must act as a hydrogen desorption barrier and its barrier properties must not degrade under ion bombardment during the operation of the neutron tube. However, the features of alteration of barrier properties of protective coatings by ion irradiation are not enough studied yet.

In this work, the effect of yttria and titanium nitride coatings on deuterium desorption from titanium layer on molybdenum substrate under heating and ion irradiation is investigated employing thermal desorption spectrometry (TDS).

2. Experimental

All of the coatings were deposited in the electron-induced gas discharge with a hot cathode. The residual gas pressure in the deposition chamber in all of the experiments was $<10^{-3}$ Pa, operating gas pressure – $1.3\cdot10^{-1}$ Pa. Argon was used as a main component of the operating gas. The energy of ions irradiating the surface of emerging coating did not exceed 20 eV, sample temperature – 450 K.

Titanium layer (sorbent layer) of 1 μm was deposited in Ar + 5 % D_2 plasma on molybdenum substrate of $7\times7\times0.5$ mm^3. Thus, the saturation of the sorbent layer by deuterium occurred during the deposition of this layer. After deposition of the sorbent layer some of the samples were covered by yttria (in Ar + 20 % O_2 plasma) or titanium nitride (in Ar + 25 % N_2 plasma) coatings of 50 nm.

Investigation of the hydrogen barrier properties of the coatings under heating and ion irradiation, as well as thermal desorption analysis (TDS analysis) of the samples, were conducted in MIKMA facility [3]. TDS analysis consisted of the linear heating of a sample from ≈300 to 1500 K with the rate of 5 K/s and simultaneous measuring of desorption flux from this sample by the quadrupole mass-spectrometer QMG220 (Pfeiffer).

For testing of the ability of a coating to prevent the deuterium desorption at the operating temperature of the neutron tube target a coated sample was heated at the temperature 623 K in vacuum ambient for 4 h. Desorption flux from the sample was measured during this procedure. After that, the sample was cooled down to ≈300 K and then underwent TDS analysis in order to measure the amount of deuterium atoms remaining after the heating.

For testing of the resistance of a coating to deuterium desorption under ion irradiation, a coated sample was irradiated with 650 eV/at. ions (mainly H_2^+) of H_2 + 2% O_2 and H_2 + 30 % O_2 plasma delivered by the flux of $5\cdot10^{19}$ $m^{-2}s^{-1}$ to the fluence of 1.8×10^{23} m^{-2} for 1 h. The sample temperature during the irradiation was equal to 623 K. After that procedure, TDS analysis of the sample was performed.

Hydrogen barrier properties of the genuine oxide films on the uncoated sorbent layers were tested the same way as those of the investigated coatings. Before and after the irradiation the content of the sample's surface layer were analyzed by the energy-dispersive X-ray spectrometry (EDX) module INCA-xAct (Oxford Instruments).

3. Results and discussion

The results of TDS analysis showed that the deposition of either yttria or titanium nitride on the sorbent layer's surface led to the increase of the temperature of deuterium desorption from this layer under linear heating (figure 1). Shift of the maximum of thermal desorption of deuterium from the yttria-coated sample was ≈180 K, from the one coated by titanium nitride – ≈240 K.

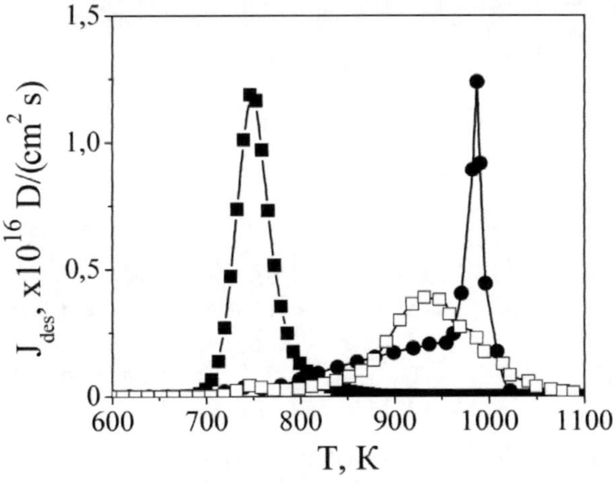

Figure 1. TDS spectra of deuterium from the samples: ■ – Mo + sorbent layer (Ti); □ – Mo + sorbent layer (Ti) + yttria; ● – Mo + sorbent layer (Ti) + titanium nitride.

Keeping of the uncoated sample (Mo + sorbent layer) at the constant temperature of 623 K for 4 h resulted in the release of 67 % of deuterium atoms contained in the sorbent layer, with 18 % of deuterium atoms desorbed during the first hour of the heating (table 1). Desorption of deuterium from the sorbent layer coated by either yttria or titanium nitride remained negligible (<5 %) during the 4 h of the similar tests. The increase of the temperature of deuterium desorption under linear heating and significant decrease of deuterium losses from the sorbent layer under the prolonged keeping at the constant temperature are the evidence that both yttria and titanium nitride coatings have better (in comparison to genuine titanium oxide film) barrier properties for the thermal desorption of deuterium contained in the sorbent layer.

Table 1. The results of tests of the coatings on ability to prevent deuterium desorption from the sorbent layer under heating and ion irradiation.

	Mo + sorbent layer (Ti)	Mo + sorbent layer (Ti) + titanium nitride	Mo + sorbent layer (Ti) + yttria
Deuterium concentration in the sorbent layer before the tests	$1.2 \cdot 10^{18}$ cm^{-3}	$1.1 \cdot 10^{18}$ cm^{-3}	$1.1 \cdot 10^{18}$ cm^{-3}
Deuterium desorption during keeping at 623 K for 1 h (4 h)	18 % (67 %)	– (<5 %)	– (<5 %)
Deuterium desorption during irradiation with ions of H_2 + 2 % O_2 plasma at 623 K for 1 h	<5 %	17 %	<5 %
Deuterium desorption during irradiation with ions of H_2 + 30 % O_2 plasma at 623 K for 1 h	<5 %	70 %	<5 %

If the keeping of the uncoated sample (Mo + sorbent layer) at 623 K for 1 h was accompanied by the irradiation with 650 eV/at. ions of either H_2 + 2 % O_2 or H_2 + 30 % O_2 plasma, noticeable deuterium losses from the sorbent layer were not observed (table 1). Moreover, during subsequent TDS analysis of the irradiated sample deuterium desorbed at higher temperatures (figure 2). Thus, ion irradiation of the sorbent layer covered only by the genuine oxide film mitigates thermal desorption of deuterium. Analogous result was obtained by us earlier [1, 2] when titanium deuteride underwent the similar radiation test. As a possible reason of this phenomenon, it was proposed that low-energy protium ions captured in the surface region may hinder deuterium diffusion from the bulk and subsequent desorption [1, 2].

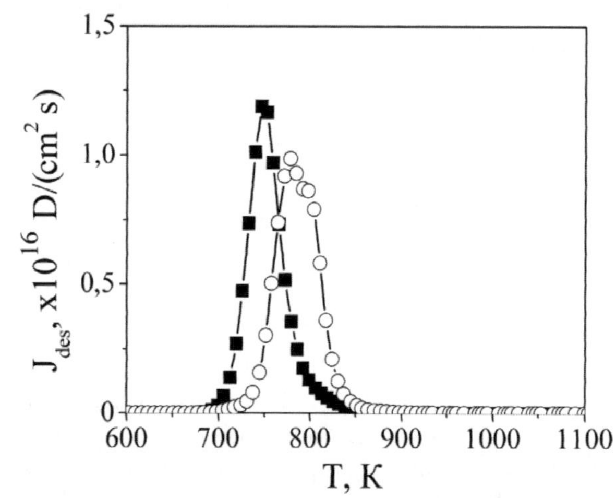

Figure 2. TDS spectra of deuterium from the sample "Mo + sorbent layer (Ti)": ■ – before the ion irradiation; ○ – after the irradiation with 650 eV/at. ions of H_2 + 30 % O_2 plasma.

Irradiation of the yttria-coated samples by ions of hydrogen plasma with both oxygen concentrations did not cause deuterium desorption from the sorbent layer (table 1). At the same time, under irradiation of the nitride-coated sample with ions of H_2 + 30 % O_2 plasma, 17 % of deuterium atoms desorbed, and irradiation in H_2 + 30 % O_2 plasma provoked desorption of 70 % of deuterium (table 1). Experimental result with titanium nitride coating is supposed to be the consequence of the degradation of the coating's barrier properties under ion irradiation. The failure of the barrier properties develops with the increase of oxygen contamination of plasma; therefore, one can believe that the interaction of oxygen with the surface of titanium nitride is a possible reason of the degradation.

After the irradiation of both types of coatings a peak at the temperature ≈780 K appeared in the deuterium TDS spectrum (figure 3). This maximum is inherent for TDS spectra of deuterium from the irradiated uncoated sorbent layer (figure 2) and that allows for suggesting that the partial degradation of the coatings occurs, apparently, due to their sputtering by the impinging ions. EDX analysis confirmed the absence of the protective coatings on particular regions of the surface. This result indicates the necessity of application of protective coatings recovery system providing the protection of the neutron tube target during its service lifetime.

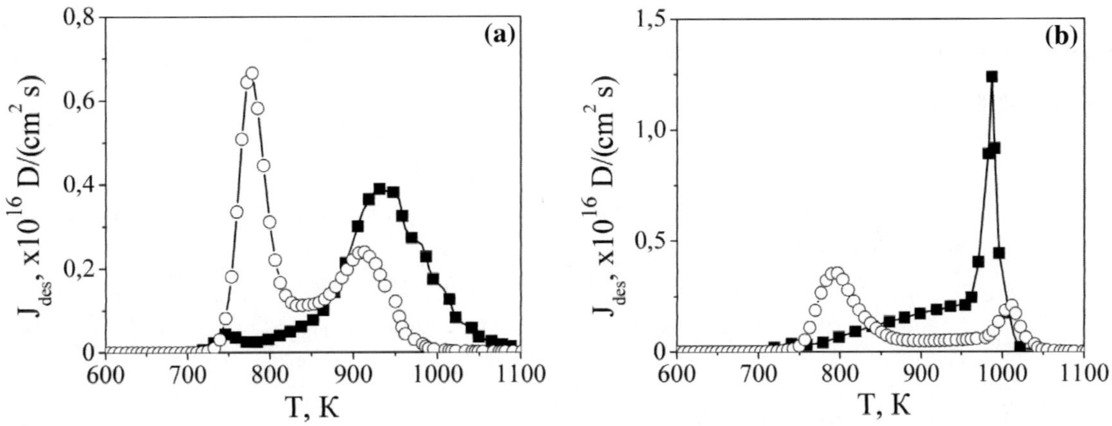

Figure 3. TDS spectra of deuterium from the samples "Mo + sorbent layer (Ti) + yttria" (**a**) and "Mo + sorbent layer (Ti) + titanium nitride" (**b**): ■ – before the ion irradiation; ○ – after the irradiation with 650 eV/at. ions of H_2 + 30 % O_2 plasma.

4. Conclusions

It is found that the temperature of deuterium desorption from titanium layer (sorbent layer) under linear heating is significantly increased if this layer is coated by a thin film of yttria or titanium nitride. It is shown that deuterium desorption from the samples with both types of coatings during prolonged keeping at the operating temperature of the neutron tube is very low. Yttria coating retain its barrier properties under irradiation by ions of both H_2 + 2 % O_2 and H_2 + 30 % O_2 plasma, whereas the irradiation of titanium nitride by ions of oxygen-contaminated plasma causes the degradation of its barrier properties and promotes the deuterium desorption from the sorbent layer. Sputtering of the protective coatings by the impinging ions requires the development of protective coatings recovery system providing the protection of the neutron tube target during its service lifetime.

Acknowledgment

This work was supported by Competitiveness Growth Program of the Federal Autonomous Educational Institution of Higher Education National Research Nuclear University MEPhI.

References

[1] Zakharov A M, Dvoichenkova O A and Evsin A E 2015 *Physics of Atomic Nuclei* **78**(14) 1643–5
[2] Begrambekov L B, Dvoichenkova O A, Evsin A E, Zakharov A M, Kaplevsky A S and Shigin P A 2016 *Journal of Surface Investigation* **10**(5) 1072–6
[3] Airapetov A, Begrambekov L, Brémond S, Douai D, Kuzmin A, Sadovsky Y and Vergasov S 2011 *Journal of Nuclear Materials* **415** S1042–5

IOP Conf. Series: Journal of Physics: Conf. Series **857** (2017) 012010 doi:10.1088/1742-6596/857/1/012010

Luminescent and electrophysical properties of nanostructured thin ZnO films

L V Grigoryev[1], A I Nisgireva[1], E A Cernyh[1], S V Kulakov[2], V G Nefedov[2] and O V Shakin[2]

[1] ITMO University, 197101, Saint Petersburg, Russia
[2] Saint-Petersburg University of Aerospace Instrumentation, 190000, Saint Petersburg, Russia

E-mail: grigoryev@oi.ifmo.ru

Abstract. In article are presented the results investigation of the luminescent properties of thin ZnO films. Present the results of the investigation of the I-V characteristic in a zinc oxide film obtained by a reactive ion-plasma method. Presented the parameters of electrically active traps and quantity the mobility of charge carriers in ZnO.

1. Introduction

Thin ZnO films created by the reactive ion-plasma method are related to direct-gap semiconductors having a large band gap (more than 3.3 eV). This determines a lot of interesting properties of this semiconductors material, which are widely used in solid-state devices of optoelectronics and electronics. Also, because of low optical losses in the visible spectral range, high mobility of charge carriers and the possibility of doping, thin ZnO films are widely used as an active material for the creation of scintillation detectors, UV-photonics and transparent electronics [1].

In optoelectronics and transparent electronics are used various methods for created thin zinc oxide films: the CVD process, pulsed laser ablation of a ceramic target made from highly purified ZnO, or electron beam deposition in a pulsed mode [2]. In comparison with these methods, the method of reactive ion-plasma sputtering of a metal target makes it possible to create ZnO films with the smallest structural defects in the thin film, besides; this method allows doping of the film composition during its growth [3, 4]. The purpose of this work was investigation the luminescent properties of thin ZnO films and investigation the processes of charge carrier transport in ZnO films obtained by reactive ion-plasma sputtering.

2. Method preparation of thin films samples

The process of ion-plasma sputtering of the metallic target was carried out using a magnetron operating in a constant current mode. Process the synthesis of a zinc oxide film occurred in a gas atmosphere consisting of 30 % oxygen and 70 % argon. The pressure of the gas mixture under the cap of the sputtering installation did not exceed 101 Pa, and the substrate temperature did not exceed 423 K. The metal target was made from zinc grade extra clear. The substrate was making of quartz particularly transparent in the UV-range. Plates of quartz were polished in grade 12 cleanliness classes and were thickness of 2 mm.

The spraying of Al electrodes occurred on the magnetron installation. According to the data of microscopic studies, the thickness of the film ZnO was 1 μm. The thickness of film Al contact was 0.5 μm. X-ray diffraction investigation of synthesized zinc oxide films showed the presence of the hexagonal structure of ZnO films and its preferential orientation in the (001) direction. The dimensions of ZnO crystallites did not exceed 13 nm. The drawing of the investigated ZnO samples with the scheme by Al electrodes is shown in figure 1.

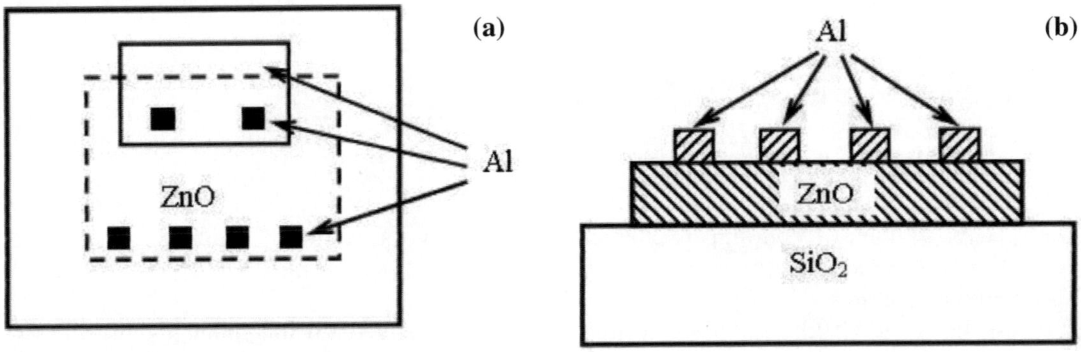

Figure 1. Appearance of the samples under study, top view (**a**) and side view (**b**).

3. Investigation of photoluminescence properties and current transport in ZnO films

Investigation of the photoluminescence spectra of ZnO films was carried out on a Perkin Elmer LM 50 fluorimeter at spectral range from 200 to 700 nm. The length of the exciting radiation was 337 nm. The measurements were carried out at a temperature of 300 K. Figure 2 shows the spectral dependence of the photoluminescence of a thin ZnO film.

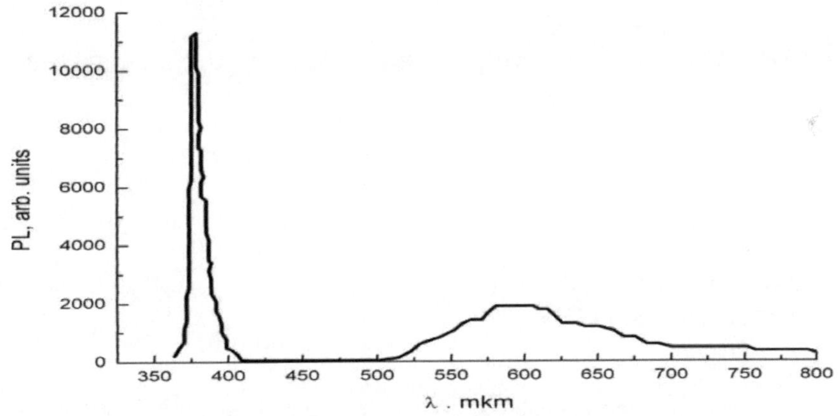

Figure 2. The photoluminescence spectrum in a ZnO film.

There are two peaks presents on the spectral dependence with maximums at 380 nm and at 600 nm. The narrow spectral peak, having a maximum at 380 nm, corresponds to the photoluminescence of excitons present in the volume of the thin ZnO film. The six fold decrease in the intensity of the broad peak relative to the narrow peak and the shift of the maximum of the spectrum from the green to the red region attest to the high structural quality of the synthesized ZnO film.

To study the current transport processes occurring in thin films ZnO, measurements of the current-voltage characteristics (CVC) at 300 K were carried out in the dark. The CVC characteristics are shown in figure 3. The CVC studies were carried out on an automated installation consisting of a programmable source voltage and a picoampermeter 6485. Al contacts sputtered on a polished quartz substrate under a ZnO layer and on a zinc oxide layer are ohmic contacts, as evidenced by the symmetry of the CVC characteristic.

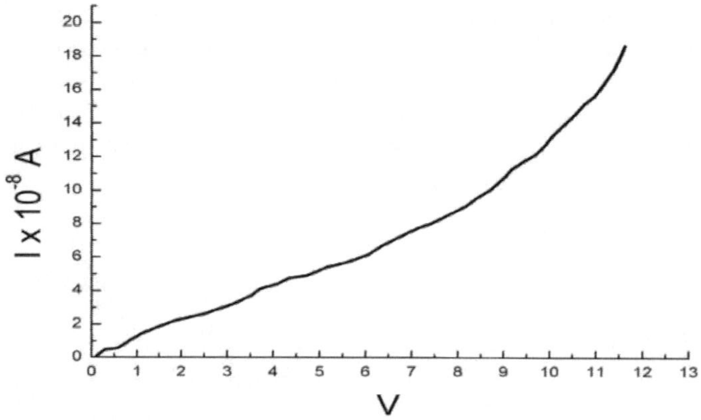

Figure 3. Volt-ampere characteristics (CVC) of the thin ZnO film.

4. Results and discussion

The processing of the CVC results with using the technique described in article [5] allows one to determine the mobility of charge carriers and the concentration of electrical activity defects.

A power-law dependence of the current is observed for the branches of the current-voltage characteristic: $I \sim U^b$. Reconstructing the CVC curve in double logarithmic coordinates, we can determine the value of the threshold voltage at which the transition to the quadratic dependence occurs current. Let us estimate the concentration n_t of small electrically active defects according to the calculated formula [5]:

$$n_t = 2.21\times 10^{20} \times U_0, \tag{1}$$

where U_0 – the threshold voltage of the transition to the current rise of the CVC characteristic, reconstructed in double logarithmic coordinates.

The value of the effective mobility of charge carriers in the zinc oxide layer at a voltage U_0 can be estimated from the calculated formula [5]:

$$\mu_{\text{eff}} = \frac{8IL^3}{9\varepsilon\varepsilon_0 U_0^2}, \tag{2}$$

where U_0 – threshold voltage of the CVC characteristic; I – current value at the inflection point; L – zinc oxide film thickness.

Concentration electrically active defects n_t make up $0.24\cdot 10^{19}$ cm^{-3}. Effective mobility of charge carriers in ZnO film make up 118 cm^2/V·s.

5. Conclusions

Synthesis of films by a reactive ion-plasma method on pre-heated quartz substrates allows obtaining nanostructured ZnO films with the presence of a hexagonal phase of zinc oxide and structural properties equivalent to the properties of ZnO epitaxial films. The photoluminescence spectra of thin ZnO films contain two peaks: narrow with a maximum at 380 nm and a broad peak with a maximum at 600 nm. The narrow spectral peak corresponds to the photoluminescence of excitons present in the bulk of a thin ZnO film. A wide spectral peak can be caused to the photoluminescence of structural defects present in the zinc oxide film. The shortwave peak has six times large the intensity compared to the intensity of the broad spectral peak. The shift to the long-wave region of a broad photoluminescence peak can be explained by the structural perfection of the synthesized zinc oxide film. Investigation of the current-voltage characteristics of thin ZnO films made it possible to estimate the mobility of charge carriers and to estimate the concentration of electrically active defects present in the volume thin film zinc oxide.

IOP Conf. Series: Journal of Physics: Conf. Series **857** (2017) 012010 doi:10.1088/1742-6596/857/1/012010

Investigation of the spectral dependences of photoluminescence showed that the maximum of the spectral peak of excitons luminescence is near 380 nm. This makes it possible to create on the synthesized films luminescent media suitable for use in scintillation devices and for use in UV photonics structures.

Acknowledgment
The work was supported by the Russian Foundation for Basic Research: № 16-07-00237.

References

[1] Endo H and Sugibuchi M 2007 *Appl Phys Lett* **90** 12196
[2] Novodvoski O A, Gorbatenko and L S Panchenko V Y 2009 *Semiconductors* **4** 439–44
[3] Ma L, Ma S and Chen H 2011 *Applied Surface Science* **257** 10036–41
[4] Jung-Shul L, Yan-Do K and Pung-Keun S 2008 *Journal Korean Physical Society* **53** 416–20
[5] Grigoryev L V, Kalmykov A E and Sorokin L M 2006 *Technical Phys Letter* **32** 33–42

IOP Conf. Series: Journal of Physics: Conf. Series **857** (2017) 012011 doi:10.1088/1742-6596/857/1/012011

Optical and photoelectrical properties of nanostructured thin ZnO films for UV-sensors

L V Grigoryev[1], S V Kulakov[2],V G Nefedov[2], O V Shakin[2], M L Grigoryeva[3] and S D Moskalenko[4]

[1] ITMO University, 197101, Saint Petersburg, Russia
[2] Saint-Petersburg University of Aerospace Instrumentation, 190000, Saint Petersburg, Russia
[3] School № 328, 192174, Saint Petersburg, Russia
[4] School № 197, 191123, Saint Petersburg, Russia

E-mail: grigoryev@oi.ifmo.ru

Abstract. The article presents the results investigations of the optical and photoelectric properties thin films zinc oxide obtained by the reactive ion-plasma method. It is shown that the optical and photoelectric properties of thin ZnO films has equivalent characteristics to the properties of single crystal zinc oxide and can be used to create UV-photoresistors.

1. Introduction

Thin ZnO films, obtained by the reactive ion-plasma method are referred to direct-gap semiconductors having a large band gap (more than 3.3 eV). This determines a lot of interesting properties of ZnO, which are widely used in solid-state devices of optoelectronics and integrated optics. Also, due to low optical losses in the visible spectral range, high mobility of charge carriers and the possibility of doping, thin ZnO films are widely used as an active material for UV-photodetector [1–3].

In optoelectronics, are used various methods to grow thin zinc oxide films such as: a CVD process, pulsed laser ablation of a ceramic target from extra clean ZnO or electron beam deposition in a pulsed mode [4]. In comparison with them, the method of reactive ion-plasma sputtering of a metal target makes it possible to create ZnO films with the smallest structural changes in the film; in addition, this method allows doping of the film composition during its growth [5]. The purpose of this work was to investigation the optical and photoelectric properties of thin ZnO films obtained by reactive ion-plasma sputtering in argon-oxygen plasma of a Zn metal target.

2. Method preparation of thin films samples

The process of ion-plasma sputtering of the metallic target was carried out using a magnetron operating in a constant current mode. Process the synthesis of a zinc oxide film occurred in a gas atmosphere consisting of 30 % oxygen and 70 % argon. The pressure of the gas mixture under the cap of the sputtering installation did not exceed 101 Pa, and the substrate temperature did not exceed 423 K. The metal target was made from zinc grade extra clear. The substrate was making of quartz particularly transparent in the UV-range. Plates of quartz were polished in grade 12 cleanliness classes and were thickness of 2 mm.

The spraying of Al electrodes occurred on the magnetron installation. According to the data of microscopic studies, the thickness of the ZnO film was 1 μm. The thickness of Al film contact was 0.5 μm. X-ray diffraction investigation of synthesized zinc oxide films showed the presence of the hexagonal structure of ZnO films and its preferential orientation in the (001) direction. The dimensions of ZnO crystallites did not exceed 13 nm. The drawing of the investigated ZnO samples with the scheme by Al electrodes is shown in figure 1.

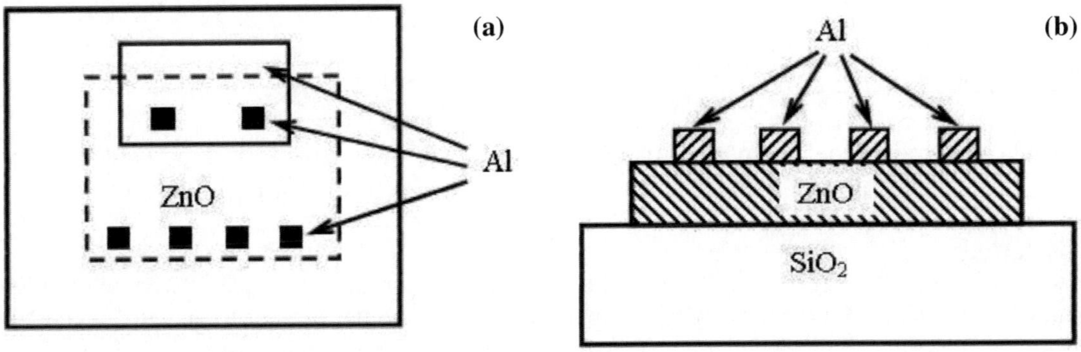

Figure 1. Appearance of the samples under study, top view (**a**) and side view (**b**).

3. Investigation of optical and photoelectrical properties in ZnO films

Investigation of the photoluminescence spectra of ZnO films was carried out on a Perkin Elmer Lambda 650 spectrometer at spectral range from 200 to 800 nm. The measurements were carried out at a temperature of 300 K. Figure 2 shows the optical properties of a thin film ZnO.

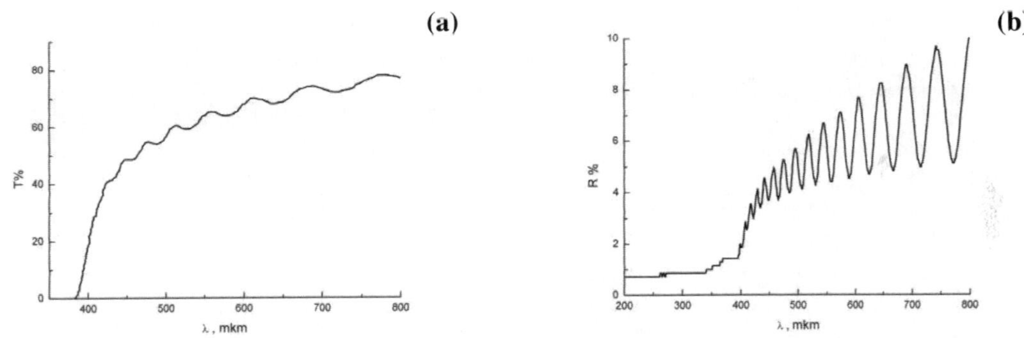

Figure 2. Transmission spectrum of ZnO film (**a**), reflection spectrum of ZnO film (**b**).

Using the experimental data of the spectral transmission and reflection dependences, the absorption coefficient were constructed using the technique described in [6] (figure 3).

To determine the width of the forbidden band of a polycrystalline ZnO film, the absorption spectrum was rearranged in the coordinates (αhν) 2-hν (figure 3). The value of the optical band gap E_g was obtained by linear interpolation on the energy axis and amounted to 3.29 eV. The obtained value of E_g differs insignificantly from the width of the forbidden band of single-crystal ZnO, which is 3.33 eV [1]. Measurements of photoconductivity currents were carried out on an automated installation consisting of a UV-source on a xenon lamp and a monochromator MDR-41. The photoconductivity current was recorded with a Keithley 6485 picoampermeter. The voltage applied to the photoresistors structure was 10 V. Figure 4 shows the spectral dependence of the photoconductivity current. It is seen that the maximum of the photoconductivity current corresponds to a wavelength of 375 nm. The increase in photosensitivity in the investigated films by a factor of 3 in the 370–380 nm spectral band, in comparison with the value of the photosensitivity in the region of the long-wave UV-spectrum, makes it possible to create photoresistors and UV-photodiodes based on these thin films.

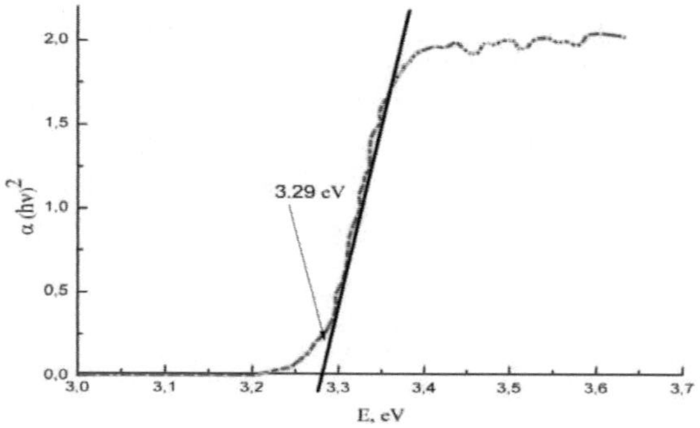

Figure 3. Absorption coefficient in film ZnO.

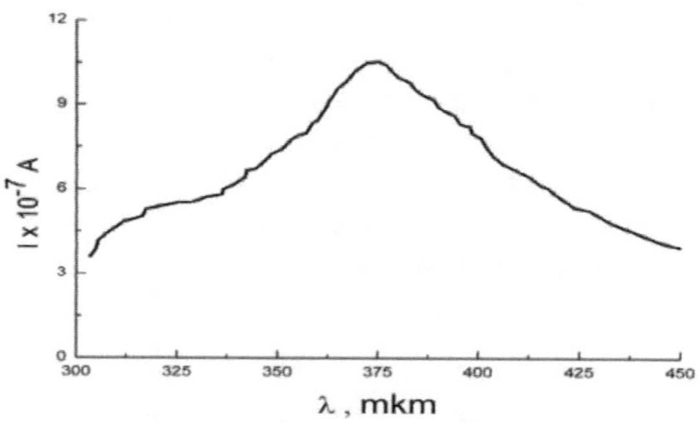

Figure 4. The spectral dependence of the photoconductivity current in a ZnO film.

4. Conclusions

Synthesis of thin films by a reactive ion-plasma method on pre-heated quartz substrates allows obtaining nanostructured ZnO films with the presence of a hexagonal phase of zinc oxide. The band gap for the synthesized ZnO film is 3.29 eV. This is small different from the value of the band gap for a single crystal or epitacsial film ZnO. Thus, zinc oxide films synthesized by the reactive ion-plasma method have low losses in the visible spectral range and have strong absorption in the UV-part of the spectrum range.

An investigation of the spectral dependences of the photoelectric current showed that the maximum of the spectral sensitivity is near 375 nm. This makes it possible to create photoresistors working in the UV spectral range on synthesized films.

Acknowledgment

The work was supported by the Russian Foundation for Basic Research: № 16-07-00237.

References

[1] Slobodchikov S I and Salikhov X M 1999 *Semiconductors* **33** 435–5
[2] Nikitin S A 2001 *Semiconductors* **7** 1329–33
[3] Mezdrogina M M and Vinogradov A Y 2016 *Semiconductors* **50** 1326–34
[4] Link M Veber J Schreiter M 2007 *Sensor Actuators B* **121** 372–5
[5] Ma L Ma S Chen H 2011 *Applied Surface Science* **257** 10036–41
[6] Yao B D Chang V F Wang F 2002 *Applied Phys Letters* **81** 757–3

Interaction of bilayer graphene with MnO(111) surface films

V V Ilyasov, I G Popova and I V Ershov

Don State Technical University, 344000, Rostov-on-Don, Russia

E-mail: viily@mail.ru, inna111109@rambler.ru, thijd@mail.ru

Abstract. Using density functional theory, we have carried out an ab initio simulation of magnetism and adsorption of Bernal-stacked (AB) bilayer graphene (BLG) on the polar manganese monoxide (111) surface terminated by oxygen. We have carried out a systemic study of the local atomic reconstructions of the bilayer graphene/MnO_x(111) interface, and their electronic properties. For the first time, the bond length and the adsorption energy have been found for different reconstructions of the surface atomic structure in BLG/MnO_x(111) systems.

1. Introduction

Bernal-stacked (AB) bilayer graphene is a most important two-dimensional carbon nanostructure for creation of photonic and spintronic devices, due its remarkable physical properties [1]. Over the last years, high-quality bilayer graphene has been synthesized on the surface of metals [1] and dielectrics [2], which opens the way for creation of real BLG-based devices. Bilayer graphene combined with magnetics, such as, e.g., manganese monoxide films, may have a potential for creation of new devices. Earlier [3], it was shown that disruption of the oxygen stoichiometry in the interface surface leads to a significant reordering of the local atomic and electronic structures of the surface atoms of carbon, oxygen and manganese in the SLG/MnO_x(111) system. However, the processes of chemisorption and magnetism in the BLG/MnO_x(111) systems have not been considered earlier. In this work we used first-principles analysis to study different variants of bilayer graphene adsorption on the oxygen-delimited substrate containing oxygen vacancies. We made use of the density functional theory (DFT) to study the adsorption of graphene, the structural and electronic properties of the BLG/MnO_x(111) systems.

2. Model and method

The theoretical model of BLG/MnO(111) system is built on the three-period slab scheme. We used a supercell of 68 atoms terminated by a monolayer of oxygen atoms and containing unit cell (2×2) of MnO in the (111) plane. Previous theoretical study shows that this surface is one of the possible thermodynamically stable forms of existence of MnO polar surfaces [4]. Figure 1(a) shows a fragment of the slab simulating the BLG/MnO(111) interface. BLG is centered on the substrate's oxygen atom. This configuration requires the minimum energy in comparison with the other positions. The MnO_x(111) slab consists of 11 nonequivalent planes in the (111) direction, including 4 manganese, 5 oxygen, and one hydrogen planes. The width of vacuum gap is 12 Å, which is allowed to exclude any interaction between the translations of the slab in the (111) direction.

We performed the self-consistent calculations of the total energy based on the DFT method with the pseudopotential approximation, which is implemented in the Quantum-Espresso package [5]. The

IOP Conf. Series: Journal of Physics: Conf. Series **857** (2017) 012012 doi:10.1088/1742-6596/857/1/012012

generalized gradient approximation (GGA) for exchange-correlation energy in PBE parameterization was used with norm-conserving pseudopotentials. The technique of calculations is similar to the work [3].

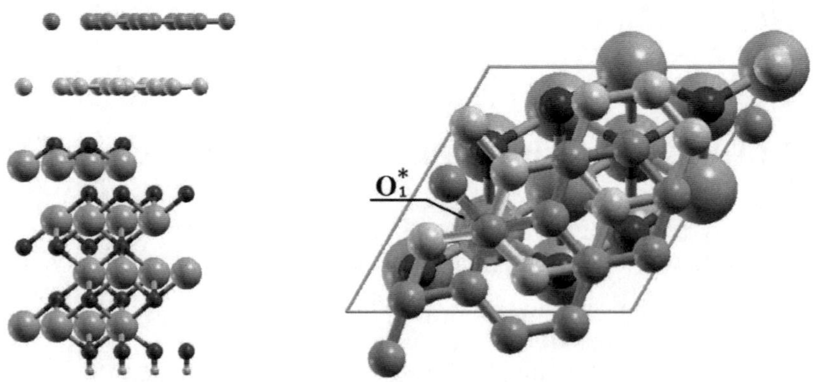

(a) **(b)**

Figure 1. Supercell model of BLG/MnO_x(111) interface with one oxygen vacancies in the interface after relaxation (side **(a)** and top **(b)** views).

3. Results and conclusions

We have performed relaxation of the two upper double atomic layers (Mn, O) of the slab consisting of manganese monoxide and two graphene layers for the atomic configurations presented in figure 1. Initially, the BLG was placed at the distance of 2 Å from the MnO_x(111) surface. The lower two double layers (Mn, O) and one layer of hydrogen below in the BLG/MnO_x(111) system were "frozen". Relaxation was continued until the sum of all forces in the system became less than 0.001 eV/Å. The equilibrium lattice constants were found, and the atomic positions of the bilayer graphene carbon atoms and of the atoms in the upper two double layers of manganese monoxide were determined. The bond lengths between the carbon atoms in BLG, and the distances between the C and O atoms in the BLG/MnO_x(111) interface after relaxation were also determined. It should be noted that the BLG adsorbed on the reconstructed surface of oxygen-delimited substrates is located at different distances $d_{C\text{–}O}$ from the oxygen atoms in the interface (see table 1). In this study the interface surface changed the oxygen stoichiometry from $x = 0.75$ to 0.5. Such choice of nonstoichiometry of the substrates is due to their wide band gaps found by preliminary calculations [3], which may be important for creation of graphene devices.

Table 1. Vertical distance between the bilayer slabs $d_{C\text{–}C}$; between the graphene lower layer and the oxygen layer $d_{C\text{–}O}$; between the first oxygen layer and the manganese layer; effective charges on the interface atoms in the considered configurations

Configurations	$d_{C\text{–}C}$	$d_{C\text{–}O}$	$d_{O_1\text{–}Mn}$	Effective charge, e		
		Å		Mn	O	C_{up}/C_{down}
A	3.145	2.846	0.964	0.88	–0.51	0.013/–0.019
B	3.157	2.830	1.120	0.82	–0.59	0.010/–0.020

In the configuration A, the interface between the BLG and the MnO_x(111) substrate undergoes a geometrical reconstruction shown in figure 1(a), the oxygen coverage rate being $\Theta = 0.75$ ML. At the oxygen coverage rate of $\Theta = 0.5$ ML in the interface (configurations B) the distance $d_{C\text{–}O}$ is reduced to the value of 2.83 Å. Simultaneously, the distances $d_{C\text{–}C}$ and $d_{O_1\text{–}Mn}$ increase (see table 1). Note that as the number of oxygen vacancies in the outer layer grows, the distance between graphene and the near-surface monolayer of manganese increases, which is possibly connected with the nature of the chemisorption mechanisms realized in each separate configuration. The hydrogen-oxygen bond length on the lower surface in the considered configurations was practically constant and equal to $d_{O_5\text{–}H}$ = 0.96 Å. The results presented in table 1 characterize the local atomic structure of the considered configurations of the atomic surface in the BLG/MnO_x(111) interface.

Adsorption of BLG on the nonstoichiometric polar surface of MnO_x(111) leads to a change in the interplanar distances between the upper layers of the oxygen and manganese atoms in the substrate $d_{O\text{–}Mn}$, between the bilayer graphene and the substrate $d_{C\text{–}O}$, and between the carbon layers $d_{C\text{–}C}$ (see table 1). The nature of the observed reordering of the atomic structure in the BLG/MnO_x(111) and O/BLG/MnO_x(111) interfaces connected with adsorption of bilayer graphene can be understood after a detailed study of the thermodynamics and the electronic structure of each of the considered interface models. Table 2 shows the adsorption energy values for BLG on substrate, in the two considered configurations differing in the oxygen coverage rate of the BLG/MnO_x(111) system interface. It also contains our DFT calculations of the adsorption energy of monolayer graphene on similar substrates [3]. Analysis of the results presented in table 2 shows that when nonstoichiometry of the oxygen interface layer is 75 %, the BLG adsorption energy value is equal to E_{ads} = 34 meV/atom. This value is 1.8 times greater than the adsorption energy of monolayer graphene on the same substrate.

Table 2. Adsorption energy E_{ads} of SLG and BLG on crystalline substrates; distance between graphene and substrate $d_{C\text{–}R}$; effective charges on the atoms of carbon δQ_C and the nearest-neighbor atoms δQ_{NNr}

Phase	$E_{ads}, \frac{\text{meV}}{\text{atom}}$	$d_{C\text{–}R}$, Å	δQ_C, e	δQ_{NNr}, e
BLG/MnO_x(111), (x = 0.75)	34	2.85	–0.019	–0.51
SLG/MnO_x(111), (x = 0.75)	19 [3]	2.83	–0.014	–0.49
SLG/MnO$(111)_O$	86 [3]	2.69	–0.029	–0.42
BLG/MnO_x(111), (x = 0.5)	26	2.83	–0.020	–0.59
SLG/MnO_x(111), (x = 0.5)	46 [3]	2.8	–0.018	–0.57
SLG/$Al_2O_3(0001)_O$	120 [6]	2.8	–	–

The vertical distance between the bilayer graphene and the substrate in the configuration A was 2.85 Å. The difference in the distance between the monolayer or bilayer graphene and the substrate was less than 1 %. When the interface symmetry is further reduced to 50 %, the distance between the BLG and the substrate decreases to 2.83 Å, while the adsorption energy also reduces to 26 meV/atom. This reduction of the BLG adsorption energy can be accounted for by charge transfer between the atoms of carbon and oxygen (see table 1). This process is influenced by reordering of the local atomic structure of the interface, in particular, the 17 % increase in the distance between the interface oxygen layer and the manganese layer, and, as a result, the 7 % decrease in charge transfer from the Mn atoms to the O atoms. In our view, the use of the exchange correlation functional in the LDA form yields an inflated value of the adsorption energy (120 meV/atom [6]). Analysis of results in table 2 shows that the use of the PBE-D2 approximation for description of the exchange and correlation energy is justified and yields comparable values for the energy of graphene adsorption on dielectric and semiconducting substrates [7, 8].

References

[1] Hao Y, Wang L, Lui Y et al. 2016 *Nature Nanotechnology* **11** 426
[2] Yan, Z, Peng Z, Sun Z et al. 2011 *ACS Nano* **5** 8187
[3] Ilyasov V V, Popova I G, Ershov I V et al. 2017 *Diamond & Related* Materials **74** 31
[4] Franchini C et al. 2006 *Phys. Rev. B* **73** 155402
[5] Giannozzi P, Baroni S, Bonini N et al. 2009 *J. Phys.: Condens. Matter.* **21** 395502
[6] Entani Sh, Antipina LYu, Avramov PV et al. 2015 *Nano Res.* **8** 1535

[7] Ilyasov V V, Nguyen C V, Ershov I V and Nguyen N H 2015 *J. Appl. Phys.* **117** 174309
[8] Ilyasov V V, Nguyen V C, Ershov I V and Nguyen N H 2015 *RSC Advances* **5** 49308

Modification of the structure and properties of commercially pure titanium through nitriding and subsequent TiN coating deposition in a single vacuum cycle

Yu F Ivanov, O V Krysina, E A Petrikova, V V Shugurov, O S Tolkachev, A D Teresov and N N Koval

Institute of High Current Electronics SB RAS, 634055, Tomsk, Russia

E-mail: yufi55@mail.ru

Abstract. The modification of titanium by ion plasma methods consisting of hard coatings deposition on a specimen surface subjected to nitriding is carried out. It is shown that complex modification of the titanium in a single vacuum cycle is followed by formation of multilayered multiphase structure which tribological properties multiply exceed the corresponding properties of the material treated on two vacuum separated setups.

1. Introduction

It is shown that the formation of hard and superhard coatings on the surfaces of material allows to increase significantly the strength and tribological properties of the material [1–10]. The main problem of such method consists in insufficiently high level of adhesion forces between coating and substrate. For the increase of the adhesion forces the different transitional layers, which form the multilayered coatings, etc., are deposited. That raises significantly the price of technology and not always results in desirable results [11, 12].

The purpose of the present work is the comparative analysis of the revealed regularities of transformation of structure and properties of commercially pure titanium surface layers which are implemented as a result of ion plasma treatment at the use of two vacuum separated setups and complex setup for electron-ion plasma finishing modification of metals and alloys in a single vacuum cycle.

2. Material and research technique

The research material was commercially pure VT1-0 titanium (0.25 Fe, 0.07 C, 0.1 Si, 0.04 N, 0.2 O, 0.01 H wt. %, the rest is Ti) [13]. The specimens had a form of plates (15×15×4.5 mm). Nitriding (650 °C, 3 h) was realized in plasma of low pressure gas discharge on the TRIO setup equipped with PINK plasma generator based on non-self-sustained arc discharge with a filament cathode. The plasma source allows to increase the plasma concentration at rather low discharge voltage, to regulate discharge current irrespective of its burning voltage and gas pressure in the working camera [14–16]. The synthesis of hard TiN coating of 0.5 μm thick on the nitrided specimen surface of commercially pure VT1-0 titanium was carried out on ion plasma QUINTA setup [17]. The complex modification of commercially pure VT1-0 titanium (nitriding and subsequent coating deposition) in a single vacuum cycle was realized on the laboratory COMPLEX setup for electron-ion plasma finishing treatment of metals and alloys which was developed and created in IHCE SB RAS (Tomsk, Russia) as a result of execution of RSF grant (project No. 14-29-00091). The treatment modes in a single vacuum cycle were identical to that used above for two vacuum separated setups.

The researches of modified material structure and properties were carried out by methods of scanning (SEM-515 Philips) and transmission diffraction (JEM-2100F) electron microscopies; X-ray diffraction analysis (XRD-7000s Shimadzu diffractometer). The microhardness was measured by Vikers method (PMT-3 microhardness tester, loads: 0.2 N and 0.5 N); wear resistance and friction coefficient were revealed by "Tribotechnic" tribometer (dry friction at the room temperature according to the "ball-on-disk" scheme (WC-8% Co, 6 mm; load of 5 N).

3. Results and discussion

The combined treatment on the above cycle, carried out on the laboratory COMPLEX setup, leads to the formation of a modified layer. The wear resistance of the last one exceeds that of the initial material more than 30 times, the friction coefficient decreases by 1.3 times at the same time. Modification of titanium on two vacuum separated setups leads to decrease of the material wear resistance in ≈0.7 times with a constant friction coefficient. The microhardness values of the surface layer of the modified commercially pure VT1-0 titanium are close and vary in the range of 5500–6000 MPa in the both cases.

The formation of titanium nitride with TiN structure and a face-centered cubic crystal lattice (space B1 group) and titanium nitride of Ti_2N composition with tetragonal crystal lattice (space $P4_2/mnm$ group) after nitriding on a TRIO setup at a temperature of 650° C for 3 h was found using X-ray phase analysis. The subsequent deposition of the nitride coating is accompanied by the formation of a multiphase state. The main phases formed after modification on the vacuum separated setups are α-Ti (≈35 vol. %) and TiN (≈65 vol. %). The main phases after nitriding and subsequent nitride coating deposition obtained on the laboratory COMPLEX setup are α-Ti (≈27 vol. %), β-Ti (≈20 vol. %), TiN (≈39 vol. %) and Ti_2N (≈14 vol. %). As a result of the researches carried out by X-ray diffraction analysis it was shown that microdistortions of the α-Ti crystal lattice of the "coating/substrate" system formed on the COMPLEX setup are ≈1.8 times lower compared with that formed on the two vacuum separated setups. In addition, the value of the microdistortions of TiN coating crystal lattice is practically independent on the number of setups used for titanium modification.

The defective substructure and phase composition of the modified titanium layer were analyzed by transmission electron diffraction microscopy. It is established that a multilayer structure is formed as a result of nitriding. The typical image of that is shown in figure 1. There are the thin (40–50 nm) surface layer (figure 1(b), (c), the layer is indicated by an arrow), the transition layer (figure 1(a), the layer is indicated by arrows) with thickness of 450–500 nm, and the bulk of the material. The surface layer has a nanocrystalline structure (figure 1(c)). The dimensions of the crystallites forming the surface layer vary in the range of 5–10 nm. The transition layer has a columnar structure (figure (1)); the cross dimensions of the columns vary from 150 to 500 nm.

The analysis of microelectron pattern obtained from the modified layer revealed the presence of titanium nitrides reflexes of TiN composition. The surface layer with a nanocrystalline structure is formed by titanium nitride of TiN composition. The transition layer with a columnar structure is formed by titanium nitride of $TiN_{0.3}$ composition.

The subsequent deposition of a nitride coating on the nitrided specimen which is carried out on the COMPLEX setup leads to formation of the "coating/substrate" system. This system keeps a multilayer structure and consists of the surface layer with thickness of 120–150 nm (figure 2, layer 1), thin intermediate layer with thickness of 15–25 nm (figure 2, the layer is specified by an arrow) and the transition layer ≈400 nm thick (figure 2, layer 2). The surface layer, as well as the intermediate, has nanocrystalline structure with the crystal grains average sizes of 8.8 nm. The transitional layer has the columnar structure. The analysis of the electron-microscopic diffraction patterns obtained from the modified specimen shows that the surface layer is titanium nitride of TiN structure, the transition layer is titanium nitride of Ti_2N structure.

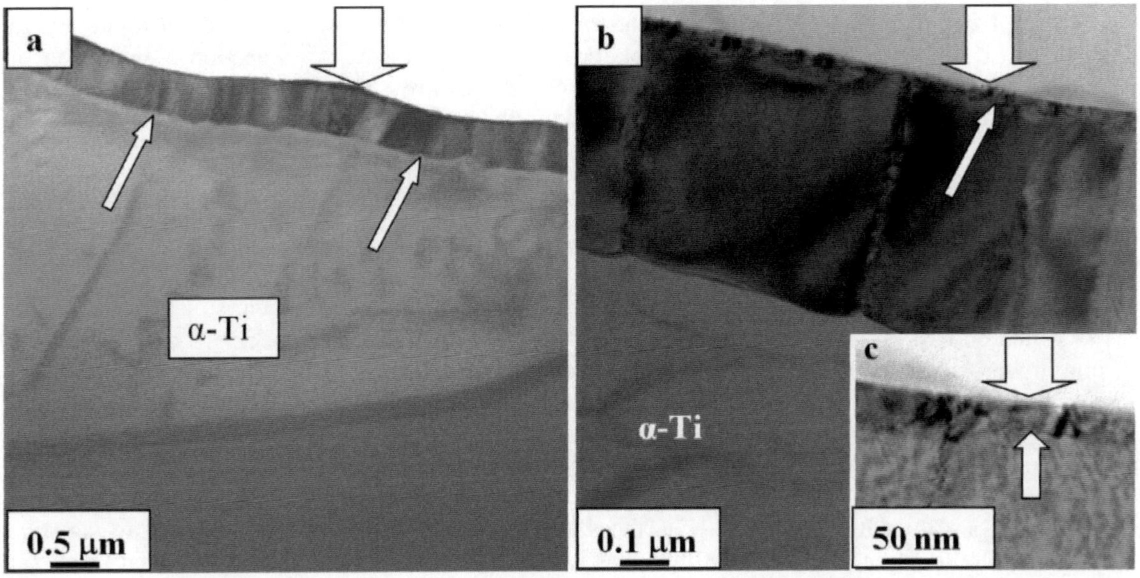

Figure 1. The electron microscopic image of the structure of the surface layer of commercially pure titanium subjected to nitriding in a plasma of a low-pressure gas discharge. A large arrow indicates the nitriding surface, small arrows on **(a)** indicate a modified layer with a columnar structure; on **(b)** and **(c)** – a thin nanocrystalline layer.

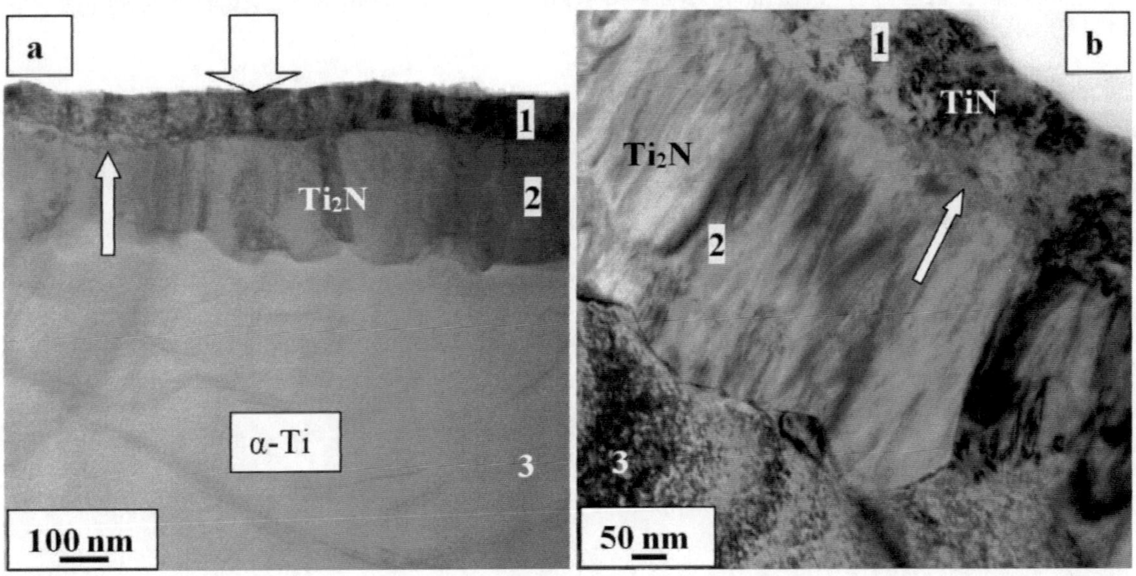

Figure 2. The electron microscopic image of a cross-section of the «hard TiN coating/VT1-0 substrate after nitriding» system formed on the COMPLEX setup. The top arrow on **(a)** specifies the surface of modification; the lower arrow on **(a)** and **(b)** – the interlayer. Transmission electron diffraction microscopy.

Comparing the results of the electron-microscopic diffraction analysis of structure of the modified layer formed as a result of nitriding and the subsequent deposition of a nitride coating it is possible to conclude that the coating deposition is followed by saturation of a titanium with nitrogen. That led to transformation of titanium nitride of $TiN_{0.3}$ structure to titanium nitride of Ti_2N structure.

The use of two vacuum separated setups at formation of the "coating/substrate" system leads to formation of a TiN structure coating with columnar structure and the cross sectional dimension of columns in the range of 100 nm. The columns have nanocrystalline structure with a crystal grains size of 50–80 nm. That significantly distinguishes them from the coating deposited on COMPLEX setup in a single vacuum cycle.

4. Conclusions

It is established that modification of commercially pure VT1-0 titanium specimens on the COMPLEX setup is followed by formation of the multilayer multiphase nanocrystalline system. That has the tribological parameters which are multiply exceeding the corresponding parameters of material which was modified on two vacuum separated setups. It is possible to assume that the main external reasons causing higher values of tribological properties of commercially pure VT1-0 titanium, modified in a single vacuum cycle on the COMPLEX setup compared with the titanium treated on two vacuum separated setups is the following: (1) better vacuum conditions in the COMPLEX setup, which provides an ultimate vacuum of $\approx 8\cdot 10^{-4}$ Pa against $\approx 5\cdot 10^{-3}$ Pa in separate setups and suppresses the formation of undesired oxide and oxynitride films on the material surface; (2) possibility to avoid contamination and oxidation of nitrided surfaces in the COMPLEX setup, as opposed to separate setups in which the probability of these processes is high as specimens are to be transferred from one setup to the other.

Acknowledgments

The work was supported by the Russian Science Foundation (project No. 14-29-00091). Modification of titanium specimens on two vacuum separated setups was carried out with partial financial support of the RFBR grant (project No. 16-58-00075-Bel_a).

References

[1] Veprek S, Veprek-Heijman M G J, Karvankova P and Prochazka J 2005 *Thin solid films* **476** 1–29
[2] Musil J, Baroch P and Zeman P 2008 *Research Singpost: India* 1–34
[3] Musil J 2000 *Surface and coatings technology* **125** 322–30
[4] Veprek S and Reiprich S 1995 *Thin Solid Films* **268** 64–71
[5] PalDey S and Deevi S C 2003 *Materials Science and Engineering* **A361** 1–8
[6] Ivanov Yu F, Koval N N, Krysina O V, Baumbach T, Doyle S, Slobodsky T, Timchenko N A, Galimov R M and Shmakov A N 2012 *Surface and Coatings Technology* **207** 430–4
[7] Li Z G, Miyake S, Kumagai M, Saito H and Muramatsu Y 2004 *Surface and Coatings Technology* **183** 62–8
[8] He J L, Setsuhara Y, Shimizu I and Miyake S 2001 *Surface and Coatings Technology* **137** 38–42
[9] Matenoglou G M, Evangelakis G A, Kosmidis C and Patsalas P 2007 *Rev. Adv. Mater. Sci.* **15** 38–43
[10] Pinakidou F, Paloura E C, Matenoglou G M and Patsalas P 2010 *Surface and Coatings Technology* **204** 1933–6
[11] Berlin E V and Seidman L A 2010 *Ion plasma processes in thin-film technology* (Moscow: Technosphera) 528 [in Russian]
[12] Andreev A A, Sablev L P, Shulaev V M and Grigoriev S N 2005 *Vacuum-Arc Devices and Coatings* (Kharkov: NSC KIPT Press) 236 [in Russian]
[13] Borisova E A, Bochvar G A and Brun M Ya 1980 *Titanium alloys. Metallography of titanium alloys* (Moscow: Metallurgy) 464 [in Russian]
[14] Lopatin I V, Akhmadeev Yu H and Koval N N 2015 *Review of Scientific Instruments* **86** 103301
[15] Vintizenko L G, Grigoriev S V, Koval N N, Tolkachev V S, Lopatin I V and Schanin P M 2001 *Russian Physics Journal* **44** 9 927–36

[16] Borisov D P, Goncharenko I M, Koval N N and Schanin P M 1998 *IEEE Transactions on Plasma Science* **26** 6 1680–4
[17] Koval N N, Ivanov Yu F, Lopatin I V, Akhmadeev Yu H, Shugurov V V, Krysina O V and Denisov V V 2015 *Russian Journal of General Chemistry* **85** 5 1326–38

IOP Conf. Series: Journal of Physics: Conf. Series **857** (2017) 012014 doi:10.1088/1742-6596/857/1/012014

Research of the texture and residual stress in E110 after surface modification and hydrogen exposure

S V Ivanova[1], S Ya Betsofen[2], A A Lozovan[2] and D E Molostov[2]

[1] National Research Nuclear University, MEPhI, 115409, Moscow, Russia
[2] Moscow Aviation Institute (National Research University), 121552, Moscow, Russia

E-mail: s.betsofen@gmail.com

Abstract. The effect of modifying the surface of components from the zirconium alloy E110 by pulsed carbon ions, irradiating electrons and titanium ions in the oxygen medium on the formation of a structural-phase and stress state in them and the patterns of interaction of surface modification with hydrogen were studied. Treatment by carbon and titanium ions does not change the basal texture of the initial E110 sheet. Irradiation by electrons leads to a significant weakening of the basal texture of the initial sheet. Holding in hydrogen leads to the formation of $ZrH_{1.5}$ hydride in samples treated by carbon and titanium ions, but in samples, irradiated by electrons even traces of hydride were not detected. It is shown that surface modifications are accompanied by a transition from the compressive stresses of the initial product (100–120 MPa) to tensile stresses from 100 to 150 MPa. It was shown that after carbon and titanium ion modification the hydrogen exposure leads to a change in the stress state from tension to compression (20–70 MPa). In samples irradiated by electrons tensile stresses (70 MPa) were preserved.

1. Introduction

Toughening the working conditions of modern power plants increases the rate of corrosion processes in zirconium products and increases the level of their hydrogenation [1]. Crystallographic texture has a significant effect on the performance of zirconium components [2–7]. It was also shown [8–10] that during the formation of oxide layers on zirconium products at the interface with the metal (Me/O), a dense oxide layer is formed that contains, in addition to the monoclinic 8 % tetragonal oxide, the amount of which decreases as you move away from the interface and go to the loose outer oxide layer. In this case, the tetragonal oxide has a pronounced texture (100) $[001]_t$, which is related to the monoclinic oxide (m) by the orientation relationship (OR): $(10\bar{3})_m//(100)_t$ or $(10\bar{3})_m//(101)_t$ and $[100]_t//[010]_m$ [9].

In addition to the texture, the properties of the oxide layer depend on the level of the compressive stresses, which decrease as the sections are removed from the boundary Me/O. It is known that the morphology and texture of hydrides depends on the texture of the substrate. Most often between them is realized OR: $(0001)_{\alpha\text{-Zr}}//\{111\}_{\delta\text{-ZrH1.5}}$ [11, 12]. However, under the influence of tangential tensile stresses, other ORs can arise, especially $\{10\bar{1}1\}_{\alpha\text{-Zr}}//\{111\}_{\delta\text{-ZrH1.5}}$. It is important to note that the change in the OR is accompanied by a reorientation of the hydrides from the tangential to the radial direction of the tubular products. In the present work we investigated the effect of surface modification and the subsequent hydrogen saturation of the alloy E110 on the formation of texture and residual macro stresses therein.

2. Texture and hydride formation in E110

Three variants of modifying the surface of the alloy E110 (Zr – 1 % Nb) were realized: treatment with carbon ions with a pulse energy of 1 J/cm^2, n = 3 (iC); electron irradiation (El) and irradiation with titanium ions in an oxygen medium (Ti). The texture and residual stresses were determined by X-ray diffraction on a DRON-4 diffractometer in filtered CuKα radiation. Residual stresses were determined by the "sin$^2\psi$" method for the reflex $(10\bar{1}5)$ with "turn" angles ψ = 0, –20, –40, –50°.

Figures 1–3 show the X-ray diffraction patterns of zirconium specimens after various variants of surface modification and hydrogen saturation. Treatment by carbon ions according to the iC regime (figure 1(a)) and irradiation with titanium ions in the oxygen medium according to the Ti regime (figure 2(a)) practically does not change the basal texture of the initial sheet from the E110 alloy. Irradiation with electrons leads to a significant weakening of the basal texture of the initial sheet (figure 3(a)).

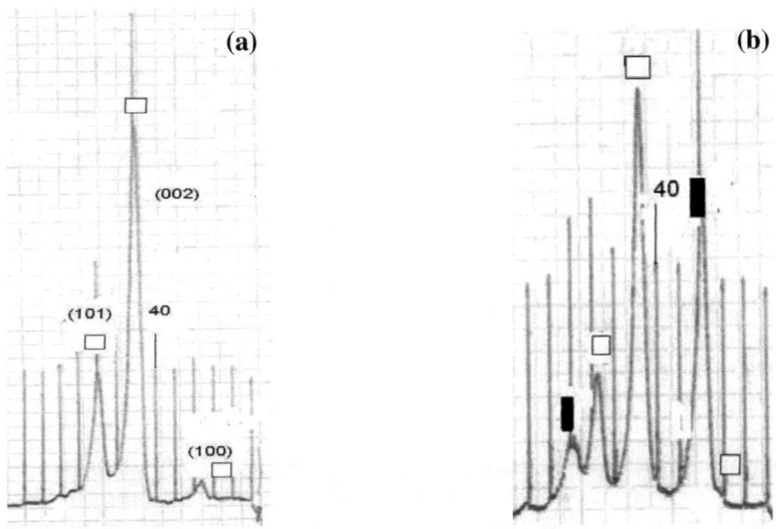

Figure 1. The X-ray patterns for the E110 product after surface treatment by carbon ions (iC) **(a)** and hydrogenation **(b)**: □ – α-Zr; ▌ – δ-$ZrH_{1.5}$.

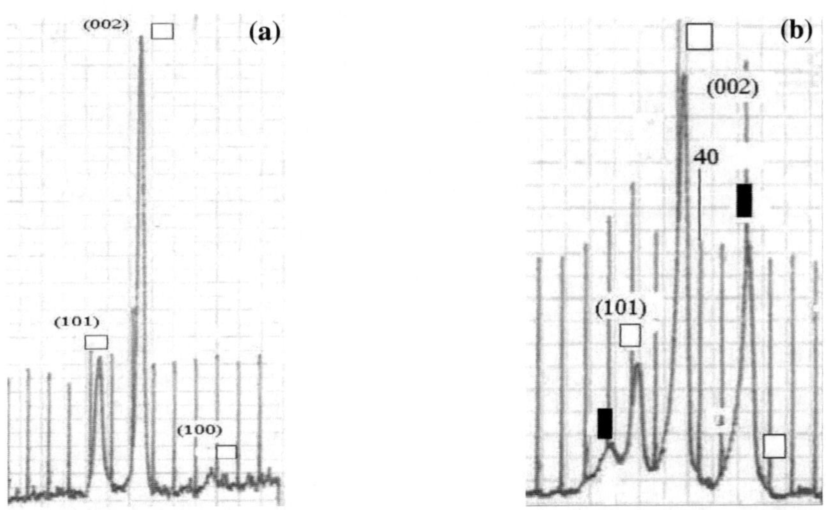

Figure 2. The X-ray patterns for the E110 product after surface treatment by titanium ions in the oxygen medium (Ti) **(a)** and hydrogenation **(b)**: □ – α-Zr; ▌ – δ-$ZrH_{1.5}$.

Hydrogen exposure leads to the formation of $ZrH_{1.5}$ hydride in samples treated in iC and Ti modes, in which the texture of the original material has not changed (figures 1(b), 2(b)). At the same time, the intensive hydride reflex (111) it is observed (figures 1(b), 2(b)), which agrees with the basal texture of zirconium and the orientation relationship between the metal and hydride $(0001)_{\alpha\text{-}Zr}//\{111\}_{\delta\text{-}ZrH1.5}$. In samples irradiated by electrons in which weakening of the initial basal texture was observed, even traces of hydride were not detected (figure 3(b)).

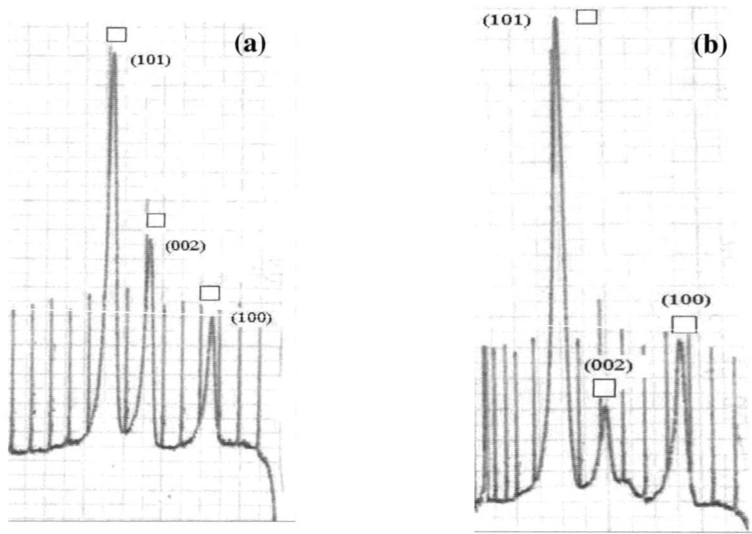

Figure 3. The X-ray patterns for the E110 product after electron irradiation (El) **(a)** and hydrogenation **(b)**: □ – α-Zr.

3. Residual stress formation after surface modification and hydrogen exposure

Figure 4 shows the results of measuring residual stresses in zirconium products. It is shown that all three methods of surface modification are accompanied by a transition from the compressive stresses of the initial product (100–120 MPa) to tensile stresses, the magnitude of which varies from 100 to 150 MPa (figure 4(a)).

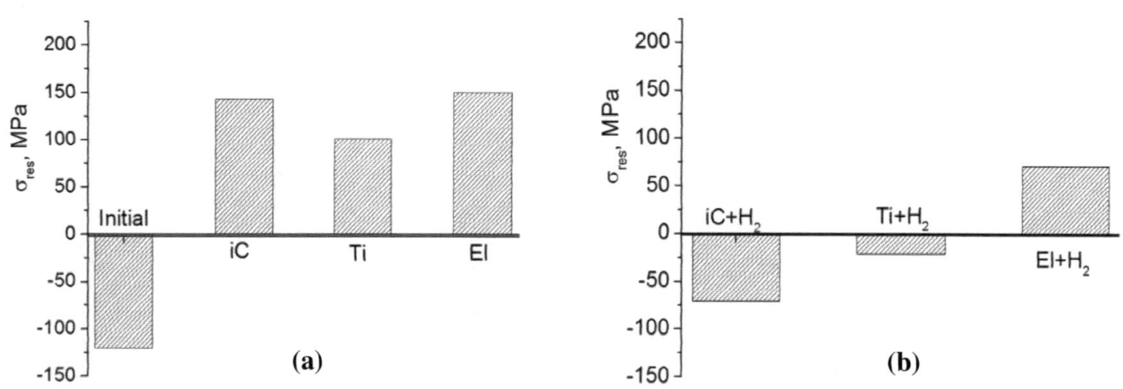

Figure 4. The magnitude of residual stresses for various variants of modifying zirconium products **(a)** and subsequent hydrogenation **(b)**.

It was found that in specimens treated in the iC and Ti regimes in which the original basal texture does not change, holding in hydrogen leads to the formation of hydride, which is accompanied by a change in the stress state from tension to compression (20–70 MPa), figure 4(b), which is a natural consequence of the volume increase as a result of hydride formation. At the same time, tensile stresses

(70 MPa) were preserved in samples irradiated by electrons (figure 4(b)) in which the basal texture was weakened and hydrogen exposure did not lead to the hydride formation.

It is likely that the formation of hydrides in samples modified by titanium and carbon ions indicates a lower permeability of their surface layers by hydrogen, possibly in connection with the formation of low-permeability oxide layers, which is facilitated by the basal texture of the metal substrate that survived after treatment. In addition, the presence of compressive stresses in the surface layers after these treatments prevents the penetration of hydrogen, figure 4(b).

4. Conclusions

It was experimentally discovered that treatment by carbon ions and titanium ions in an oxygen medium does not change the basal texture of the initial E110 sheet. Irradiation by electrons leads to a significant weakening of the basal texture of the initial sheet.

It was shown that a hydrogen exposure leads to the formation of $ZrH_{1.5}$ hydride in samples treated in iC and Ti ions, but in samples irradiated by electrons even traces of hydride were not detected.

It was shown that surface modification is accompanied by a transition from the compressive stresses of the initial product (100–120 MPa) to tensile stresses from 100 to 150 MPa. It was found that in specimens treated by the iC and Ti ions in which the original basal texture does not change, hydrogen exposure leads to the formation of hydride, which is accompanied by a change in the stress state from tension to compression (20–70 MPa). However, tensile stresses (70 MPa) were preserved in samples irradiated by electrons, in which the basal texture was weakened and hydrogen exposure did not lead to the hydride formation.

Acknowledgments

This study was supported by the Russian Science Foundation (project No. 15-19-00241).

References

[1] Zinkle S J and Was G S 2013 *Acta Materialia* **61** 735–58
[2] Kearns J J 2001 *J. Nucl. Mater.* **299** 2171–4
[3] Murty K L and Charit I 2006 *Progress in Nuclear Energy* **48** 325–59
[4] Romero J, Preuss M and Quinta da Fonseca J 2009 *Acta Mater.* **57** 5501–11
[5] Gruber J A, Brown S A and Lucadamo G A 2011 *J. Nuclear Mater.* **408** 176–82
[6] Gey N, Humbert M, Gautier E and Béchade J L 2004 *J. Nucl. Mater.* **328** 137–45
[7] Daymond M R, Holt R A, Cai S, Mosbrucker P and Vogel S C 2010 *Acta Mater.* **58** 4053–66
[8] Szpunar J A, Qin W, Li H and Kumar K 2012 *J. Nucl. Mater.* **427** 343–9
[9] Lin J, Li H, Nam C and Szpunar J A 2004 *J. Nucl. Mater.* **334** 200–6
[10] Akhiani H and Szpunar J A 2013 *Appl. Surf. Sci.* **285** 832–9
[11] Santisteban J R, Vicente-Alvarez M A, Vizcaíno P, Banchik A D and Almer J D 2010 *Acta Mater.* **58** 6609–18
[12] Vicente Alvarez M A, Santisteban J R, Domizzi G and Almer J 2011 *Acta Mater.* **59** 2210–20

IOP Conf. Series: Journal of Physics: Conf. Series **857** (2017) 012015 doi:10.1088/1742-6596/857/1/012015

Computer experiment on studying the properties of thermally sprayed alumina coatings with lamellar structure

V I Jordan[1], O P Solonenko[2] and V A Blednov[2]

[1] Altai State University, Barnaul, Russia
[2] Khristianovich Institute of Theoretical and Applied Mechanics, Siberian Branch of RAS, Novosibirsk, Russia

E-mail: jordan@phys.asu.ru, solo@itam.nsc.ru

Abstract. Results of stochastic modeling of lamellar structure formation and characteristics (microstructure, porosity, roughness) of ceramic coatings made of corundum powders (α-phase of Al_2O_3) are given in conditions typical for atmospheric plasma spraying (APS), and also detonation and supersonic flame spraying (HVOF). Comparison of characteristics of coatings for two model splats morphologies at their random laying on the base surface is presented.

Thermal spraying of coatings made of metal oxides powders (Al_2O_3, ZrO_2, TiO_2, Cr_2O_3, etc.) is widely used in industry for application of protective surface layers of various functionality. With that, increase in quality and improvement of structure of the sprayed materials in many ways depend on comprehension of the processes taking place at stochastic laying of splats as a result of which lamellar structure of coating is formed.

Despite the fact that a considerable amount of experimental data about the functional characteristics of thermally sprayed coatings is gathered so far, this data is insufficiently systematized. Thereof, there is no comprehensive approved idea of interrelation between spraying conditions and properties of sprayed coatings. In this regard, the computational technology, including program code, for modeling of lamellar structure formation of thermally sprayed coatings and their functional characteristics are presented in paper [1], which use, in our opinion, will promote more targeted design of the required coatings. Besides the problem-oriented core, the software includes several subsystems, in particular, the visualization subsystem of structure and surface morphology of coatings, and also the expandable referral database on properties of materials.

The “SPLAT” subsystem allows calculating geometrical characteristics of splats (melt droplets, spread and solidified on the base surface) at the preset values of key physical parameters (KPPs): particles velocity, temperature and size prior to impact, temperature of base and morphology of its surface. Besides, with the help of this subsystem it is possible to define the subspace of initial space of KPPs in which the given set of requirements to splat is fulfilled. It is the basis for computer-aided design of coatings and it allows to physically justify formulated requirements to equipment and spraying mode. For calculation of thickness of individual splats in the vicinity of the stagnation point at impact with the base of metals droplets, oxide ceramics and cermets droplets at the preset values of KPPs, the experimentally approved models [2–4] are used. Cylinder approximation for final splat morphology, as well as equality of volumes of droplet and splat, allow finding the final diameter of splat (figure 1(a)). Besides, in the course of modeling of coating formation an opportunity for carrying out the deposition of equivalent splats with smooth periphery (further named “smooth splat”) is realized in the software

IOP Conf. Series: Journal of Physics: Conf. Series **857** (2017) 012015 doi:10.1088/1742-6596/857/1/012015

(figure 1(b)). For this purpose, as well as in the previous case, thickness of splats in the vicinity of stagnation point at spreading and solidification of droplet is calculated according to [2–4].

The "COATING" subsystem is based on an algorithm of stochastic laying of splats on the base with the surface geometry changing at spraying (Monte-Carlo method), therefore lamellar structure of coating is formed.

Figure 1. Microsections of model splats: **(a)** – cylindrical; **(b)** – equivalent smooth.

The use of specified two geometrical representations of splat (figure 1) allows estimating their influence on properties and structure of coatings on micro- and macrolevel (see figures 2–4 and figure 10). As it was noted above, in the absence of systematization of the experimental data and comprehensive approved idea of interrelation between the spraying modes and functional characteristics of the obtained coatings, the developed computational technology gives the chance to design them, estimate their functional characteristics (porosity structure, roughness, bonding strength with the substrate, cohesive interlayer bonding strength, etc.). Results of modeling the lamellar structure of coatings sprayed from corundum powder (α-phase of Al_2O_3) which is widely used at thermal spraying of wear-resistant, heat-protective, insulating, etc. are given below.

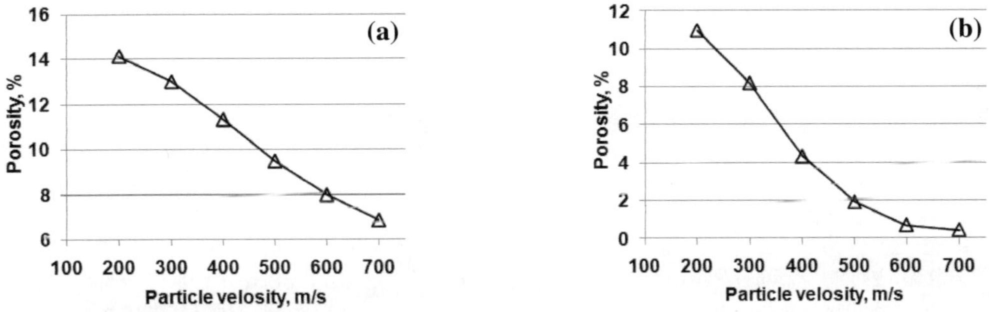

Figure 2. Porosity *P* of corundum coatings vs. velocity of particles U_p: **(a)** – cylinder approximation; **(b)** – smooth splat approximation.

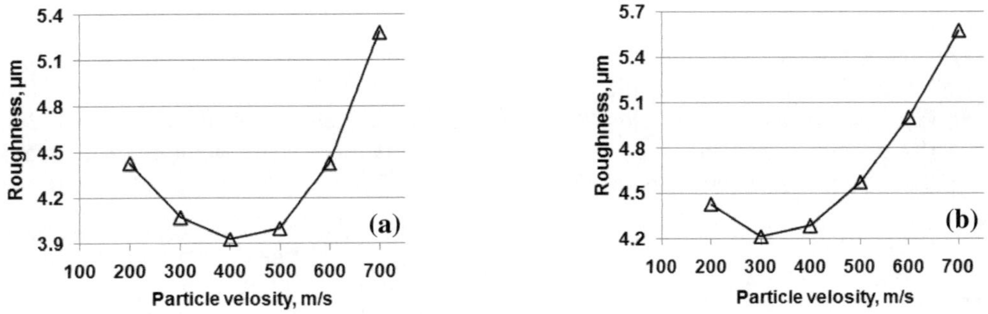

Figure 3. Roughness R_a of corundum coatings vs. velocity of particles U_p: **(a)** – cylinder approximation; **(b)** – smooth splat approximation.

Figures 2, 4 confirm the fact that when using splat geometry in the form of thin cylinder, in comparison with smooth splat approximation, coatings porosity is always overestimated in comparison with the experiment [5, 6]. Therefore, results of calculations only for smooth splat morphology are given bellow in figures 5–8. Calculations of roughness of coatings carried out, with other things being equal, for two model morphologies of splats agree both qualitatively and quantitatively (figure 3). The results of modeling presented in figures 2, 3 are obtained at the following KPPs values: temperature of particles

T_{p0} = 2330 K, diameter of particles D_p = 24 μm, temperature of steel substrate (St. 45) T_{b0} = 400 K, thickness of coatings H = 300 μm. The following KPPs values correspond to the results presented in figures 4, 5: diameter of particles D_p = 24 μm, temperature of steel substrate (St. 45) T_{b0} = 400 K, thickness of coatings H = 300 μm. For plasma spraying (APS) the velocity of particles is U_p = 200 m/s, for detonation and supersonic flame spraying (HVOF) – U_p = 627 m/s. The following KPPs values correspond to the results presented in figures 6, 7: temperature of particles T_{p0} = 2330 K, temperature of substrate (St. 45) T_{b0} = 400 K, thickness of coatings H = 300 μm. For plasma spraying the velocity of particles is U_p = 200 m/s, for a high-velocity spraying – U_p = 627 m/s.

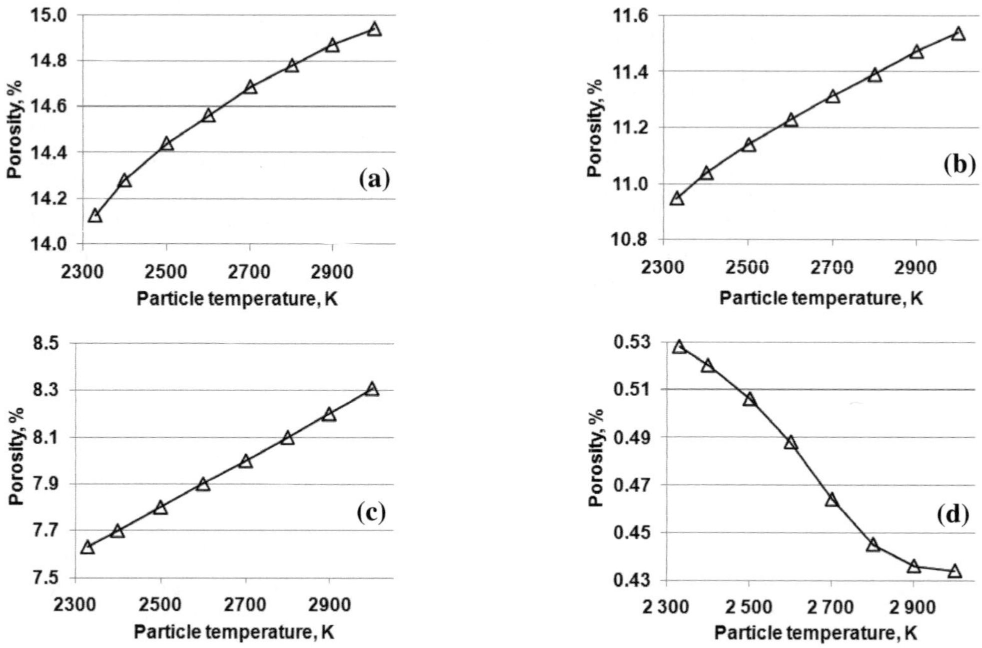

Figure 4. Porosity P of corundum coatings vs. temperature of particles T_{p0}. APS (U_p = 200 m/s): **(a)** – cylinder approximation; **(b)** – smooth splat approximation. HVOF (U_p = 627 m/s): **(c)** – cylinder approximation; **(d)** – smooth splat approximation.

The following conditions of spraying correspond to the results of modeling presented in figure 8: temperature of particles T_{p0} = 2330 K, diameter of particles D_p = 24 μm, substrate – steel (St. 45), thickness of coatings H = 300 μm. For the APS plasma mode, the velocity of particles is U_p = 200 m/s, and for the HVOF mode – U_p = 627 m/s. Under the same conditions, the roughness of coatings also changes slightly and its mean value, respectively, equals: R_a ~ 4.55 μm (APS) and R_a ~ 6.1 μm (HVOF).

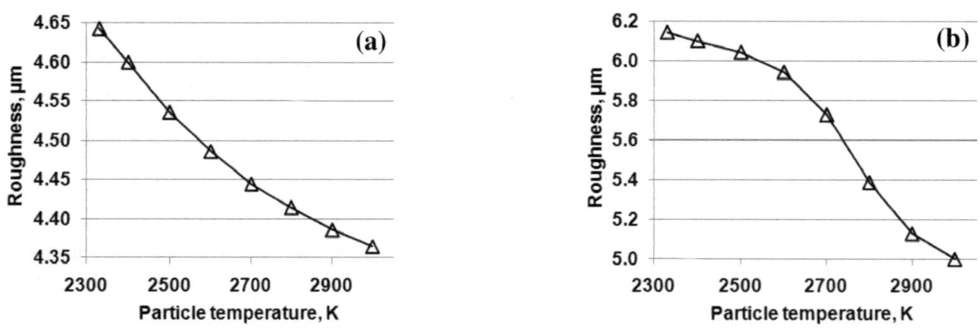

Figure 5. Roughness R_a of corundum coatings vs. temperature of particles T_{p0} (smooth splat): **(a)** – APS (U_p = 200 m/s); **(b)** – HVOF (U_p = 627 m/s).

The microsections of two plasma coatings illustrating, all other things being equal, distinctions in porosity structure when using two model morphologies of splats (figure 1) are given in figure 9. The porosity of coating (figure 9(a)) obtained with the use of cylinder approximation equals 14.7 % according to figure 4(a) and is overestimated in comparison with the experimental value of 11 % given in [5]. In turn, the porosity of coating (figure 9(b)) obtained with the use of smooth splat approximation equals 11.3 % (figure 4(b)), which agrees with the experiment [5].

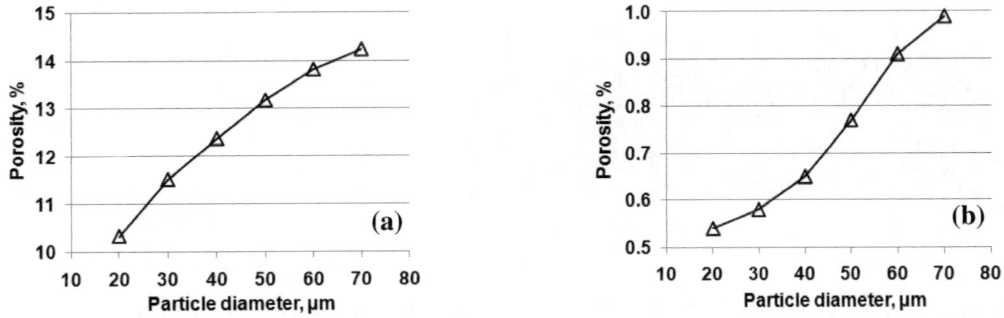

Figure 6. Porosity P of corundum coatings vs. diameter of particles D_p (smooth splat): **(a)** – APS ($U_p = 200$ m/s); **(b)** – HVOF ($U_p = 627$ m/s).

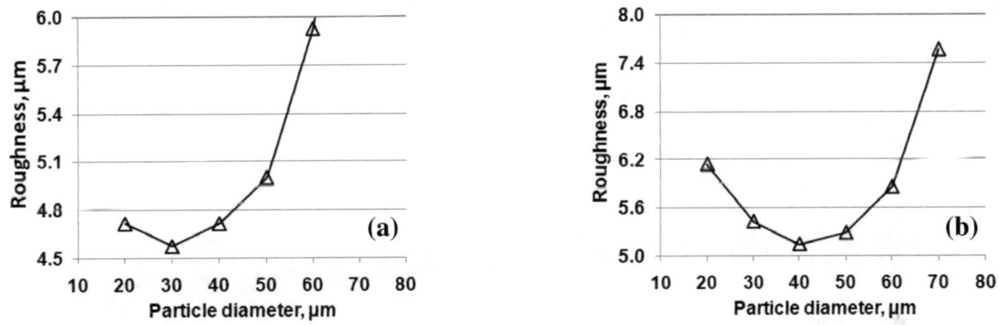

Figure 7. Roughness R_a of corundum coatings vs. diameter of particles D_p (smooth splat): **(a)** – APS ($U_p = 200$ m/s); **(b)** – HVOF ($U_p = 627$ m/s).

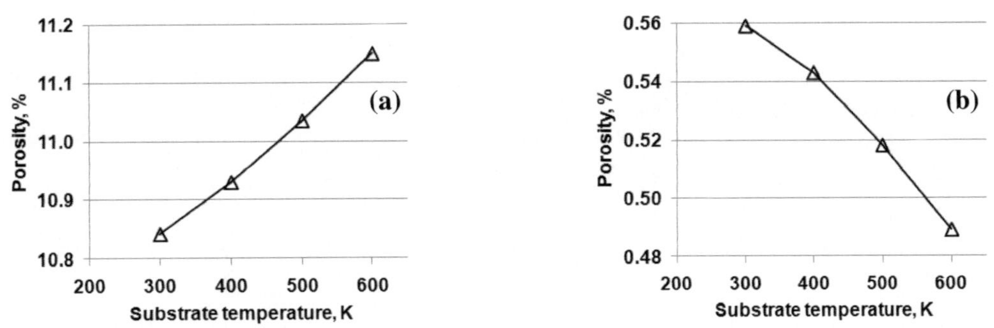

Figure 8. Porosity P of corundum coatings vs. substrate temperature T_{b0} (smooth splat): **(a)** – APS ($U_p = 200$ m/s); **(b)** – HVOF ($U_p = 627$ m/s).

The confirmation of validity of coatings modeling with use of the second model morphology of splat for detonation spraying conditions ($T_{p0} = 2330$ K, $U_p = 627$ m/s, $D_p = 24$ μm, $T_{b0} = 400$ K, thickness of coatings $H = 300$ μm) is the calculated value of porosity $P = 0.53$ % which almost agrees with the experimental value $P = 0.6$ % of the paper [6].

IOP Conf. Series: Journal of Physics: Conf. Series **857** (2017) 012015 doi:10.1088/1742-6596/857/1/012015

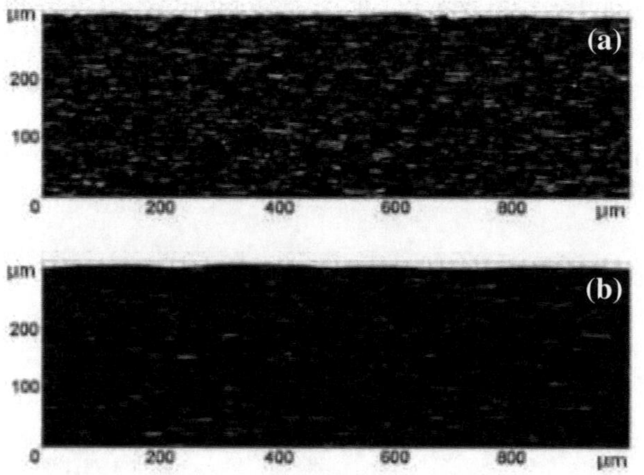

Figure 9. Microsections of plasma coatings from the corundum powder at KPPs values: U_p = 200 m/s, T_{p0} = 2700 K, D_p = 24 μm, temperature of steel substrate (St. 45) T_{b0} = 400 K: **(a)** – cylinder approximation; **(b)** – smooth splat approximation.

References

[1] Solonenko O P, Jordan V I and Blednov V A 2015 *Advances in Materials Science and Engineering* **2015** 396427
[2] Solonenko O P and Smirnov A V 1998 *Doklady Physics* **43** 689
[3] Solonenko O P 2001 *Journal of Physical Mesomechanics* **4** 45
[4] Solonenko O P 2015 *Surf. Coat. Techn.* **277** 128
[5] Vardelle A, Vardelle M and Fauchais P 1982 *Plasma Chem Plasma Process* **2** 255
[6] Ulianitsky V, Shtertser A, Sadykov V and Smurov I 2016 *Materials and Manufacturing Processes* **31** 1433

IOP Conf. Series: Journal of Physics: Conf. Series **857** (2017) 012016 doi:10.1088/1742-6596/857/1/012016

Study of the wear resistance of ion-plasma coatings based on titanium and aluminum and obtained by magnetron sputtering

G V Kachalin, A F Mednikov, A B Tkhabisimov and S V Sidorov

National Research University «MPEI», 111250, Moscow, Russian Federation

E-mail: kachalingv@mpei.ru

Abstract. The paper presents the results of metallographic researches and erosion tests of ion-plasma coatings (based on titanium, aluminum and their nitrides), which were formed on samples of 12Kh13 and EI961 blade steels. Erosion tests and studies of characteristics of obtained by magnetron sputtering coatings were carried out by using a set of research equipment UNU "Erosion-M" NRU "MPEI". It was found that the formed Ti/Al-TiN/AlN coatings increase the duration of blade steels erosion wear incubation period by at least in 1.5 times and have a layered structure with thicknesses of nitride layers 1.3–1.6 μm and intermediate metallic layers 0.3–0.5 μm, with a total thickness of coatings of 10–14 μm for 12Kh13steel samples and 19–21 μm for EI961 steel samples.

1. Introduction

It is possible to enhance the functional properties of traditional ion-plasma nitride coatings by alloying them with other elements and applying technological approaches that reduce the size of individual structural elements of the coating down to several nanometers.

Forming TiN coatings with aluminum and it's nitrides to create TiN/AlN nanostructured compositions allows to increase the operating temperature, reduce the coefficient of friction and, in general, increase the wear resistance of such coatings [1, 2].

It is in practical interest to apply such approaches to improving the functional characteristics of titanium nitride coatings, which are used to protect rotating blades of powerful steam turbines low-pressure part [3].

In the present work, the task was to obtain and to study the characteristics and erosion resistance of Ti/Al-TiN/AlN multilayer nanostructured coatings formed on 12Kh13 and EI961 steels, which are widely used for manufacturing rotating blades of the steam turbines last stages.

2. Techniques for the formation of ion-plasma coating and conducting metallographic researches and erosion tests

Ti/Al-TiN/AlN coatings were formed in the "Hephaestus+" experimental-industrial installation [4] by using planar unbalanced magnetrons. Titanium grade VT1-0 and aluminum grade A99 were used as the targets-cathodes. The coatings were applied according to the scheme shown in figure 1 on special 12Kh13 and EI961 steel specimens for erosion tests [5]. During the formation of coatings, the samples made a planetary motion. The rates of axial and planetary movement of the samples were 3 and 30 rpm, respectively. Argon and nitrogen of high purity were used as the plasma-forming and reaction gases. Preliminary treatment of the samples surface was carried out by electrolyte-plasma polishing in an aqueous solution of ammonium sulfate (6 %).

The process of ion-plasma formation of Ti/Al-TiN/AlN coatings included the following steps:

– vacuum chamber pumping out with preliminary heating of samples and technological equipment;

– ionic cleaning;

– formation of metal and nitride coating layers.

The regime parameters of the coating formation technological process main stages on the "Hephaestus+" installation are given below (see tables 1–3).

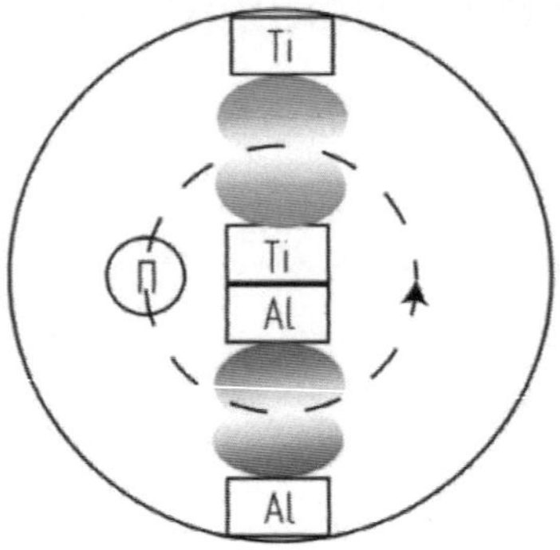

(a)

(b)

Figure 1. Nanostructured coating formation scheme **(a)** and "Hephaestus+" installation general view **(b)**.

Table 1. Basic operating parameters of the pumping and preheating phase.

Stage duration, min	Voltage on the substrate, V	Total power of magnetrons, kW	Substrate temperature, °C	Pressure in the vacuum chamber, Pa
45–60	–	–	100–150	no more $8 \cdot 10^{-3}$

Table 2. Basic regime parameters of the ionic cleaning phase.

Stage duration, min	Voltage on the substrate, V	Total power of magnetrons, kW	Substrate temperature, °C	Pressure in the vacuum chamber, Pa
15–30	700–1200	–	no more 350	no more 0.2

Table 3. Basic regime parameters of the coatings formation stage.

Stage duration, min	Voltage on the substrate, V	Nitrogen consumption, l/h	Total power of magnetrons, kW	Substrate temperature, °C	Pressure in the vacuum chamber, Pa
300–400	60–100	0–5.4	16–24	no more 350	0.25–0.30

Determination of the formed coatings characteristics was carried out by using the complex of research equipment of the UNU "Erosion-M" of the NRU "MPEI", which allows carrying out the researches of the protective coatings composition, structure, thickness, microhardness and erosion resistance.

By using the TESCAN MIRA 3 LMU scanning electron microscope equipped with X Max 50 spectrometer (Oxford Instruments), the composition, structure and thickness of the coatings were studied. In addition to the general energy-dispersive microanalysis of the coating composition, its layer-by-layer analysis was performed on the GD Profiler 2 glow discharge optical spectrometer (Horiba Jobin Yvon) to obtain the distribution profiles of the elements in the coatings.

The metallographic sections were made by using a complex of preparation equipment, including: abrasive cutting machine with linear movable cutting system POWERMET 3000 (Buehler GmbH), electro-hydraulic press SIMPLIMET 1000 (Buehler GmbH), grinding-polishing machine BETA/1 (Buehler GmbH).

The microhardness of the coatings, as well as of the uncoated samples, was determined by using a DuraScan 20 (Emco-Test) hardness tester. The measurements were carried out according to the Vickers method at a load of 0.05 kgf (0.491 N).

Tests of the samples to the high-speed impact of liquid droplets resistance were carried out on the erosion test rig "Erosion-M". "Erosion-M" is a rotary-type testing rig that works as follows: two test samples are attached to the ends of a rod rotating in a vacuum chamber and cross each vertical rotation of liquid droplets emerging from a special droplet generator. At the "Erosion-M" rig it is possible to simulate various conditions for the interaction of liquid droplets with the surface of experimental samples: impact velocity, the size (diameter) of liquid droplets, the number of droplet flows, the attack angle. Within the framework of present research, erosion tests were carried out at a sample velocity of 300 m/s and a droplet diameter (distilled water) of 800 μm. To perform a comparative analysis of the erosion tests results, were plotted the dependences of the erosion wear average depth E, (m) on the liquid mass deposited on the eroded surface m (kg/m^2).

3. Results and discussion

As a result of the carried out experiments series were obtained Ti/Al-TiN/AlN coatings with a thickness of 7 to 21 μm. For all these coatings erosion tests were carried out and in this work presented the coatings with better erosion resistance.

Figures 2 and 3 show images of transverse sections and a characteristic view of the EI961 and 12Kh13 blade steels specimens surface with coatings.

Figure 2. Image of the surface morphology and microstructure of the Ti/Al-TiN/AlN coating based on the EI961 specimen.

Figure 3. Image of the surface morphology and microstructure of the Ti/Al-TiN/AlN coating based on the 12Kh13 specimen.

It was revealed that the coatings have a layered structure, the thickness of the nitride layers is 1.3–1.6 μm, and the intermediate metallic layers 0.3–0.5 μm. The total thickness of the coating is 10–13 μm for 12Kh13 steel specimens and 19–21 μm for EI961steel specimens. The average thickness values of the obtained coatings are shown in table 4. The results of the elemental composition energy-dispersive microanalysis with the determination of the average content of elements in the coating are presented in table 5.

Table 4. Thickness of the formed coatings.

Sample	Sample description	h, μm
1	Ti/Al-TiN/AlN coating on EI961 steel	20.1 ± 1.3
2	Ti/Al-TiN/AlN coating on 12Kh13 steel	11.8 ± 1.7

Table 5. Results of elemental composition of coatings analysis.

Sample	Average content of elements in the coating, weight %						
	N	O	Al	Si	Ti	Cr	Fe
Ti/Al-TiN/AlN coating on EI961 steel	4.2 ± 1.3	–	31.2 ± 0.5	–	62.2 ± 0.8	0.3 ± 0.1	1.3 ± 0.2
Ti/Al-TiN/AlN coating on 12Kh13 steel	4.0 ± 1.1	–	29.6 ± 0.8	–	63.9 ± 0.9	0.4 ± 0.1	1.5 ± 0.2

It is not possible to determine the thickness of individual Ti (TiN) or Al (AlN) nanolayers, and the structure of coatings obtained under the conditions of current experiments is more similar to coatings obtained from targets (cathodes) based on Ti-Al alloys. The surface of the coatings has a granular structure with a characteristic grain size of 200–300 nm.

Figure 4 shows the profiles of the elements distribution along the thickness of coatings obtained on an optic-emission spectrometer of a glow discharge. In coatings, the aluminum content is about 30 %, titanium is about 63 %, nitrogen is about 4 %. In the upper layers of coatings, the nitrogen concentration reaches 7–9 % by weight. It also noted the presence of iron (1.3–1.5 %) and chromium impurities, which, possibly, fall into the coatings due to the spraying of the tooling certain elements.

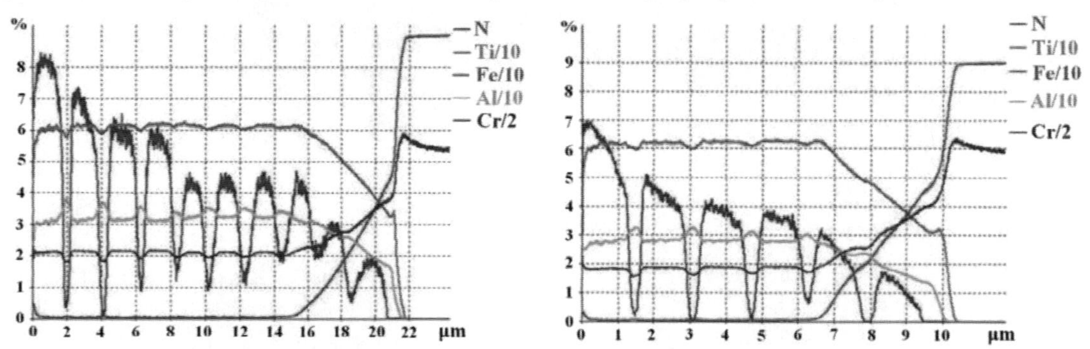

(a) (b)

Figure 4. Profiles of the elements distribution in coatings on of EI961 **(a)** and 12Kh13 **(b)** steels specimens.

Table 6 presents the mean values of the microhardness of the surface of the EI961 and 12Kh13 blade steel specimens with and without coatings. The analysis of the data in table 6 shows that the formed coatings increase the microhardness of the surface up to 4–5 times compared with the reference level.

The results of erosion tests in the form of kinetic erosion curves ($E = f(m)$) are presented in figures 5 and 6. Analysis of the obtained dependences shows that Ti/Al-TiN/AlN coatings increase the incubation period of the erosive wear process of 12Kh13 and EI961 steels not less than in 1.5 times.

4. Conclusions

The Ti/Al-TiN/AlN coatings considered in present work are promising for solving the problem of improving titanium nitride coatings used to protect the powerful steam turbines wet steam stages rotating blades from water droplet erosion, but they will not solve the wear problem during all periods of this process. The solution of this problem, according to the authors' opinion, is possible in a combination of surface modification and the formation of ion-plasma coatings.

Table 6. Results of the mechanical properties study of 12Kh13 and EI961 steel specimens with Ti/Al-TiN/AlN coating and without coating.

Sample	Sample description	Vickers microhardness H0,05, HV
1	12Kh13 without coating	271 ± 5
2	EI961 without coating	280 ± 7
3	12Kh13 with coating based on TiAlN	1150 ± 30
4	EI961 with coating based on TiAlN	1140 ± 30

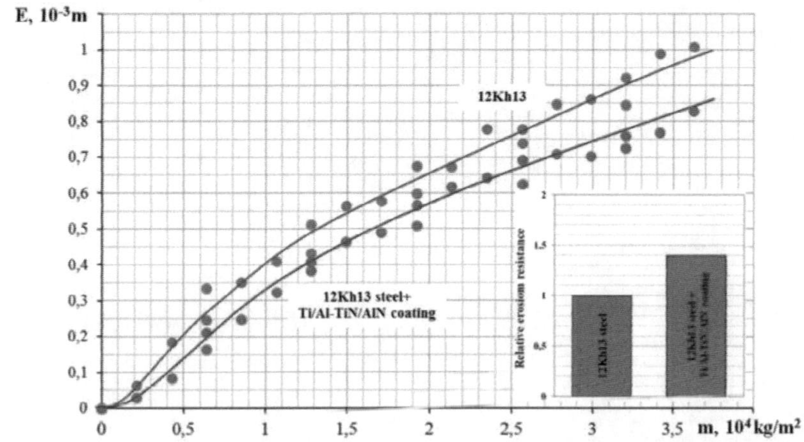

Figure 5. Curves of erosion wear of 12Kh13 steel uncoated and coated with Ti/Al-TiN/AlN at a collision velocity C_{imp} = 300 m/s and a droplet diameter d_d = 800 μm.

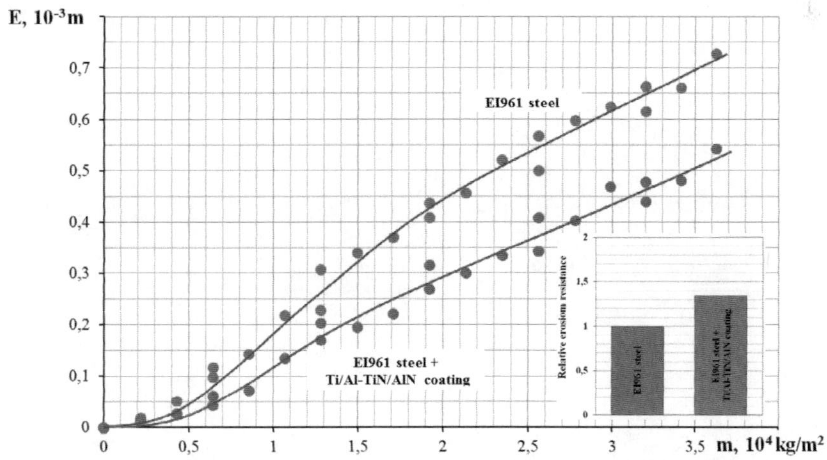

Figure 6. Curves of erosion wear of EI961 steel uncoated and coated with Ti/Al-TiN/AlN at a collision velocity C_{imp} = 300 m/s and a droplet diameter d_d = 800 μm.

Acknowledgements
This study was financially supported by the Ministry of Education and Science of the Russian Federation under agreement № 14.577.21.0180 dated October 27, 2015 (unique identification number RFMEFI57715X0180).

References

[1] Muboyadzhyan S A, Alexandrov D A and Gorlov D S 2010 *Metals* **5** 39–51
[2] Sobol O V, Dmitriyak V V, Pogrebnoi N A, Pinchuk N V and Meilikhov A A 2015 *East-European Journal of Advanced Technologies* **2/5** 53–9
[3] Kachalin G V, Mednikov A F and Tkhabisimov A B 2014 *WIT Transactions on Ecology and the Environment* **186** 729–37
[4] Kachalin G V, Ryzhenkov A V, Medvedev K S, Bychkov A I and Parthenenok M A 2014 *Reliability and safety of power engineering* **2** 8–12
[5] Ryzhenkov V A, Seleznev L I, Mednikov A F and Tkhabisimov A B 2014 *Thermal Engineering* **61**(8) 594–9

IOP Conf. Series: Journal of Physics: Conf. Series **857** (2017) 012017 doi:10.1088/1742-6596/857/1/012017

BN coatings deposition by magnetron sputtering of B and BN targets in electron beam generated plasma

A S Kamenetskikh[1,2], **N V Gavrilov**[1], **O V Koryakova**[3] **and S O Cholakh**[2]

[1] Institute of Electrophysics, 620016, 106 Amundsen str., Yekaterinburg,Russia
[2] Ural Federal University, 620002, 19 Mira str., 620002 Yekaterinburg, Russia
[3] Institute of Organic Synthesis, 620137, 22 Sofia Kovalevskayastr., Yekaterinburg, Russia

E-mail: alx@iep.uran.ru

Abstract. Boron nitride coatings were deposited by reactive pulsed magnetron sputtering of B and BN targets (50 kHz, 10 μs for B; 13.56 MHz for BN) at 2–20 mA/cm^2 ion current density on the substrate. The effect of electron beam generated plasma on characteristics of magnetron discharge and phase composition of coatings was studied.

1. Introduction

Boron nitride (BN) coatings have been used in various fields of science and technique due to their unique features. It is known there are several polymorphic modifications of BN, including superhard (up to 72 GPa) cubic c-BN phase [1]. RF magnetron sputtering of dielectric targets made of hexagonal h-BN is mostly used for deposition of BN coatings by PVD methods. Coatings with high c-BN content are formed when conducting targets of B_4C are used in DC or DC pulsed mode, however this coatings contains more than 5 % of C contamination [2]. The specific resistance of boron drops lower than 1 Ω·cm when target are heated till ~600 °C that allows using B-target in DC sputtering systems [3].

Intensive ion assistance is a necessary condition for deposition of coating with high c-BN content by vacuum-plasma methods. Coatings containing ~80–90 % of c-BN are formed in narrow range of ion current density and energy [4]. To increase the ion current density during BN coating deposition by magnetron sputtering, unbalanced magnetic field [5] or ion-beam assistance are often used [6]. In the present work ion current density on the samples was regulated in the range 2–20 mA/cm^2 by adjustment of the current of low-energy (~10^2 eV) electron beam injected into the working chamber. In previous work the authors showed high efficiency of such method of plasma parameters control [7]. The aim of the work is the study of the effect of ion component of electron beam plasma on the characteristics of magnetron discharge with B and h-BN targets and on the phase composition of coatings.

2. Experimental technique

The sketch of experimental system is shown in figure 1. Plasma generation system included a plane balanced magnetron *1* and a source of broad (50 cm^2) electron beam *2*, placed on the vacuum chamber *3* 330 mm in diameter. Discs 80 mm in diameter and 4 mm thick made of B and h-BN (99.9 % purity) were used as sputtering targets. B target was sputtered in DC pulsed mode (50 kHz, 10 μs) with amplitude of discharge current up to 1 A; h-BN target was sputtered in RF mode (13.56 MHz) with discharge power 200 W. Discharge with self-heated hollow cathode made of titanium nitride [7] was used in the electron source. Discharge current was regulated in the range of 2–30 A and discharge voltage changed from 300 V to 25 V. Energy of electron beam during the coating deposition was 100 eV.

Coatings were deposed on polished surfaces of samples 15×15 mm made of silicon and steel AISI430. Samples were ultrasonically cleaned in acetone solution before placing into the chamber. The working chamber was pumped off till 10^{-2} Pa by turbomolecular pump. Samples were cleaned by ion etching in Ar plasma generated by electron beam. Pulse bias of samples during ion etching was 500 V (50 kHz, 10 μs), ion current density – 2 mA/cm^2. The bias voltage during coatings deposition was 200 V.

Gas mixture Ar/N_2 was feed through electron source. Coatings were deposited under constant gas flow rates Q_{Ar}/Q_{N_2} – 30/7 (sccm) in the mode of B target sputtering and 3/1 in the mode of h-BN target sputtering; total gas pressure was 0.12 and 0.04 Pa, correspondingly.

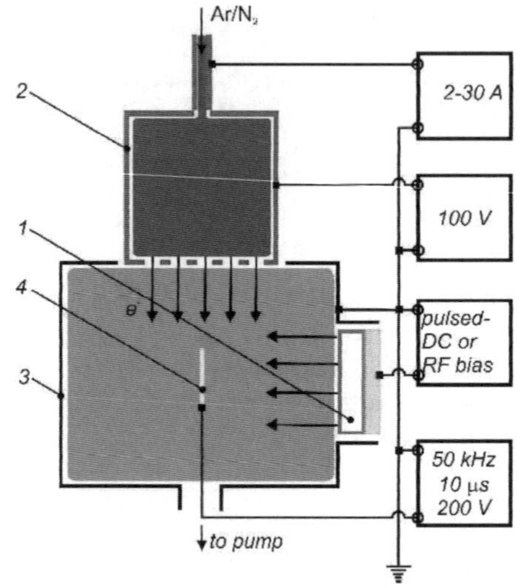

Figure 1. The diagram of experiment: *1* – magnetron, 2 – electron source, *3* – working chamber, *4* – samples.

The coatings was analyzed by IR-Fourier spectroscopy. The spectra of frustrated total internal reflection of the coatings were registered on IR-Fourier spectrometer "Nicolet 6700" (Thermo Electron Corporation) using "Smart orbit" add-in device with diamond crystal. Intensity of peaks and squares under optic density curve was calculated using "Omnic" program.

3. The effect of electron beam generated plasma on characteristics of magnetron discharge with B cathode

Characteristics of magnetron discharge with B target are determined significantly by temperature dependence of boron electric resistance. Heating of B target in discharge plasma at constant current amplitude is accompanied by discharge voltage decrease by several hundred V during ~10^2 s. Current voltage characteristic for steady state discharge (figure 2) has non-monotonic character. Increase of discharge current from 0.1 to 0.5 A leads to rise of target temperature by 450 °C and to corresponding decrease of specific resistance by ~10^5 Ω·cm^2; the voltage across the discharge gap is reduced by 110 V. Specific resistance of B target does not significantly affect the magnitude of discharge voltage at further growth of discharge current. The right branch of dependence curve corresponds to typical current voltage characteristic of magnetron discharge [8].

The increase of electron beam current till 25 A leads to decrease of magnetron discharge voltage from 580 to 60 V (figure 3). The drop of discharge voltage is determined by ions, which arrived from beam plasma. Ion current density (j_i) grows monotonically along with beam current increase and reaches ~23 mA/cm^2 at 25 A. Figure 4 shows the dependencies of magnetron discharge voltage (obtained at beam current 5 A) and frequency of ionizing collisions of electrons with gas atoms, calculated from the ratio $\nu_i \sim \sigma_i(\upsilon)\cdot\upsilon$ (where σ_i is the cross section of electron impact ionization; υ – electron velocity), on accelerating voltage of electron source. It is seen that frequency of ionizing collisions reaches maximum value at electron energy ~250 eV; magnetron discharge voltage is minimum (~440 V) in this case.

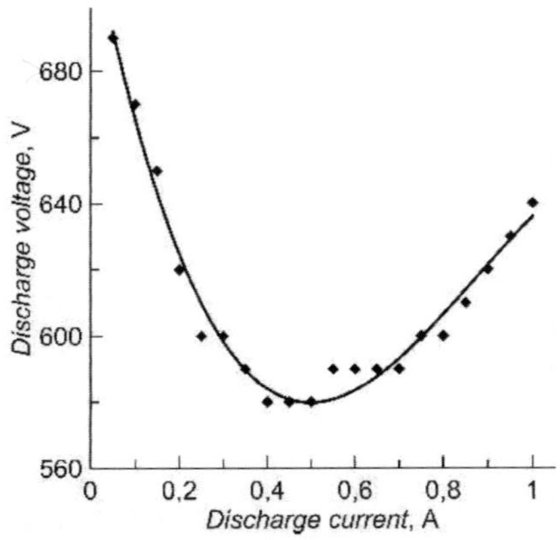

Figure 2. The voltage current characteristic of magnetron discharge with B target.

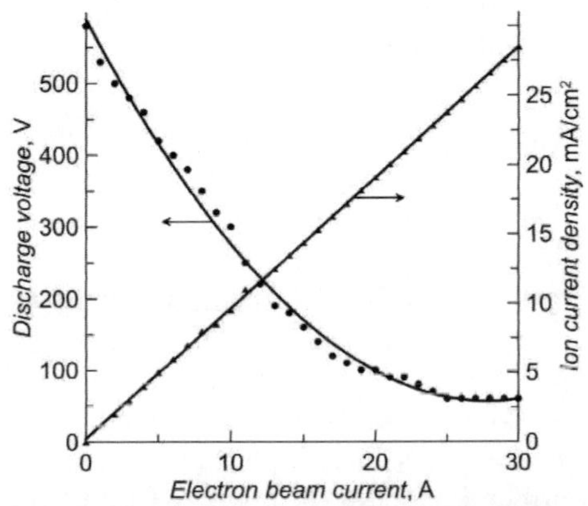

Figure 3. The dependencies of magnetron discharge voltage and density of ion current from the beam plasma on the beam current.

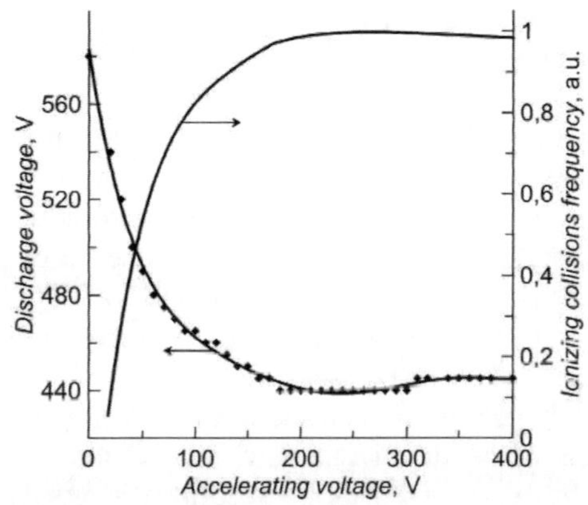

Figure 4. The dependencies of magnetron discharge voltage and frequency of ionizing impact of electrons with gas atoms on accelerating voltage of the electron source.

4. The effect of beam plasma on the phase composition of BN coating

The IR spectra of BN coatings deposed on Si substrate in B target sputtering mode are shown at figure 5. At all IR spectra a line with wavenumber ~1080 cm^{-1} corresponding to c-BN phase can be seen. It is known that wavenumber 1004 cm^{-1} corresponds to c-BN peak in the stress-free samples [9]. Observed shift of the peak towards large values of wavenumber indicates the presence of intrinsic stresses in the coating. Their value estimated from the ratio obtained in [10] is 13 GPa. Such a high level of intrinsic stresses impairs the coatings adhesion, so the maximum thickness of adhesive BN coatings obtained in B target sputtering mode did not exceed ~1 μm.

According to [4], relative content of c-BN in coatings (n_c) can be estimated from the ratio of intensities of peak at 1080 cm^{-1} and peak at 1366 cm^{-1}, corresponding to h-BN phase. The increase of electron beam current from 5 to 20 A, providing the change of ion fluence ratio to the number of atoms per square unit of the coating Φ/N from 3 to 24, is accompanied by n_c growth from 0.67 to 0.83.

Besides c-BN and h-BN phases, a phase having wurtzite structure (w-BN), corresponding to the line with wave number 1171 cm^{-1} [11] (figure 5) as well as fullerene-like (e-BN) phase with lines 907, 1260, 1409 and 1580 cm^{-1} [12] presents in the coating. The e-BN phase was first revealed after exposing of high dynamic loads achieved by explosive method on h-BN powder [13]. In work [11] e-BN phase was observed in coatings deposited by RF sputtering of h-BN target.

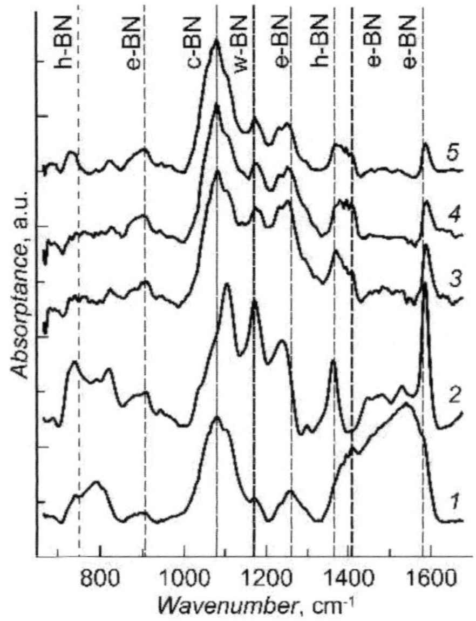

Figure 5. IR spectra of BN coatings deposited in B target sputtering mode. Electron beam current: *1* – 2; *2* – 5; *3* – 10; *4* – 15; *5* – 20 A.

The authors of the work [11] describe the appearance of e-BN phase in the coating as follows. The h-BN clusters (which are in state of sp^2-hybridization) are sputtered from target surface, acquired enough energy to form sp^3 bonds during interaction with plasma particles and form structure with ratio of sp^2 and sp^3 bonds 1:1 on coating surface. In coatings deposited by reactive magnetron sputtering of one-component target in DC pulsed mode, e-BN phase was first revealed. The intensity of e-BN phase peaks decreases under the increase of electron beam current. According to [11], the growth of energy transmitted by plasma particles to atoms of the coating facilitates formation of a large number of sp^3 bonds, hence the concentration of phases h-, e-, w- and c-BN in the coatings is increases along with the growth of beam current. In our experiment conditions for e-BN formation are created already at electron beam current 2 A providing Φ/N ratio ~2.4. Further j_i increase promotes preferential c-BN formation.

IR the spectra of coatings obtained by RF sputtering of h-BN target are shown at figure 6. The coatings with main phase h-BN are formed when the ratio Φ/N changes from ~3 to 14 in the range of beam current 2–8 A. In the spectra the peaks corresponding to e-BN (1580, 1409 cm^{-1}) are also observed, however their intensity at beam currents below 14 A is significantly lower than intensity of h-BN peak (~1340–1360 cm^{-1}). Phase c-BN appears in coatings at electron beam currents more than 8 A. Relative content of c-BN reaches maximum 0.58 at beam current 12 A and decrease sharply to 0.05 at further current increase.

The peak of c-BN phase lays in the range of wavenumbers 1022–1038 cm^{-1}. The intrinsic stress of coatings estimated from the peak shift value is ~5 GPa, that is significantly lower than in coatings deposited in the mode of B target sputtering. According to the model of static stresses [14], the minimum level of intrinsic stresses necessary for c-BN formation makes ~4 GPa, and the content of c-BN at stress value ~5 GPa reaches 55–60 % that correlates with obtained results.

Thus, at comparable values of Φ/N the level of intrinsic stress in the coatings deposited by RF sputtering of h-BN target is ~2.5 times lower than in coatings obtained by DC pulsed sputtering of B target. This improves adhesion of coatings and allows to make them thicker. The main advantage of B target sputtering mode (under conditions of Ar-N_2 plasma generation by broad electron beam) consists in higher content of c-BN (~0.6–0.8) in coatings that is achieved in wide range of Φ/N ratio (from 3 to 24).

5. Conclusions

Two variants of BN coating deposition – by RF magnetron sputtering of h-BN target and DC pulsed sputtering of B target – were compared under conditions of the coating exposure in the plasma of low-energy (100 eV) electron beam. Ions from electron beam plasma facilitate initial heating of B target and reduce magnetron discharge voltage. Changing of electron beam current allows controll the ratio of ion fluence to the number of atoms in the coating in the range of 2–24 and optimization of the conditions of c-BN phase formation. Maximum content of c-BN phase in coatings obtained by B and h-BN targets sputtering, made 83 and 58 %, and the levels of intrinsic stresses are 13 and 5 GPa, correspondingly.

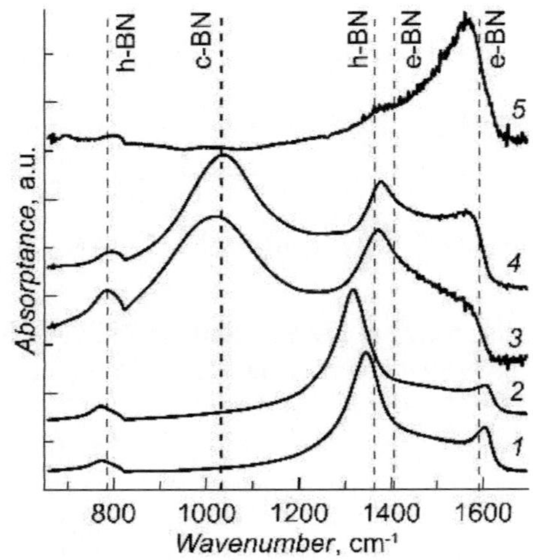

Figure 6. IR spectra of BN coatings deposited in h-BN target sputtering mode. Electron beam current: *1* – 0; *2* – 6; *3* – 10; *4* – 12; *5* – 14 A.

Acknowledgment
The work was accomplished with support of Complex Program of the Ural Department of RAS (project № 15-17-2-28).

References

[1] Mirkarimi P B, McCarty K F and Medlin D L 1997 *Mater. Sci. Eng. R.* **21** 47–100
[2] Kouptisidis S et al. 1998 *Diamond and Related Materials* **7** 26–31
[3] Hahn J et al. 1996 *Diamond and Related Materials* **5** 1103–12
[4] Inagawa K et al. 1987 *J. Vac. Sci. Technol. A* **5**(4) 2696–700
[5] Otano-Rivera W 1998 *J. Vac. Sci. Technol. A* **16**(3) 1331–5
[6] Djouadi M A et al. 2004 *Surf. Coat. Technol.* **180–1** 174–7
[7] Gavrilov N V et al. 2015 *Journal of Physics: Conference Series* **652** 012024
[8] Westwood W D, Maniv S and Scanlon P J 1983 *J. Appl. Phys* **54**(12) 6841–6
[9] Wentzcovitch R M, Chang K J and Coben M L 1986 *Phys. Rev. B* **34** 1071–9
[10] Sanjurjo J A et al. 1983 *Phys. Rev. B* **28**(8) 4579–84
[11] Zhu P W et al. 2002 *J. Solid State Chem.* **167** 420–4
[12] Olszyna A, Konwerska-Hrabowska J and Lisicki M 1997 *Diamond and Related Materials* **6** 617–20
[13] Batsanov S S, Blokhina G E and Deribas A A 1965 *J. Struct. Chem.* **6** 209–13
[14] McKenzie D R et al. 1996 *Surf. Coat. Technol.* **78** 255–62

Improvement of manufacture technology and research of actuators based on ionic polymer-metal composites

I K Khmelnitskiy, L O Vereshagina, V E Kalyonov, A P Broyko, A V Lagosh, V V Luchinin and D O Testov

Research and Education Center "Nanotechnologies", Saint Petersburg Electrotechnical University "LETI", 197376, Saint Petersburg, Russia Federation

E-mail: khmelnitskiy@gmail.com

Abstract. The purpose of this work is to improve the operational characteristics of actuators based on ionic polymer-metal composites (IPMC) by improving their manufacture technology. The optimal thickness of the ion-exchange membrane was determined. The technology of metal electrode deposition, the composition of actuator electrolyte solution and the storage method of actuators were optimized.

The creation of power-efficient, inexpensive actuators and micro robotic systems based on them made by micro- and nanosystem technology is an actual problem [1].

Recently, special attention is paid to actuators based on IPMC, due to its low driving voltage and large deflections [2]. Ionic polymer-metal composite (IPMC) is synthetic composite nanomaterial that consists of ion-exchange polymer coated with a conductive metal. Mechanical deformation of the actuator is caused by the applied electric field. And on the contrary, mechanical deformation of the membrane induces electrical signal. So, IPMC actuator could be used as sensor and as actuator. The operation actuator parameters depend on many factors [3–5].

The principle of IPMC actuator operation is following. After applying of external electric field, ions of the same charge begin to move from the anode to the cathode. So, this electroosmotic fluid flow results in excessive fluid pressure near an electrode. Thus, the actuator deformation is caused by the pressure difference between the upper and bottom electrodes [5] (figure 1).

The purpose of this work is to improve the operational characteristics of IPMC actuators by improving their manufacture technology. So, it was necessary to solve the following tasks for attainment of this object: the choice of an optimal thickness of the ion-exchange membrane, the technology improvement of metal electrode deposition, the selection of a proper actuator electrolyte solution and the research of storage methods for actuators.

The thickness of the commercially available Nafion® film varies from 50 to 180 μm. The efficiency of IPMC actuators changes with the thickness because some parameters such as bending stiffness, driving voltage, etc., depend on the thickness value.

When one's chooses a thickness of actuator membrane the crucial parameters are the developed force and the created displacement. The actuator made of the membrane with a thickness 175 μm develops larger force than the one made of the membrane with a thickness 50 μm (figure 2), but the values of displacement are similar for both (figure 3). Thus, the membrane with a thickness of 175 μm was chosen for further investigations.

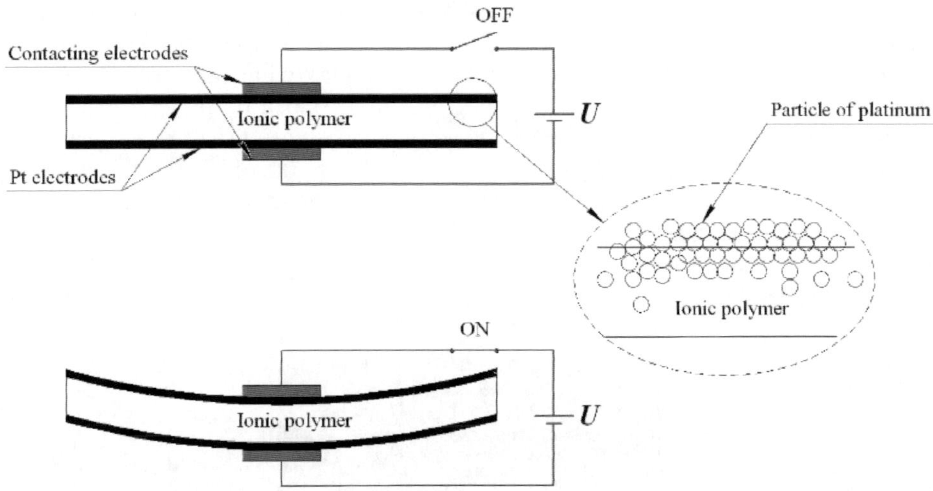

Figure 1. The IPMC actuator principle of operation.

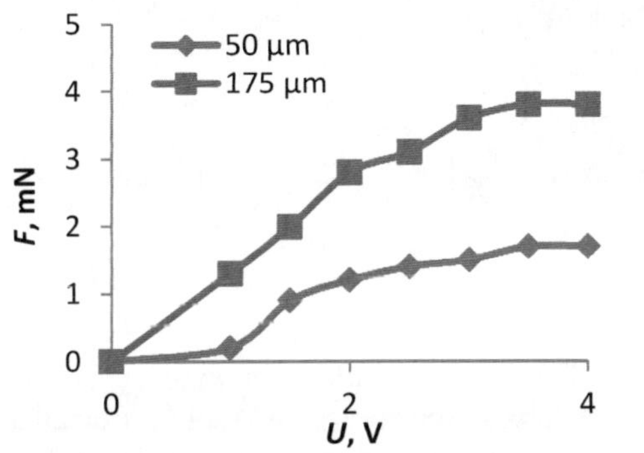

Figure 2. Developed force vs. driving voltage for actuators based on IPMC with different thickness.

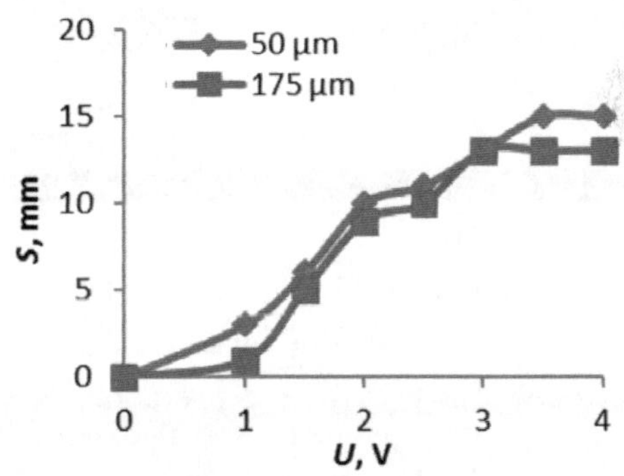

Figure 3. Created displacement vs. driving voltage for actuators based on IPMC with different thickness.

In order to reduce a driving voltage and increase the transformer effectivity, it is necessary to produce wear-resistant, flexible and high conductive electrodes on the membrane surface. One of most effective methods for electrode manufacturing is deposition by chemical reduction from a platinum solution. This process consists of two steps. At first, platinum particles are precipitated in the near-surface layer. Next, platinum electrodes are deposited directly on the membrane.

During the first step, membrane is soaked in salt solution of tetraammineplatinum chloride $Pt(NH_3)_4Cl_2$. Thus, cations containing platinum diffuse into the membrane by ion-exchange process. Then, platinum is deposited by reduction with sodium borohydride $NaBH_4$ solution. Highly dispersed metal layer is formed inside the membrane.

At the second stage, platinum precipitates from the solution, containing a platinum salt $Pt(NH_3)_4Cl_2$, hydrazine H_2NNH_2 and hydroxylamine NH_2OH. Thus, the electrode thickness on the surface increases. Decrease of electrode resistance could be obtained by the thickness increase of the deposited platinum layer. This could be achieved by repeating the second stage twice.

The samples of actuator were produced with electrodes deposited on the membrane by three methods:

- the first method: apply the first and then the second stage (1 + 2).
- the second method: repeat first stage twice and then the second once (1×2 + 2).
- the third method: apply the first and then the second stages, each twice (1×2 + 2×2).

The focused ion beam method (Helios NanoLab 400 FIB station) was used to investigate the IPMC structure. Sample sections of IPMC actuator were obtained by the focused ion beam (figure 4(a)). In the figure 4(a) one can see two layers of the platinum particles which differ in morphology. The bottom layer – diffusive, obtained at the first step of platinum deposition process when the platinum particles penetrate into the membrane. Thickness of this layer for different samples varies from 6 to 10 μm and the sizes of the dispersed particles are within the range of 30–90 nm (figure 4(b)). The concentration of platinum particles gradually decreases with the increase of distance from membrane surface. The upper layer thickness obtained after additional platinum deposition at the second step depends on the time of deposition and the number of iteration. Thus, thickness varies from 2 to 10 μm. The layer consists of granular platinum particles with a size within the range of 40–60 nm (figure 4(c)).

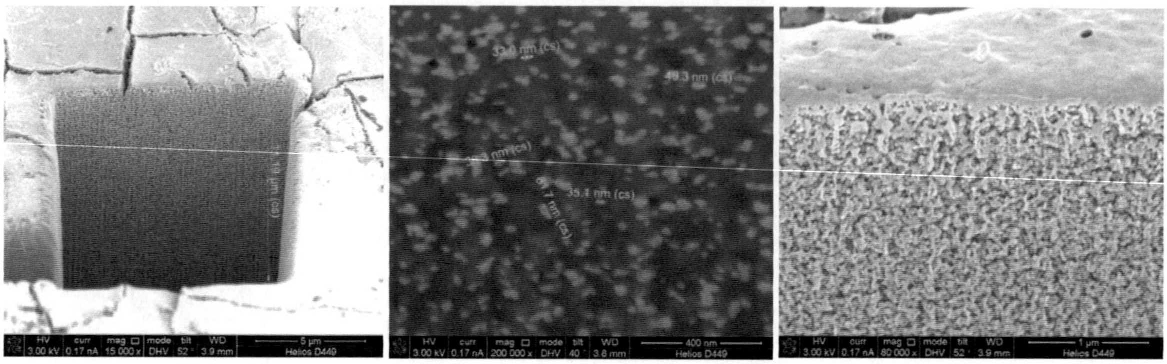

(a) (b) (c)

Figure 4. Images of IPMC actuator section obtained by FIB station: whole section **(a)**, diffusive **(b)** and surface **(c)** platinum layers.

The next step was to study the operational characteristics of actuator samples obtained by different electrode deposition methods. The largest thickness and the lowest resistance of platinum electrode layer were obtained by the third method. The maximum value of developed force is about 10 mN (figure 5). The highest displacement of 15 mm is shown by actuator produced by the second method (figure 6).

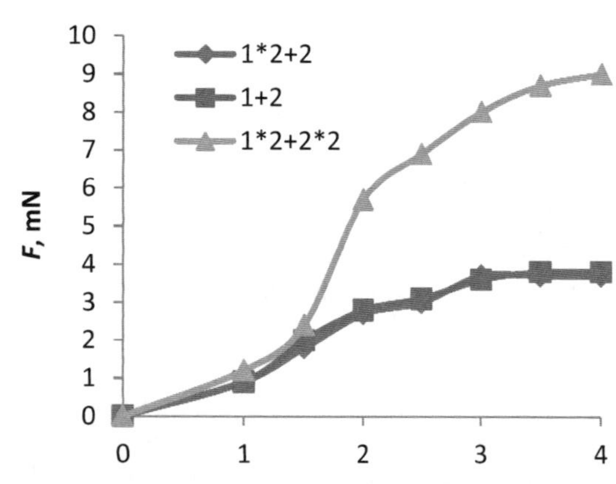

Figure 5. Developed force vs. driving voltage for actuators based on IPMC with electrodes produced by different methods.

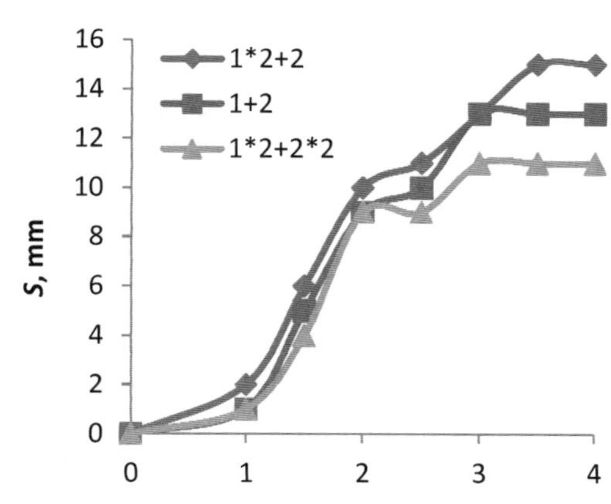

Figure 6. Created displacement vs. driving voltage for actuators based on IPMC with electrodes produced by different methods.

The actuator characteristics depend on the electrolyte composition dissolved in water. And the nature of cations which freely move within the membrane play the key role. Therefore, the characteristics of actuators soaked with different salt solutions were investigated. The highest displacements (figure 7) and forces (figure 8) are observed for actuator soaked with 0.1M $CuSO_4$ solution.

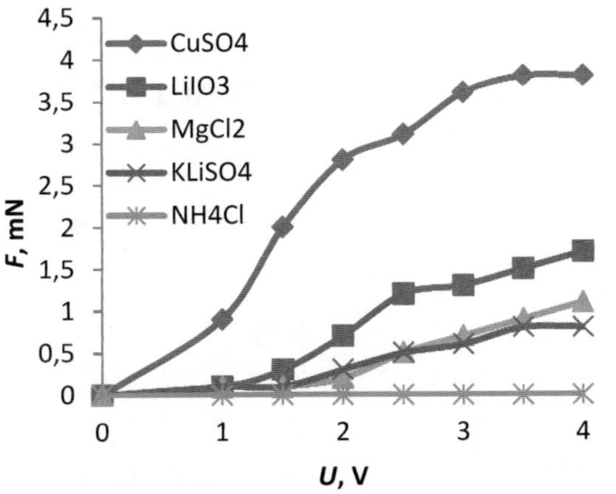

Figure 7. Developed force vs. driving voltage for actuators based on IPMC soaked with different electrolytes

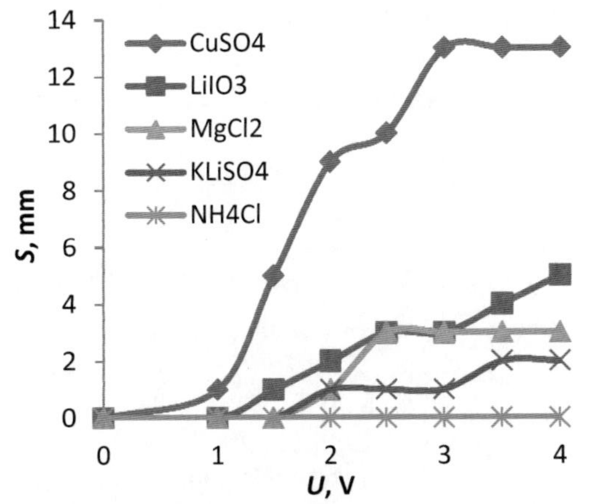

Figure 8. Created displacement vs. driving voltage for actuators based on IPMC soaked with different electrolytes.

The method of actuator storage between usage has a great influence on operational characteristics. The actuator storage in a dry mode is not acceptable, because the Nafion® membrane volume changes during drying and soaking, which can lead to a change in the electrode structure. Therefore, the actuator must be stored in solution., The characteristics deteriorate almost twofold, when the actuator is stored in a 0.1M $CuSO_4$ solution, and no more than 20 %, when it is stored in water.

Acknowledgments
The work has been funded by Russian Science Foundation (Grant No. 16-19-00107).

References
[1] Kalyonov V E, Broyko A P, Korlyakov A V, Khmelnitskiy I K et al. 2013 *Journal of nano and microsystem technique* **12** 41–6
[2] Shahinpoor M and Kim K J 2001 *Smart Mater. Struct.* **10** 819–33
[3] Korlyakov A V, Khmelnitskiy I K, Luchinin V V, Kalyonov V E et al. 2016 *Journal of nano and microsystem technique* **5** 277–85
[4] Korlyakov A V, Khmelnitskiy I K, Broyko A P, Kalyonov V E et al. 2015 *Biotekhnosphere* **6** 41–7
[5] Khmelnitskiy I, Vereschagina L, Kalyonov V, Krot A et al. 2016 *Proceedings of the 2016 IEEE North West Russia Section Young Researchers in Electrical and Electronic Engineering Conference* 58–61

Frictional transfer and the self-organization phenomenon in the friction

I V Kolesnikov and D S Manturov

Rostov State Transport University, 344038, Rostov-on-Don, Russia

E-mail: kvi@rgups.ru, manturovds@rgups.ru

Abstract. The paper is devoted to the investigation of the mechanism and kinetics of the surface structures formation in the process of metal-polymer frictional contact. IR spectroscopy methods have showed that the formation kinetics of a frictionally transferred film is determined by the adhesion of the composite components and the direction of the electric field at the contact.

The phenomenon of energy and matter transfer plays a determinant role in many technological and operational processes, including friction units of rolling stock. All polymer self-lubricating materials form a transferred film having the properties of a lubricant. The lubricating action of such a film is defined by the polymer nature, its structure, components, counterbody material, and roughness.

As is known, in closed systems processes pass in the direction of increasing entropy and establishing an equilibrium state.

As for tribosystems, they are open systems, exchanging matter and energy with each other and the environment. Being nonequilibrium, tribosystems can develop unpredictably even under the same operating conditions. If there is no possibility to impose a development path for closed systems, then such a path can be selected for an open tribological conjugation system taking into account the inherent elements of self-regulation.

Structural self-organization is carried out due to the formation of secondary structures. The composition and properties of surface structures are determined primarily by the material of the frictional pair [1, 2]. As for metal-polymer tribosystems, there is no doubt that a film transferred from a polymer sample to a metal counterbody has a significant effect on the friction and wear of the metal-polymer conjugation. However, the kinetics and mechanism of friction transfer have not been studied sufficiently. In particular, the role of temperature, triboEMF in the formation of the frictionally transferred film has not been clarified. Besides, the qualitative and quantitative composition of transferred films has not been evaluated as well. A set of methods was used to investigate the mechanism of the surface structures formation of in the process of metal-polymer frictional contact. The methods are characterized with a high sensitivity to changes both on the friction surface and in the structure of the materials selected for the study.

As a result of a purposeful search, such thermoplastic polymers as Polycaproamide PCA and Phenylone C2 were chosen as polymeric binders for antifriction composite materials. F4MBP polytetrafluoroethylene powder in an amount of 5–10 wt. % was selected as antifriction filler. Nanodimensional additions of manganese spinel (40–100 μm) with a concentration not exceeding 1.5–3 wt. %. were introduced into the composite in order to improve its characteristics. After obtaining the nanodimensional components on a MicroCer bead mill, a size analysis of particles in the range of 0.01 to 40 μm was carried out using a CPS24000 measuring system.

The process of samples formation consists of several stages. The briquetting of Phenylone C-2 powder with fillers in a mold is carried out without heating at a pressure of 500 kg/cm^2. Drying of the obtained briquette of Phenylone with fillers is carried out at 200 °C for one hour. Hot briquettes are placed in a preheated mold and maintained for 1 min. Then they are pressed at temperature of 320–340 °C and at a pressure of 500 kg/cm^2 for 1 min. Final operation is the sample cooling together with the mold to 100 °C, followed by pressing out the finished sample.

Tribological tests were carried out at a fixed speed of sliding and a stepped loading on the friction machine AI-5018 by a «finger-roller» scheme. The finger was made of composite material, and the roller was made of steel 45.

The tests were carried out at a roller speed of 3 rev/s and a discretely varying load (in the ranges of 209, 310, 420, 520, 620, 720, 820, 920, 1020 N) with a constant registration of the friction coefficient (figure 1). After the end of the test, the amount of wear of the polymer sample and the roller was determined as the difference between the weight before and after the tests.

Besides, tribological tests were carried out on the end type friction machine «finger-flat disk» constructed in the laboratory at loads P = 2 MPa and sliding speed V = 0.5 m/s (figure 2). Such construction of the machine allows controlling the condition of tribomating surface, the presence, the kinetics of formation and stability of the friction transfer film or a lubricating film in any time intervals, without disturbing the surface.

We investigated the mechanism of the transferred film formation on the counterbody's metal surface to determine the causes of low friction and high wear resistance of composite polymeric materials with antifriction fillers and nanodimensional additives. The investigations of the elemental composition and chemical bonding of transferred films on a surface with a metal counterbody, as well as the depth of films were carried out using IR spectroscopy.

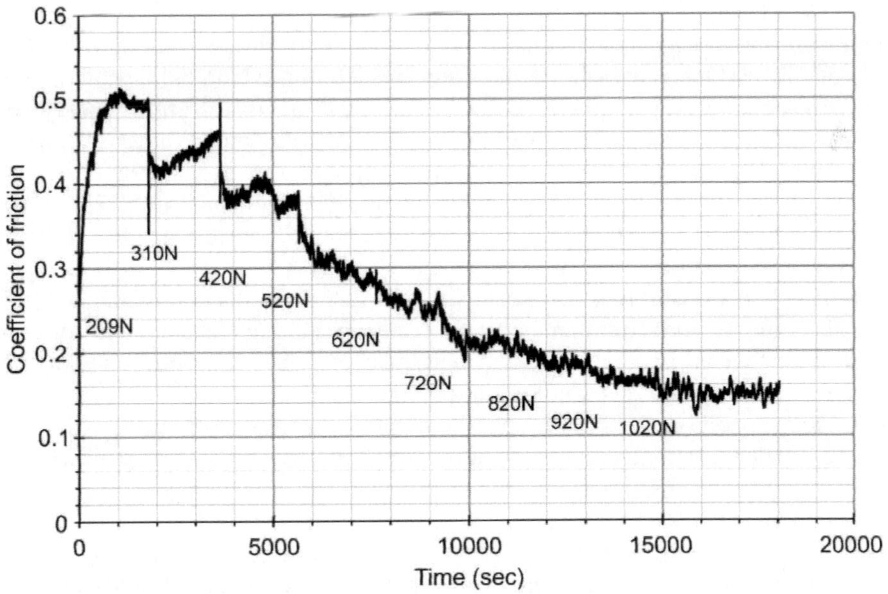

Figure 1. Dependence of the friction coefficient of Phenylone C2 + 10 % F4MB composite on the time at different load values («finger-roller» friction scheme).

As a result of tribological researches, it was established that some polymers like Polycaproamide PCA, Phenylone C2 possess rather strong adhesion at friction to the metal counterbody and do not form a highly oriented film. While the other polymers like polyethylene HDPE and fluoropolymer F4MB have weak adhesion to the metal, but differ in their ability to form transferred particles of friction polymer with a small tangential drag and a thickness of 0.1...0.3 μm.

For the research of tribocontact surfaces by FTIR spectroscopy of MFTIR we got flat samples with the test surface is not less than 10×30 mm. The design of the end friction machine allows controlling the status of the tribomating surface, i.e. the presence, the kinetics of the formation and stability of the frictionally transferred film in any time intervals without disturbing the surface. To study the surface is necessary to record the reading of FTIR spectroscopy spectra of tribomating surface at different times of tribocontact and determine the ratio of the line intensities in the spectra in different frequency ranges, related to the various components.

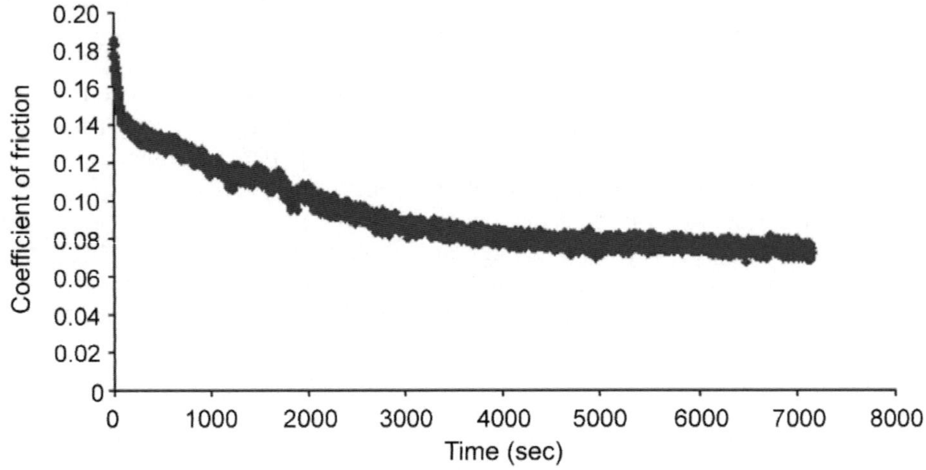

Figure 2. Dependence of the friction coefficient of Phenylone C2 + 10 % F4MB composite on the time obtained on the end type friction machine.

Typically, when recording spectra for the qualitative analysis and for determining the presence of the film on the tribomating surface, most intense lines are selected, for example, lines of valence vibration of C–H polymer molecule (2800–3200 cm^{-1}). The appearance of the absorption bands of symmetric (2900–2870 cm^{-1}) and an asymmetric valence vibration (2990–2950 cm^{-1}) in MFTIR spectrum is the criterion for the presence of a frictionally transferred film in the tribomating area. Deformation vibrations (1000–1350 cm^{-1}) and bands of adjoint oxidizing C=O bonds (1450–1750 cm^{-1}) are more informative while studying the destruction process of the polymer molecule and the structural changes in the surface layer of the sample.

The intensity and energy position of these bands is sensitive to breaks in polymer chain, adjoint end groups, chemical elements and compounds formed at tribomating. The conclusions about these processes can be made by comparing the intensity and the energy position of the spectra lines obtained from the friction surfaces with the spectra of initial polymer compositions.

The presence and condition of the frictionally transferred film at friction with various tested composites studied under different loads and tribomating times were evaluated based on analytical lines in the FTIR spectra of the friction surfaces

As a result of conducted researches it was also established that the formation process of transferred film takes place in several stages. At first mainly due to products of binding friction transfer, as at the initial stage (20–40 min) only phenylone bands 1300–1750 cm^{-1} are recorded in MFTIR spectrum and their intensity is relatively weak (figure 3, curve 1). Then an increase of friction transfer products of F4MB filler takes place (figure 3, curve 2). In the spectrum the bands of valence bond CF (1150–1200 cm^{-1}) appear. Their intensity is increasing with time of tribocontact until the process stabilization (60–90 min).

We carried out a range of studies determining the triboEMF on the transferred film during its formation to confirm that the antifriction layer of F4MB products is retained on the first PCA by electrostatic forces. It was proved that at the first stage of the film formation the charge sign is positive due to products of frictional transfer of the binder composite.

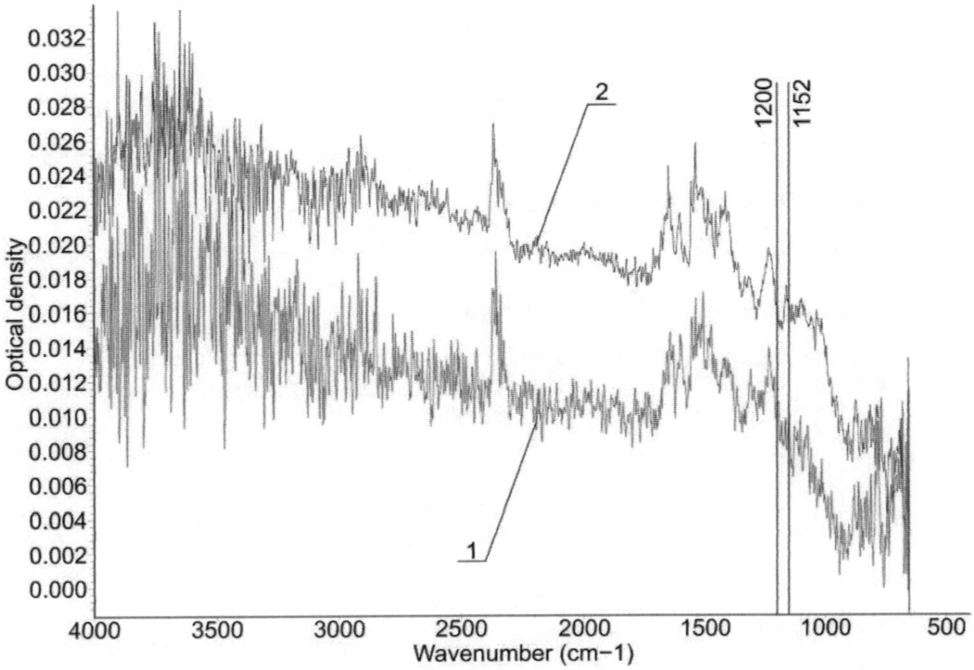

Figure 3. FTIR spectra of the frictionally transferred films on a steel disk while tribomating with composites Phenylone C2 + 20 % F4MB at the load 2 MPa and the sliding velocity of 0.5 m/s for different times of tribocontact: (1 – 30 min; 2 – 60 min).

At the second stage, when there is an increase of F4MB products, the charge sign is negative. The obtained result suggests the following mechanism of the polymeric PFP formation. First, the wear products of the binder are firmly retained by adhesion on the metal counterbody's surface at friction. Then, the wear products of the F4MB filler form a second layer, which is retained by the electrical forces on the binder due to their heteropolarity. As a result, a highly oriented transferred film is formed on the metal counterbody, which makes it possible to provide a small coefficient of friction.

Acknowledgment
This work is supported by the Russian Science Foundation under grant 14-29-00116 provided by the Rostov State Transport University.

References

[1] Mashkov Ju K, Kropotin O V, Shil'ko S V and Pleskachevskij Ju M 2013 *Self-organization and structural modification in Metropoli-dimensional tribosystem* (Omsk: OmSTU Press) 232
[2] Kolesnikov I V 2014 *Engineering Journal of Don* **4** 8

Electro-physical and mechanical properties of films based on zirconium oxide in sensitive elements of MEMS

N E Korobova and S P Timoshenkov

Department of microelectronics, National Research University of Electronic Technology "MIET", 124498, Zelenograd, Moscow, Russia

E-mail: korobova3@mail.ru

Abstract. Studies have been carried out to obtain functional films based on zirconium oxide (ZrO_2, PZT) and to ensure the stability of their metrological characteristics under harsh operating conditions – over a wide range of temperatures, thermal shock and vibration for use in microsystem technology. The structural features of film structures for establishing the relationship "structure-property", as well as assessing the role of defects have been investigated.

Indispensable materials for electrical engineering, radio engineering, instrumentation and other fields use are oxide materials [1]. The search and development of new methods for synthesizing oxide materials capable to high-efficiency, non-energy-consuming and environmentally friendly technological processes are always relevant [2]. This paper presents the study results of thin zirconium oxide films for functional purposes. Yttria-stabilized zirconia films deposited by sol-gel technology on platinum substrates and heat treated in air at temperatures from 600 to 1500 °C from 0.5 to 3.0 h were the study objects. Samples heating and cooling was carried out together with the furnace. Specific volume resistivity (ρ_V), angle tangent of the dielectric loss (tg δ), dielectric constant (ε), compressive and bending strengths (σ_{com}, σ_{ben}) were measured for initial and heat-treated samples. Measurements of electro-physical and mechanical properties were carried out in accordance with GOST 5458-75. The phase composition and samples microstructure were studied by X-ray diffraction (diffractometer DRF-2, CuK_α-radiation). The initial samples had a light gray color. Heat treatment at temperatures below 1000 °C did not lead to a noticeable change in the properties and samples color. Lighter sections appeared on the surface of the films after heat treatment at (1100 ± 100) °C, and at temperatures above 1300 °C the films were painted white. There was a significant decrease in strength properties and an increase in tg δ films with increasing temperature and time of heat treatment. Also small changes were in ρ_V and ε.

Carbon materials with protective films of stabilized zirconium dioxide have been tested under high-temperature irradiation. Irradiation was carried out in a helium atmosphere in a high-temperature reactor channel at 2000 °C. We provided fluency set from $5 \cdot 10^{19}$ to $2 \cdot 10^{20}$ neutrons/cm^2 controlling distribution of the flux density over the cell height and the irradiation time.

Irradiated samples have been used for determination by non-destructive methods change in dimensions, mass, density, electrical resistivity, modulus of elasticity. Then, flexural strength, thermal conductivity, and compressive strength were measured breaking the sample, so the adopted methods allowed obtaining maximum information from each irradiated sample.

Stabilized zirconium dioxide coating on graphite allowed the crystal lattice parameter *c* either not change, or slightly decrease ($\Delta c/c$, % = 0.1–0.3), parameter *a* did not change. Physical characteristics (specific electrical resistance ρ from 10–18 μΩ·m, thermal conductivity λ = (100 ± 5) W/mK, coefficient of linear thermal expansion (5.2 ± 0.3)·10^{-6} K^{-1}, tensile strength under compression σ_{com} = 30–50 MPa, dynamic modulus of elasticity E = (6.2 ± 0.3) GPa.

The crystal lattice parameter c_{002} = 0.6712–0.7728 nm increased upon irradiation, but the relative change of these properties decreased exponentially with the irradiation temperature. Mass loss due to oxidation and corrosion, cracking due to different forming speeds, films peeling from the substrate due to insufficient adhesion have been observed when using as a precursor metal alkoxide with an aging period more than 48 h (figure 1).

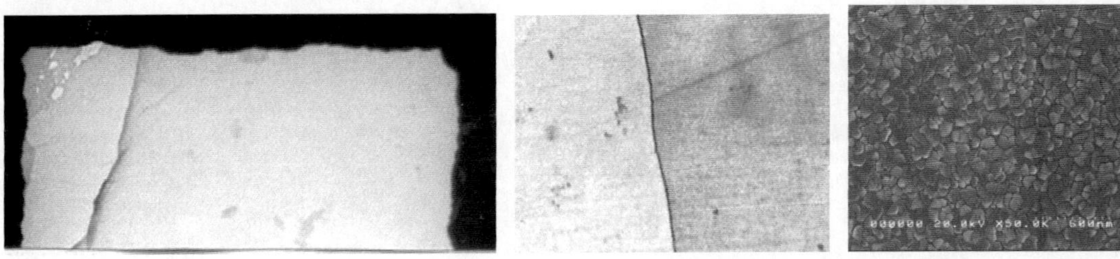

Figure 1. Film peeling from the substrate when using precursor metal alkoxide with an aging period more than 48 h.

The initial crystallization in the precursor alkoxides reduced the strength characteristics and also reduced thermal and electrical resistivity. The phase composition and mechanical properties of films based on zirconium dioxide have been studied after firing in vacuum. It was found that ZrO_2 – 3 % (mol.) Y_2O_3 films obtained by heat treatment at 1200 °C were dense samples of black color with an average grain size 50 nm containing T and T' phases. Oxygen vacancies in such film created as a result of annealing were caused a reduced volume of the unit cell of the tetragonal phase and also caused its stability under various mechanical influences until destruction, and at cooling up to –180 °C.

Piezo-ceramics is a promising material for the production of sensitive sensors and high-precision actuators. At present, lead zirconate titanate (PZT – $Pb(Zr_xTi_{1-x})O_3$) is widely used in the electronics industry. The advantages of devices based on PZT materials include high sensitivity and accuracy, wide operating temperature range [3–4]. Precursor – ceramic PZT powder – was applied to silicon plates covered with platinum electrode, by electrophoresis in isopropyl alcohol. Under the applied electric field (E = 100–300 V/cm), charged PZT particles in suspension with (0.01–10 mass %) concentration were deposited to one of the electrodes within 0.5–10 min. After electrophoretic deposition PZT film was subjected to drying and annealing. The phase and structural film non-equilibrium have been studied with the help of XRD and (SEM) taking into account the multi-component ferroelectrics nature and complex structure of formation processes. Optimum regimes of electrophoretic deposition and high-temperature annealing were found. It was found that films with a thickness from 6 to 16 μm, which had been annealed at 880 °C during 20 min, had porous structure, and tuberous surface with micro-cracks (figure 2).

Amorphous phase presence was noted in the films. Film samples with a thickness from 7 to 10 μm obtained under the same conditions, but having annealing at 950 °C for 1 h, were solid, nonporous, and had a homogeneous polycrystalline structure. Using samples with Si-SiO_2-Ti/TiO_2-Pt-PZT hetero structures, it was shown that the perovskite texture was determined by two factors. {111} determines platinum layer orientation, {100} appears in the presence of PbO and/or TiO_2 oxide phases at the interface between PZT and Pt in the activation case of diffusion processes by high-temperature annealing. Structural imperfections that arise during the films formation lead to micro stresses.

The elastic substrate interaction with the film can be considered as one of the reasons for the appearance of micro-stresses. The "film-substrate" system is a complex heterogeneous system with different mechanical characteristics (as it is observed in MEMS). The thinner film, the more substrate influence has on the state of film lattice. Due to the vacancies diffusion to the surface and the "healing" of micro-cavities, the film volume decreases during annealing, and tensile stresses arise in it, which relax by the formation of cracks. Compressive macro-stresses, which relaxation leads to the surface folds were found in the films obtained from aged sols. Films made from fresh sols were not so susceptible to "folding." Significant differences in the parameters of massive and film ferroelectrics include a wide scatter of characteristics. Two factors are responsible for this phenomenon- structural imperfections and dimensional effects. Since the true dimensional effects essentially manifest themselves only at thicknesses less than 0.05 μm or at the sizes of their constituent particles less than 0.05 μm, the main influence on the properties of the resulting films was exerted by various structural defects.

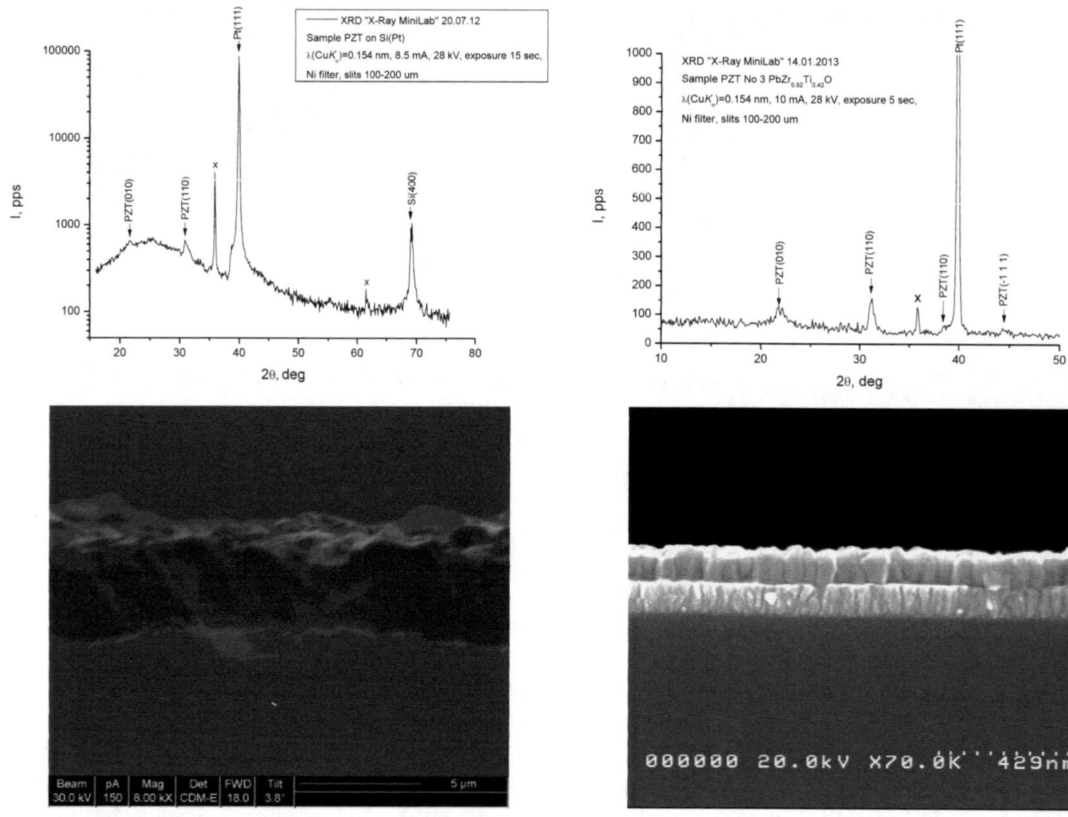

Figure 2. XRD and SEM of PZT films after annealing at: **(a)** – 880 °C during 20 min; **(b)** – 950 °C during 1 h.

Thus, from the above results it follows that new approaches to thin film formation technology have provided coatings on complex profile large-sized substrates with an effective "price-property" ratio. A high efficiency of thin films structure optimizing for a particular application was shown by using metal alkoxides with different aging times.

Acknowledgements

This work was financially supported by the Russian Science Foundation with unique identifier for applied research is 16-19-00177.

References

[1] Chaplygin Yu A ed. 2015 *Nanotechnology in electronics* **3** (Moscow: Technosphere) 480

[2] Uchino K ed. 2010 *Advanced Piezoelectric Materials: Science and Technology* (Woodhead Publishing Series in Electronic and Optical Materials) 687
[3] Korobova N 2015 *Science of Advanced Materials* **7** 1–5
[4] Korobova N, Timoshenkov S and Vodopyanov V 2013 *Materials Science and Technology Conference and Exhibition 2013 (MS&T)* 2656–62
[5] Korobova N, Timoshenkov S and Mukimov D *2014 Materials Science and Technology (MS&T)* 127–31

Laser cladding technological machine. Investigation of efficiency of various nozzles design

R S Korsmik[1], G A Turichin[1,2] and K D Babkin[1]

[1] Peter the Great Saint-Petersburg Polytechnic University, 195251, Saint-Petersburg, Russia
[2] Saint-Petersburg State Marine Technical University, 190121, Saint-Petersburg, Russia

E-mail: r.korsmik@ltc.ru

Abstract. The paper shows the correlation between nozzle design and efficiency of laser cladding process. The nozzles developed by authors are characterized by uniform distribution of the powder in the jet transection. The stable formation of the cladding beads is achieved when the gas-powder jet focused below the processing surface.

1. Introduction

An integral part of increasing the life process of turbo-fan engines is their maintenance and repair. In particular, this applies the main and most expensive mass-use structure elements – working blades. The causes of their deterioration or breakage are vibrations, shocks, abrasive deterioration, high temperatures and other factors during operation [1]. The cladding processes are used for restoration of damaged turbine blades.

TIG [2, 3] and EB [4, 5] cladding technology does not provide the same quality as laser method [6]. Both wire [7] and powder [8] are used as filler material. In case of wire-filler, the whole amount of material transfers to the cladded track. It is unquestionable advantage of this type of filler material from the economic side of technology. However, the fact that all amount of wire should be melted in the action zone causes the number of technological difficulties during automation process. In view of this, filler wire fed manually and cladding duration of one blade could be up to 1 h [8]. In case of powder-filler, material transfers to cladded track partly. Amount of cladded powder depends on number of technological parameters of process and design of equipment, which influence to powder transportation and melting processes [9]. In the case of large areas cladding, it is advisable to use a laser beam scanning to increase the area of the melting zone. In this case, weight of cladded material could be up to 80 % of fed powder [10]. As applied to restoration cladding of turbine blades, scanning of laser beam is not expediently, because of narrow width of processing area (approximately 1 mm). In this way, there is an issue to increase amount of powder, which transfers to cladding, with help of better focusing ability of gas-powder jet. This paper shows powder transfer coefficient investigation in depending on nozzles design, which are working tool of laser cladding technological machines.

2. Methods, materials and equipment

Experimental research was done on laser cladding technological complex based on industrial robot LRM-200iD_7L, Fanuc; laser irradiation source LK-700, IPG Photonics; laser head FLW D30, IPG

IOP Conf. Series: Journal of Physics: Conf. Series **857** (2017) 012021 doi:10.1088/1742-6596/857/1/012021

Photonics with removable cladding nozzles (table 1); powder feeder Oerlikon Metco Powder Feeder Twin 150 with track transection of metering disk of 5×0.6 mm^2.

Table 1. General geometrical investigating nozzles parameters.

Nozzle	Cone angle, deg	Gap between cones, mm
ILWT-N2	36.0	0.26
ILWT-N3	40.0	0.24

Powder of Co-alloy EuTroLoy16006-04 (Stellite 6) Castolin Eutectic was used for experiments. The fracture grain size of powder is 20–53 and 50–150 μm. For calculating powder transfer coefficient was used formula:

$$K = \frac{M/T}{G}, \quad (1)$$

where M – is powder mass, transferred through hole in container per time unit T; G – nominal powder rate. Nominal powder rate was determined by weighting of powder, transferred to container in time unit T. Experiments for measuring of powder transfer coefficient were done with use of laboratory stand, which models melting pools of various size. Stand consists of plate with holes of 0.8; 1.5; 2.0; 3.0; 4.0 mm diameter, which is parallel to nozzle edge, and container that is placed on the weighting-machine under the plate. Powder transfer coefficient was studied for each of holes with gap between nozzle edge and plate from 4 to 16 mm.

The series of cladding beads were done to actualize results of modeling experiment. The gap between nozzle edge and plate was from 6 to 15 mm. For the actual powder transfer coefficient were calculated duration of 50 layers. Cladding parameters: laser power – 300 W, beam diameter on the substrate – 0.8 mm, cladding speed – 20 mm/s.

3. Results and discussion

Nominal powder rate determined in result of experiment is 12.9 g/min. Results of measurement transferred powder mass through the holes shows that powder transfer coefficient decreases with decreasing of hole diameter. The figure 1(a), (b) shows depending powder transfer coefficient for nozzles "ILWT-N2" and "ILWT-N3".

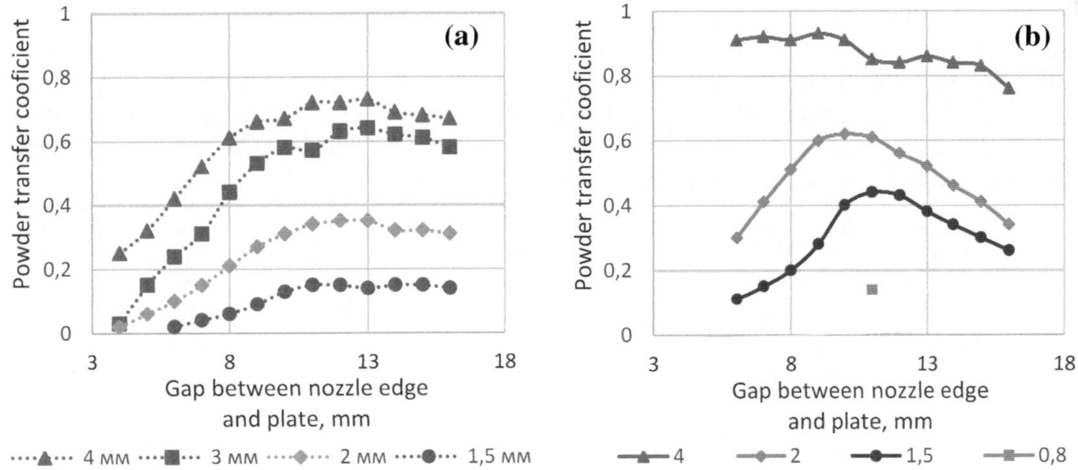

Figure 1. The powder transfer coefficient in depending of distance to plate and hole diameter: **(a)** – nozzle ILWT-N2"; **(b)** – nozzle "ILWT-N3".

IOP Conf. Series: Journal of Physics: Conf. Series **857** (2017) 012021 doi:10.1088/1742-6596/857/1/012021

During the influence investigation of gap between nozzle edge and plate, was found a maximum value, which means about focus of gas-powder jet. The focus distance for nozzle "ILWT-N2" is 13 mm. With decreasing of hole diameter the curve inflection becomes more expressed. The focus distance of gas-powder jet for nozzle "ILWT-N3" is 11–12 mm from the bottom edge. The curve inflection is more expressed for diameters up to 2 mm. For hole diameter of 4 mm dependence character becomes more smooth. The powder transfer coefficient at least does not change with gap form 6 to 10 mm, and saves maximum value.

The powder transfer coefficients in the focuses of gas-powder jets were compared to determine efficiency of nozzles. Comparing curves are showed in figure 2.

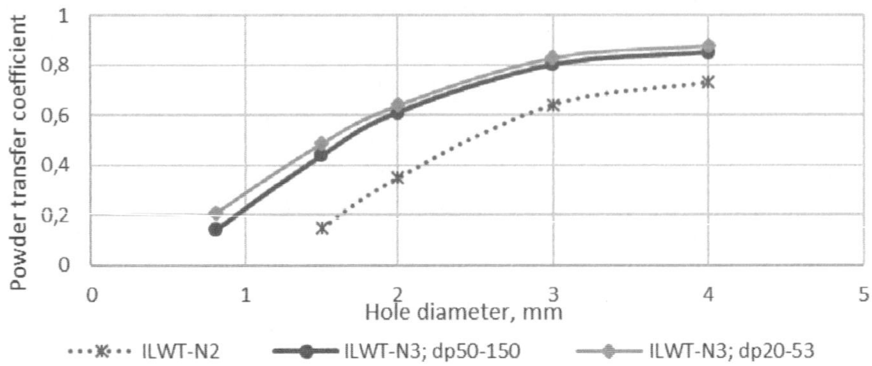

Figure 2. Powder transfer coefficient in the focus point of gas-powder jet.

As shown in diagram, the nozzle "ILWT-N3" provides higher powder transfer coefficient then nozzle "ILWT-N2". In case of nozzle "ILWT-N3", the powder transfer coefficient increases in 1.2–3 times for holes 1.5–4 mm in comparison with nozzle "ILWT-N2". For hole diameter of 0.8 mm the powder transfer coefficient is 0.14 whereas nozzle "ILWT-N2" cannot focuses gas-powder jet to such size.

Investigation of powder fracture influence reveals that decreasing of grain size insignificantly increases powder transfer coefficient in 1.03–1.5 times.

For further cladding investigation were solved to use nozzle "ILWT-N3", which provides higher powder transfer coefficient, and powder with fracture of 50–150 μm, which costs lower. In the results of cladding, 10 built-up "walls" were obtained. The width of each wall is 1.15 mm.

The height of walls is not equal. It changes from 5.03 to 15.83 mm. Obtained walls could be separated to two groups. The first one is walls with stable formation (the focus of gas-powder jet is below of substrate surface). The second group is walls with unstable formation (the focus of gas-powder jet is above of substrate surface) [11]. Maximum height of wall is achieved during cladding with gap between nozzle and substrate surface is 12 mm. The figure 3 shows dependence of cladded height and distance between nozzle edge and substrate surface.

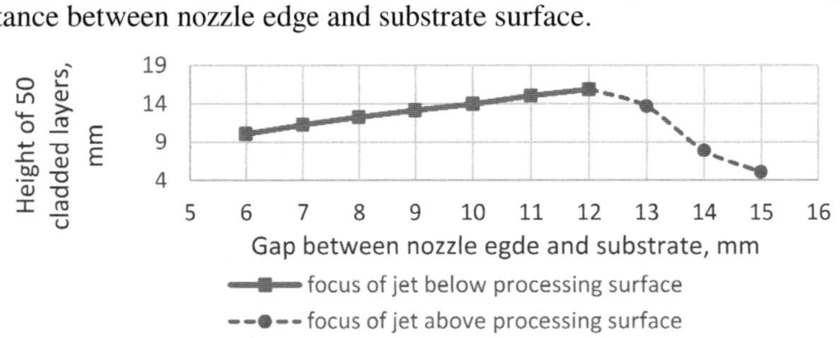

Figure 3. Dependence of cladded height and gap between nozzle edge and substrate surface.

Cladding duration of 50 layers is 75 s. It means that 16.13 g of powder was fed (powder rate is 12.9 g/min). Cladded mass is 4.18 g. In this way, actual powder transfer coefficient is 0.26.

4. Conclusions

Based on conducted experiments it can be seen, that various designs of cladding nozzles provide different values of powder transfer coefficient. The nozzle "ILWT-N3" has better focusing ability than the nozzle "ILWT-N2" and increases the material utilization during the processing of small areas.

The proposed modeling method makes it possible to compare the different nozzles without conducting cladding experiments. Comparing the powder transfer coefficient, measured after cladding the samples, with the powder transfer coefficient, measured after the experiments at the stand, which simulates melting pools of different diameters (figure 2), it can be seen that the results are practically the same.

Surfacing must be done at the distance between the nozzle edge and the cladding surface, which does not exceed the focal length of the gas-powder jet provided by the nozzle. Investigation have shown a monotonous increasing of the deposited height (due to increasing of powder transfer coefficient), up to the focus point of the gas-powder jet. When the focus point of gas-powder jet sets above of cladding surface, the powder transfer coefficient decreases because of the low and unstable concentration of particles in the jet. This leads to excessive losses of filler material and leaving from a stable condition of process.

Acknowledgments

The work was carried out with financial support from the Ministry of Education and Science of the Russian Federation in the framework of realization complex project "Development of restoration of gas turbine engine working blades by laser cladding method". Contract No 02.G25.31.0215 from 27.04.2016.

References

[1] Korsmik R S, Turichin G A, Klimova-Korsmik O G, Zemlyakov E V and Babkin K D 2016 *Vestnik of Samara University. Aerospace and Mechanical engineering* **15** 60–9
[2] Sorokin L I 2004 *Svarochnoe Proizvodstvo* **7** 36–8
[3] Klimov V G 2016 *Vestnik Moskovskogo aviatsionnogo instituta* **23**/1 86–97
[4] Shulov V A, Paikin A G, Bytsenko O A, Teryaev D A, Engelko V I and Tkachenko K I 2010 *Strenghtening Technologies and Coatings* **3** 34–8
[5] Serkov A V, Lonshakova O N, Tikhomirov A E, Babich I I, Gejkin V A, Puzanov S G, Fokin G A, Kronev S A and Matveev A N 2015 *Patent RU 2419526 Method of repairing surface defects of gas turbine blade body*
[6] Zemlyakov E V, Babkin K D, Korsmik R S, Sklyar M O and Kuznetsov M V 2016 *Photonics* **4** 10–25
[7] Shitarev I L, Smelov V G and Sotov A V 2014 *Proceedings of V International scientific-practical conference "Innovation technologies and economy of mechanical engineering"* **1** 119–23
[8] Klimov V G, Zhatkin S S, Schedrin E Yu and Kogteva A V 2015 *Proceedings of the Samara Scientific Center of the Russian Academy of Sciences* **17** 782–8
[9] Turichin G A, Somonov V V and Klimova O G 2014 *Applied Mechanics and Materials* **682** 160–5
[10] Turichin G A, Zemlyakov E V, Pozdeeva E Yu, Tuominen J and Vuoristo P 2012 *Metal Science and Heat Treatment* **54** 139–44
[11] Turichin G A, Zemlyakov E V, Klimova O G and Babkin K D 2016 *Physics Procedia* **83** 674–83

IOP Conf. Series: Journal of Physics: Conf. Series **857** (2017) 012022 doi:10.1088/1742-6596/857/1/012022

Spectral analysis of the charge and elemental composition of the vacuum arc discharge plasma flux during deposition of carbon coatings

D K Kostrin[1], S A Trifonov[1] and A A Lisenkov[2]

[1] Department of electronic instruments and devices, Saint Petersburg Electrotechnical University "LETI", 197376, Saint Petersburg, Russia
[2] Institute of Problems of Mechanical Engineering, Russian Academy of Sciences, 199178, Saint Petersburg, Russia

E-mail: dkkostrin@mail.ru

Abstract. In the present work spectral analysis of the charge and elemental composition of the plasma flux of a vacuum arc discharge on example of clear dispersion of the graphite cathode and also when adding benzene or argon as a working gas is studied. It is shown that the intensity of the main spectral lines of the atomic and molecular carbon in the emission spectrum of the vacuum arc discharge plasma is almost constant for all considered modes of operation, while the intensity of the spectral lines of ionized carbon changes significantly when adding a working gas.

1. Introduction

Carbon materials and coatings present a great interest in the various fields of science and technology. The unique ability of carbon is an almost infinite number of its condensed forms. One of the most promising technological methods of receiving carbon films and coatings is a vacuum arc discharge that is a high current, self-maintained discharge, growing in a closed volume in the cathode material vapors [1, 2]. The emission center of this discharge is a cathode spot, that is characterized by small dimensions and which is the source of the primary electrons and the base of the arc column [3]. High temperature in the spot causes intensive sputtering of the cathode material and a high concentration of a metal vapor [4–6].

Specificity of a vacuum arc discharge is the fact that the total ion current is only 6...10 % of the value of the discharge current. In the widespread vacuum arc plasma sources of coaxial structure due to the presence of an external axial symmetric magnetic field density distribution of the charged particles in the working volume is not uniform and the maximum of the distribution is located on the axis of the system (figure 1).

2. Features of the vacuum arc discharge affecting the charge composition of the plasma flux

Processes in the arc discharge are divided into:

- generating, occurring in the cathode spot (in the molten layer of a liquid metal) and which have an impact on the cathode surface and the plasma parameters;
- processes in the cathode area, determining the conditions of generation, dynamics of the development and movement of the cathode spots;

• processes in the area of transportation, in other words in the interelectrode space, starting from a region with a strongly nonideal low temperature plasma.

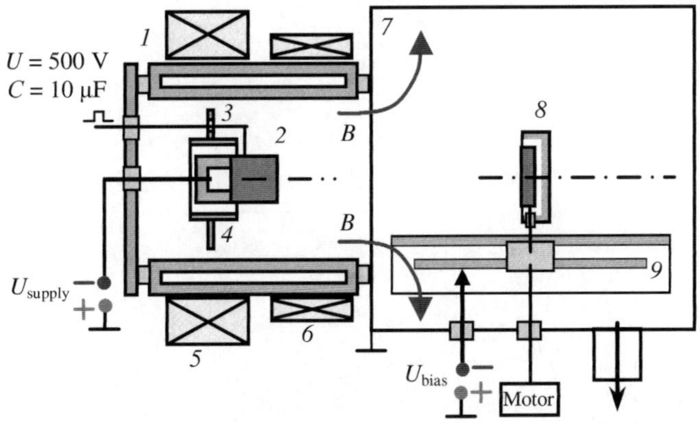

Figure 1. Design of a coaxial vacuum arc plasma source: *1* – anode; *2* – cathode; *3* – igniting electrode; *4* – shield; *5*, *6* – stabilizing and focusing solenoids; *7* – working volume; *8* – probe or workpiece; *9* – planetary gear mechanism.

Multiple ionization observed in the plasma of a vacuum arc discharge leads to an increase in the average charge of ions and growth of ion beam energy without increase of the accelerating voltage [7–9]. In the cathode area ionization of the plasma stops and there is a stabilization of its parameters. However, it should be noted that in the vacuum arc plasma sources, plasma parameters are changed both in the process of evaporator operation (cathode temperature, discharge current, magnetic field), and when using the working gas which provides the degree of ionization of the plasma flux (argon, helium) and the implementation of the plasmachemical synthesis of compounds (for carbide compounds carbon containing gas is introduced into the plasma flux). Therefore analysis of the charge and elemental composition of the plasma flux in all areas of transportation is a vital task.

Degree of ionization of the erosion products and their charge composition depends not only on the material of the cathode, but also on the mode of operation of the evaporator. The degree of ionization of the vapor of the cathode material is determined by its properties. The proportion of ionized particles in the flux emitted by the surface of the cathode for such materials as carbon, molybdenum and tungsten is 70, 80 and 90 % respectively. With decrease in boiling point of the material this percentage is reduced to 30…50 % for copper and silver, and 15...25 % for cadmium and zinc. Products of the erosion of the cathode contain ions of different degree of charge. Basically there are single, double and triple charged ions. While the more refractory material is (except graphite) the higher the average charge of the ions in the plasma of a vacuum arc discharge.

3. Spectral analysis of the charge and elemental composition of the plasma flux

For the determination of the plasma flux composition an emission spectral analyzer on the basis of a modernized one dimensional CCD array Toshiba TCD1304 with 3648 pixels (8×200 μm) was used [10–12]. The investigated range of wavelengths of the plasma was 350…950 nm (optical resolution of the spectrometer is 1 nm). To identify the components that are part of the plasma flux the emission spectra of the arc discharge in the area of the substrate were investigated. Figure 2 shows spectrum of the arc discharge in the visible range obtained in the area of transportation during sputtering of a graphite cathode PGM-6 in vacuum at a pressure of $7.8\cdot10^{-3}$ Pa and a discharge current of 80 A. Spectral lines show presence of only pure carbon in the plasma flux.

Increase of the discharge current leads to the decrease of the content of charged ions in the flux and growth of the number of singly charged ions. Thus the maximum of the ion energy distribution is shifted towards smaller energies.

For the percentage increase in the carbon component being deposited on the treated surface vapors of benzene (C_6H_6) or another carbon containing gas are introduced into the working volume. In this case to the chemical processes a hydrocarbons decomposition reaction due to the interaction with elements of the plasma flux and by pyrolysis is added. Speed of a chemical reaction is proportional to the current concentration of the reactant products (law of mass action or law of Guldberg and Waage).

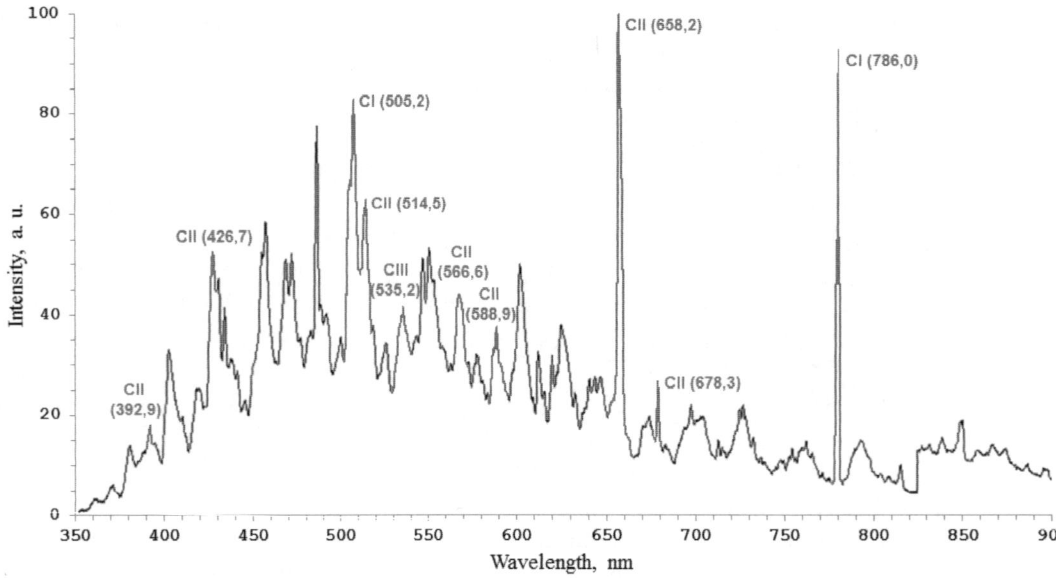

Figure 2. Spectrum of the plasma radiation in the process of pure graphite sputtering.

Noticeable decomposition of the aromatic hydrocarbons in the gas phase is observed at the temperatures of about 900 K and more. Under these conditions molecule becomes thermodynamically unstable and decomposes to fragments with a smaller number of atoms or to single atoms. In the gas discharge plasma formation of CH_x radicals is due to the process of dissociation by a direct electron impact. However the less hydrogen atoms are contained in the molecule the easier is the decomposition of hydrocarbon.

When using benzene vapor splitting of the hydrocarbons with long bonds to molecules of a shorter length with the formation of a whole range of hydrocarbons is as follows:

$$C_6H_6^{\uparrow} + e \xrightarrow{W,T} C_nH_{Xn}^{\uparrow} + C_xH_{Yn}^{\uparrow} + H_2^{\uparrow} \xrightarrow{W,T} C\downarrow + C_xH_Y^{\uparrow} + H_2^{\uparrow}.$$

Figure 3 shows spectrum of the plasma radiation during graphite sputtering with the addition of benzene at a pressure of $8.7 \cdot 10^{-2}$ Pa.

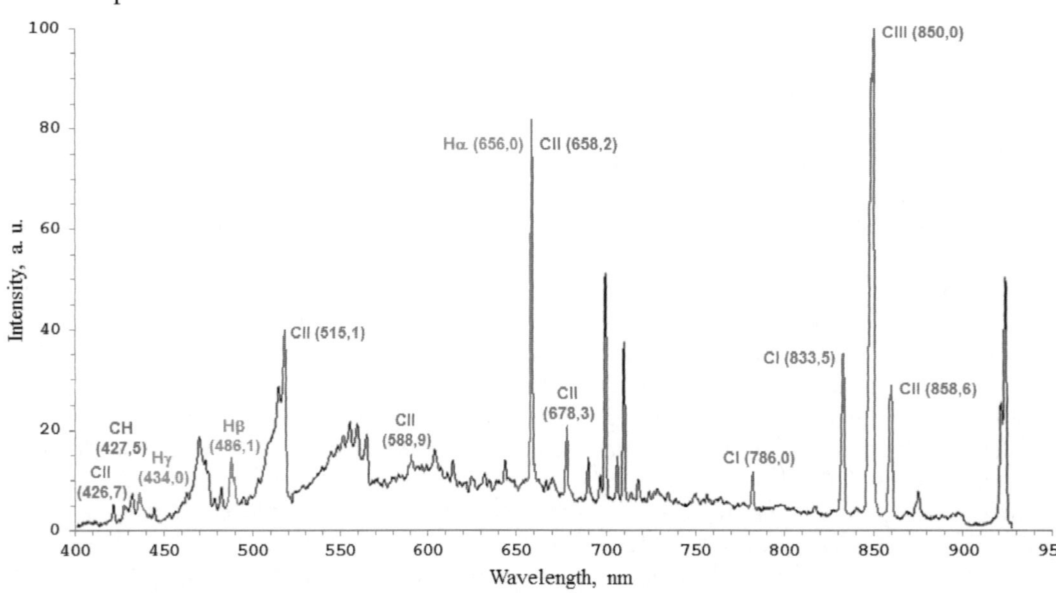

Figure 3. Spectrum of the plasma radiation in the process of graphite sputtering with added benzene.

IOP Conf. Series: Journal of Physics: Conf. Series **857** (2017) 012022 doi:10.1088/1742-6596/857/1/012022

Introduction of the benzene vapor as a working gas into a plasma flux significantly alters the radiation spectrum. For this spectrum appearance of the Balmer series lines of the atomic hydrogen (H_α and H_β), line of carbon ion CII and emission spectra of C_2 (Swan band $d^3\Pi$–$a^3\Pi$) and CH (Gere band $A^2\Delta$–$X^2\Pi$) molecules is characteristic.

At a temperature of about 2000 K the first radicals of C–H system will be atomic hydrogen and acetylene C_2H_2, concentration of which with further increasing temperature rises dramatically. After 2500 K (temperature of complete decomposition of methane) new radicals CH, C_2H, CH_3 and C_3 emerge, with their quantity increasing with further rise of the temperature. Starting from 3000 K there is a significant amount of monatomic and after 3500 K – diatomic carbon. At temperatures above 3500 K diatomic carbon C_2 and CH reach their maximum concentrations, while the content of C_2 in the system is close to the content of acetylene C_2H_2. Presence of the intense bands of C_2 and CH in the spectrum during pyrolysis (figure 3) shows that the average temperature of a gas according to the equilibrium composition of C–H system is not less than 3000–3500 K.

For maintenance of a stable discharge into the working volume a buffer inert monatomic gas argon (or helium), which do not enter into chemical reactions and form compounds, is often added. The first ionization potential for argon is 15.75 V and for helium – 24.47 V. Due to the chemical reactions the presence of a buffer gas helps to increase the degree of ionization of the plasma flux:

$$Ar+e \rightarrow Ar^+ + e + e; \quad Ar + e \rightarrow Ar^* + e; \quad Ar^* + C \rightarrow Ar + C^+ + e + h\nu; \quad Ar^+ + C \rightarrow Ar + C^+.$$

Figure 4 shows the spectrum of plasma radiation during sputtering graphite with the addition of argon, used as a buffer inert monatomic gas.

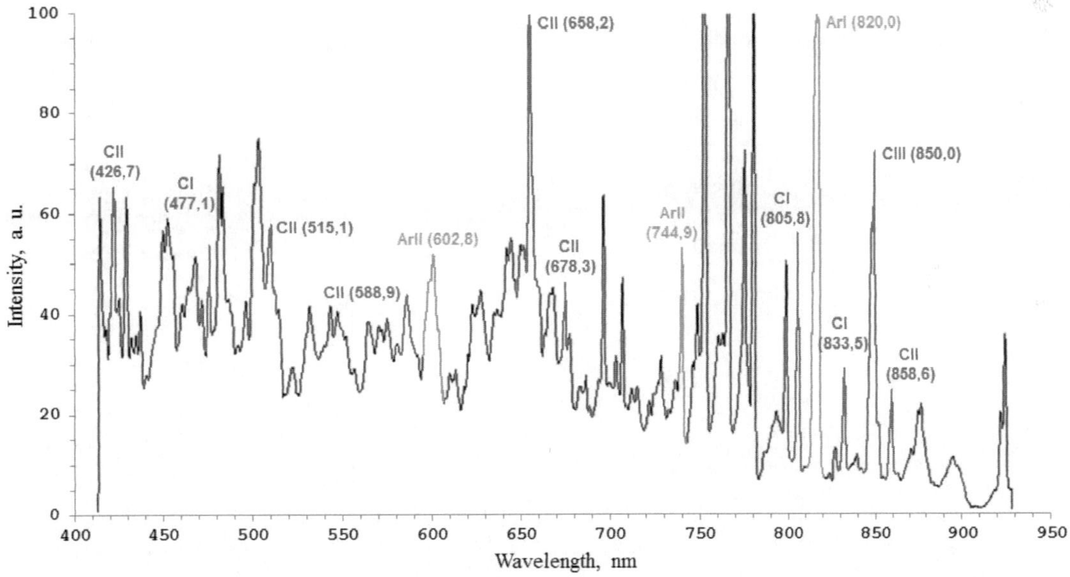

Figure 4. Spectrum of the plasma radiation in the process of graphite sputtering with added argon.

Due to the chemical reactions the presence of a buffer gas helps to increase the degree of ionization of the plasma flux that leads to the appearance on the spectrum of additional lines of carbon (CI, CII, CIII), as well as spectral lines of argon (ArI, ArII).

4. Conclusion

Obtained results show that the intensity of the atomic and molecular lines of carbon is almost constant for all investigated modes of operation, while the intensity of lines of the ionized carbon changes significantly when adding the working gas. Interaction of particles in a dense plasma of a vacuum arc discharge leads to the appearance in the plasma flux of multiply charged ions which would not otherwise emerge due to the lack of energy required for such a degree of ionization.

References

[1] Anders A 2008 *Cathodic Arcs. From Fractal Spots to Energetic Condensation* (New York: Springer)
[2] Bystrov Yu A, Vetrov N Z, Lisenkov A A and Kostrin D K 2014 *Vakuum in Forschung und Praxis* **5** 19–23
[3] Bystrov Yu A, Kostrin D K, Lisenkov A A and Vetrov N Z 2015 *Vakuum in Forschung und Praxis* **2** 22–5
[4] Bystrov Yu A, Vetrov N Z and Lisenkov A A 2014 *Technical Physics Letters* **12** 1126–9
[5] Bystrov Yu A, Vetrov N Z and Lisenkov A A 2013 *Technical Physics Letters* **10** 914–6
[6] Bystrov Yu A, Vetrov N Z and Lisenkov A A 2012 *Technical Physics Letters* **10** 938–40
[7] Anders A 2012 *Plasma Sources Science and Technology* **21** 035014
[8] Anders A 2001 *IEEE Transactions on Plasma Science* **29** 393–8
[9] Bugaev A S, Gushenets V I, Yushkov G Y, Oks E M, Anders A, Brown I, Gershkovich A and Spadke P 2001 *Russian Physics Journal* **9** 912–20
[10] Uhov A A, Gerasimov V A, Kostrin D K and Selivanov L M 2014 *Journal of Physics: Conference Series* **567** 012039
[11] Kostrin D K, Uhov A A and Lisenkov A A 2016 *Vakuum in Forschung und Praxis* **3** 34–7
[12] Kostrin D K, Lisenkov A A, Uhov A A and Ramazanov A N 2016 *Journal of Physics: Conference Series* **729** 012030

IOP Conf. Series: Journal of Physics: Conf. Series **857** (2017) 012023 doi:10.1088/1742-6596/857/1/012023

Method of correction optical spectra for absorption

E N Kotlikov and E V Iurkovetc

Institute for Innovation and Master Degree Basic Training, Department № 3, State university of aerospace instrumentation, 190000, Saint-Petersburg, Russian Federation

E-mail: ekotlikov45@mail.ru

Abstract. A new method for determining optical constants of films is proposed. The method is based on correction optical spectra for absorption. Free from absorption bands spectra can be used for determining optical constants by the known methods. The new method is illustrated by an example of barium fluoride spectrum at the range 1.5–12 µm.

There are a number of spectrophotometric methods for determining optical constants of films that are based on the measurement of reflectance and transmittance spectra [1–3]. These methods are applicable for unabsorbing films, when there exist analytical solutions for optical constants at the transmittance éxtremum points [3]. However, application of these analytical methods is impossible in the presence of absorption bands. In this case, different numerical-analytical methods are used. These methods are based on using the minimization quality function. Note that this problem is an inverse one and, as a rule, it has several solutions.

A method of correction of the optical spectra of films is proposed. It allows one to exclude absorption from the spectra. The method is based on additivity of the law of energy conservation absorbed in the film and the possibility to separate it into the energy in transmittance and the energy in reflectance. After this correction we obtain transmittance and reflectance spectra without absorption, which allows us to use the known methods for determining optical constants of unabsorbing films.

Let us consider the correction method for absorption A of the optical spectra of reflectance R and transmittance T. The value of the film's absorption A can be calculated (if we know the thickness h and the refraction coefficient n) through the extinction coefficient k (or absorption coefficient α). All these variables depend on a wavelength. R, T and A spectra are related by the law of energy conservation. Absorption A introduces some distortions into the reflectance and transmittance spectra. Since A is an additivity variable, we can present absorption A as the sum of its contributions in the reflectance A_R and in the transmittance A_T:

$$A = A_R + A_T, \tag{1}$$

here A_T and A_R are defined as:

$$A_T = T_0 - T; \quad T_0 = T + A_t; \tag{2}$$

$$A_R = R_0 - R; \quad R_0 = R + A_r, \tag{3}$$

where R_0, T_0 are the reflectance and transmittance coefficients of the unabsorbing film that has the same values of the refraction coefficient as the absorbing film. If we know A_R and A_T, we can correct the spectra for absorption by adding these variables to the reflectance and transmittance coefficients for each wavelength.

To calculate $A_R(\lambda)$ and $A_T(\lambda)$ we use the correction functions f_R and f_T that determine the contribution of the total absorption A to $A_R(\lambda)$ and $A_T(\lambda)$. Let us introduce these correction functions f_R and f_T:

$$f_R = \frac{R_0 - R}{A}, \quad f_T = \frac{T_0 - T}{A}. \tag{4}$$

We can separate the absorption A into two parts (A_R and A_T). The correction functions show the contribution of the distortions introduced by the absorption A into the reflectance and transmittance spectra. Therefore the sum $A_R + A_T = 1$.

The corrected reflectance and transmittance spectra are calculated by the equations:

$$R_{cor} = R + f_R A, \quad T_{cor} = T + f_T A. \tag{5}$$

It follows from (5) that in order to find f_R and f_T we have to know R_0 and T_0, i.e. the spectra free of absorption. Actually, it is our aim to determine these spectra. To find analytical equation for the optical constants of the spectra free of absorption is not possible because there are several solutions. To produce a model of the absorption spectrum corresponding to the real one seems to be impossible because it is an inverse problem with the same difficulties as the one of determining optical constants.

The following method is proposed to calculate the correction functions. We introduce the model of an absorbing film, with its spectra approximately corresponding to the spectra of the film considered. Then the spectra R, T and A of this model are calculated. In addition, the reflectance and transmittance spectra of the film's model without absorption, but with the same refraction coefficients as the absorbing film are also calculated and then their spectra R_0 and T_0 are calculated as well. After calculating R_0, T_0, R, T and A the correction functions f_R and f_T are to be calculated using (5) for different film models. The correction functions f_R and f_T are to be used to correct the experimental spectra of the films. The calculation of R_0, T_0, R, T, A, f_R and f_T spectra has been performed by the program FilmMgr [4], which provides the possibility to introduce both the dispersion of the refraction coefficients and the absorption of the films.

The calculated correction functions f_R and f_T depend on the film's parameters and the form of its absorption spectrum. The impact of an amplitude and an absorption spectra form on the correction function has been analyzed. Absorption spectra with different amplitudes and widths have been considered.

The correction functions for film's models with the refraction coefficient 1.35, the thickness 3.14 μm on ZnSe substrate at the range 2–4 μm are shown in figure 1. Graphs 1 correspond to the absorption amplitude 8 %, graphs 2 – 16 %, graphs 3 – 30 %.

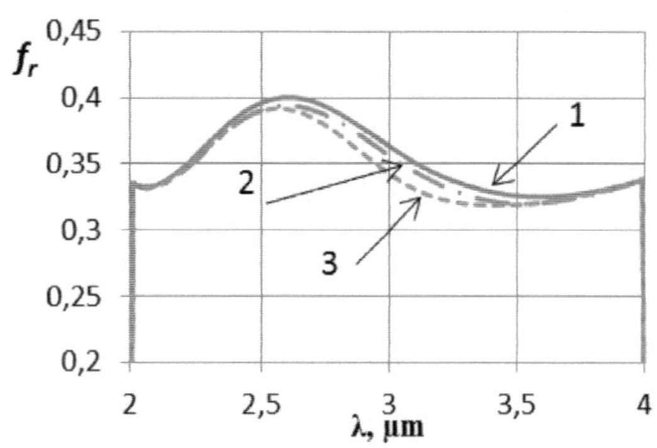

(a)

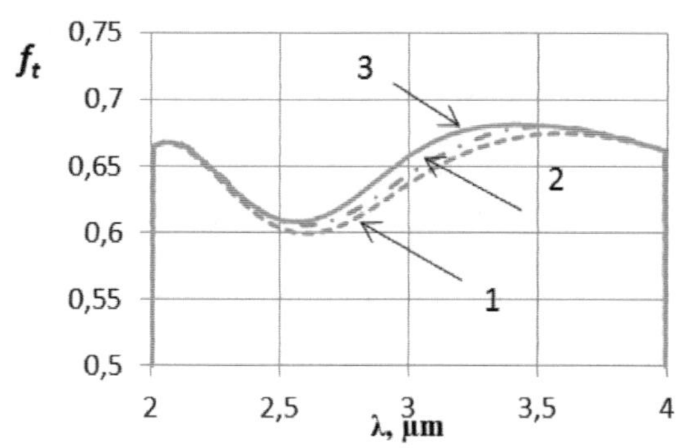

(b)

Figure 1. The correction functions f_R **(a)** and f_T **(b)** for different models of films.

Figure 1 shows that the main character of the correction functions is connected with the extrema of the film's transmittance. Note that the thickness of this film should be found with the minimal error. It follows from the figures 1(a), (b) that the errors of the spectra correction do not exceed 0.2 % when using the correction functions for 30 % absorption. This error corresponds to the accuracy of the spectra amplitude measurements. In other words, the accuracy of the spectra resulting from the correction corresponds to the accuracy of the spectra measurements regardless of the absorption spectrum form.

As an unabsorbing substrate we have used ZnSe plates of CVD make, on which the films of barium fluoride with the thickness 3.14 μm have been applied by the resistive coating method. The spectra of the films have been measured by spectrophotometer FSM-1201 with the accuracy of 0.2 %.

Figure 2 shows the reflectance (1), transmittance (2) and absorption (3) spectra of the barium fluoride on the ZnSe substrate. The corrected spectra of reflectance (4) and transmittance (5) are shown in the same figure.

The refraction and absorption coefficients are found from the corrected spectra by the method described in [4]. The results of the refraction coefficient calculations are shown in figure 3.

The reflectance and transmittance spectra of the film with absorption have been calculated by the already determined data of the refraction and absorption coefficients. They are shown in figure 2 and coincide with the corrected spectra (graphs 1 and 2) with ±0.3 % accuracy.

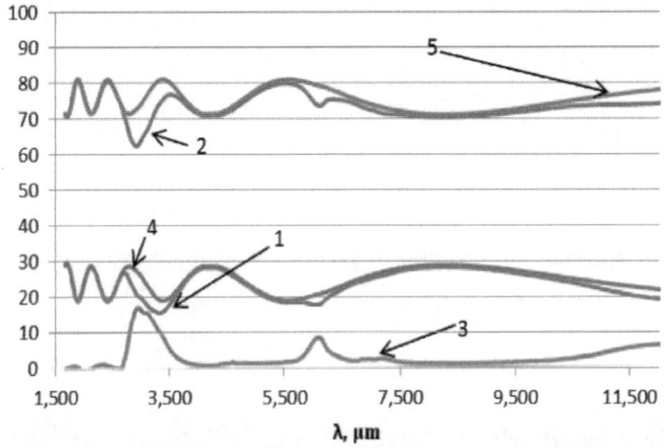

Figure 2. The spectra of reflectance (1), transmittance (2) and absorption (3). The corrected reflectance (4) and transmittance (5) spectra.

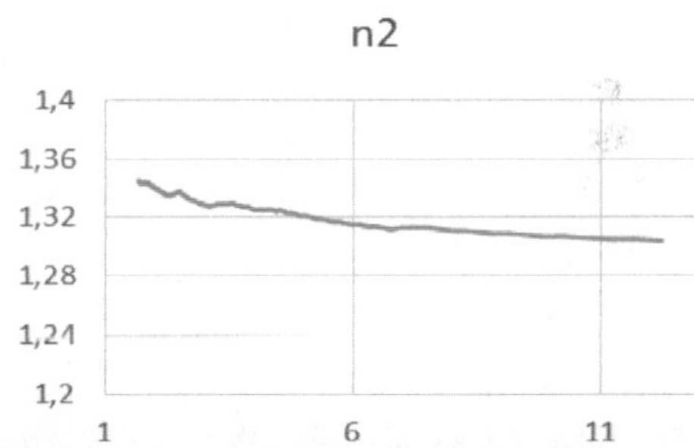

Figure 3. The dependence of the refraction coefficient of BaF_2 on a wavelength.

The proposed method of spectra correction allows one to recalculate the reflectance and transmittance spectra so as to exclude the effect of absorption from the optical spectra. This correction technique allows one to apply the methods developed for determining optical constants of unabsorbing films for the absorbing ones.

References

[1] Kotlikov E N, Novikova Y A and Tropin A N 2016 *Designing and manufacturing of interference coatings* (St. Petersburg: SUAI) 288
[2] Poelman D and Smet P F 2003 *J. Phys. D: Appl. Phys.* **36** 1850–7
[3] Konovalova O P and Shaganov I I 1988 *Journal of Optical Technology* **8** 39–41
[4] Kotlikov E N, Kotlikov A N and Iurkovetc E V 2016 *Coll. Reports Scientific Session GUAP* 253–7

IOP Conf. Series: Journal of Physics: Conf. Series **857** (2017) 012024 doi:10.1088/1742-6596/857/1/012024

The adhesion solidity, physico-mechanical and tribological properties of the coating of titanium nitride

L A Krivina, Yu P Tarasenko and Ya A Fel

Institute of problems of mechanical engineering of RAS, 603024, Nizhny Novgorod, Russia

E-mail: npktribonika@yandex.ru

Abstract. Influence of variable technological factors (arch current, fractional pressure of gas in the camera) on structure, physic-mechanical and tribological features of an ion-plasma coating of titanium nitride has been investigated. The adhesion solidity has been put to the test and the mechanism of destruction of a covering has been also researched by a skretch-test method. The optimal mode of spraying at which the formation of the nanostructured bar coating of TiN has been defined. The covering offers an optimal combination of physic-mechanical, tribological and solidity features.

1. Introduction

Nowadays, when abrasion of basic means of production and transport facilities in Russia reaches 80 %, the main task is to increase reliability and operability of various details of clusters and mechanisms in machine-building branch. In addition, increase of reliability and resource of further developed technique is also relevant.

In spite of the fact that thin coverings on the basis of refractory nitrides of the transitional metals are rather well studied and are widely used in many areas of human activity, scientific and practical interest in them does not fade away for the specific physic-mechanical properties – high hardness, a low friction coefficient, endurance, resistance to corrosion, etc. [1, 2].

2. Experimental technique

The ion-plasma coating of TiN was applied on the VU-2MBS installation on small-size thin-walled exemplars of steel on the modes given in the table 1.

Table 1. Technological modes of an ion-plasma spraying of a covering of titanium nitride (reference voltage of U = 140 V).

Technological parameters of spraying	Fractional pressure of gas in the building bag P, Pa	
	0.04	0.06
Arch current I, A	160	160
	140	140
	120	120

The spraying of coating was realized in definite periods of time of processing in one production cycle: cleaning and activation of a surface of details with method of ionic bombing Ti^{+} in a periodic mode:

5 s (processing) + 5 s (pause), etc.; drawing an underlayer α-Ti – 10 min; TiN deposition – 50 min.

The microstructure of the coating of TiN was studied with a submicroscopy method on a raster microscope of VEGA//TESKAN.

Microhardness (H_{μ}) of compositions "covering + basis" was measured on PMH-3 microhardness, theoretical calculations of the true microhardness of a coating were made according to the formula [3] that takes into account the influence of a substrate on measurement process:

$$H_c = H_{com} + (H_{com} - H_b)/[(2t/h + 3(t/h)^2],$$

where H_c – a microhardness of a coating; H_{com} – a composition microhardness; H_b – a substrate microhardness; t – thickness of a coating; h – dimpling depth.

Research of frictional features was carried out using laboratory installation of friction at dry sliding at loading 1 N on an indentor of sapphire.

Tests of endurance was carried out at the Echo installation at dry transversal contact of the rotating core (steel U10, HRC = 65) by diameter ~8 mm load of 3 N during various time. Endurance of the original material and coating of TiN was estimated in geometrical parameter – diameter of a spot of the depreciation.

Process of destruction and the adhesion solidity are analysed by a sclerometric method by Revetest RST skretch-tester. Tests were passed with use of a diamond spherical indentor. Test parameters: the growing load of the indentor (AL) – from 1 to 50 N, the speed of movement of the indentor – 1 mm/min, length of scratch (L) – 3 mm, the loading rate – 16.33 N/min, sensitivity of the acoustic emission – 5.

3. Results of the research

By method of a submicroscopy it has been ascertained that at all modes of spraying the coating with the bar structure of TiN grains (figure 1) is received. The tendency of formation of the nanostructured TiN grains with decrease of pressure of reactionary gas in the camera and current of an arch (figure 2(a)) has been revealed.

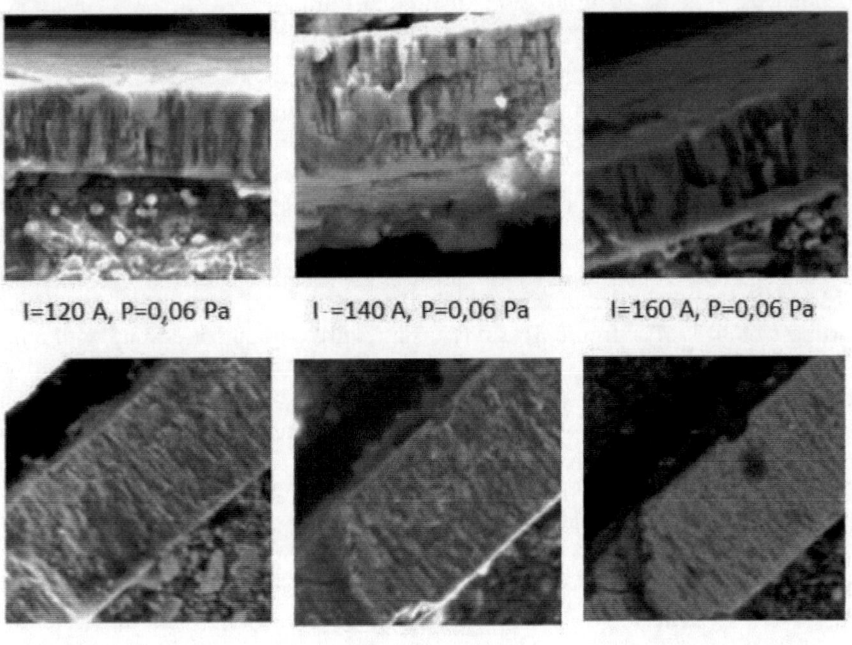

Figure 1. The nanostructured bar covering of TiN received by various modes of spraying, x10000.

The received dependences of modification of a microhardness of a coating TiN on arch current at different fractional pressure of gas in the camera showed that the maximal value H_μ^{50} ~19.4 GPa of a covering is received at arch current of I = 120 A and is caused by the finely divided microstructure. Decrease of microhardness of a coating by increasing the grain size (figure 2(b)) is installed with increase of arch current and gas pressure in the camera.

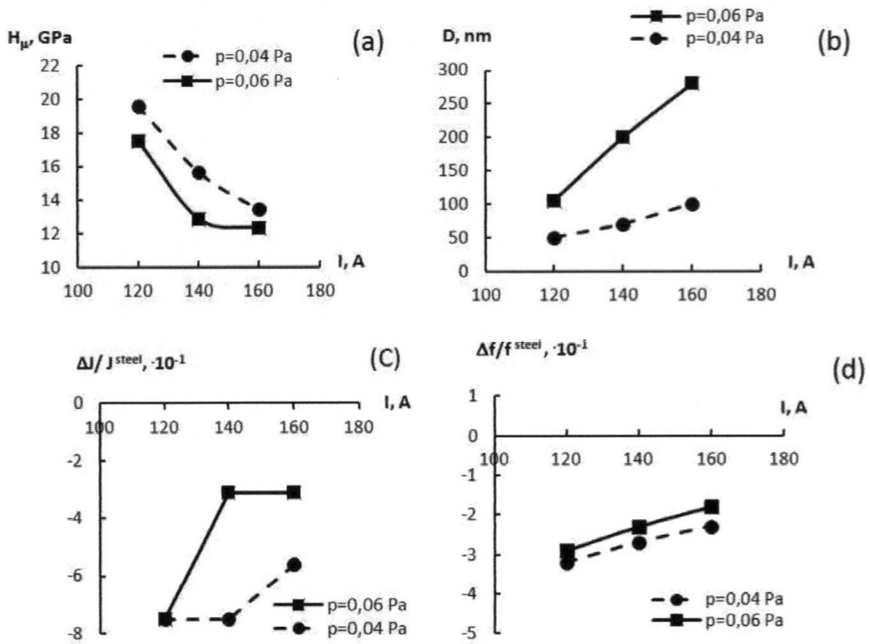

Figure 2. Dependences of cross sectional size of the bar grains TiN **(a)**, microhardness **(b)**, relative change of friction coefficients **(c)** and endurance **(d)** of coating of TiN on arch current at different fractional pressure of gas in the camera.

Application of the nitride coating promotes decrease of the friction coefficient of the processed surface. Spraying of the TiN coating at the "U = 140 V, P = 0.04 Pa, I = 120 A" mode leads to decrease of a friction coefficient of an effective area of a steel detail in ~1.4 times (figure 2(c)).

It has been ascertained that the titanium nitride coating received at arch current of 120 A (figure 2(d)) have the greatest endurance (in ~4 times).

Analyzing results of the adhesion tests it is possible to mark out various threshold values of ultimate load leading to various types of destruction. Three stages of destruction are typical for coating with a microhardness of ≤13 GPa (figure 3) and two stages are typical for coating with a microhardness ≥15 GPas (figure 4).

At the first stage (load of an indentor to ~13 N) the indentor penetrates monotonous into a coating, at the same time the frictional force poorly increases, and amplitude of an acoustic emission remains invariable. The indentor leaves smooth slight marks on a coating. Sliding of a diamond indentor on a covering takes place with very low friction coefficient.

At the second stage an increase in amplitude of an acoustic emission and change of an inclination of curves of a frictional force and friction coefficient take place. At scratching with loadings more than 13–15 N chevron cracks appear at the bottom of scratch, cleavage of separate flakes (a cohesion destruction of a covering occurs) and islet detachment is observed on edges of scratches. Increase in loading over 20 N leads at first to the local and then permanent cleaving of a covering. The permanent cleaving of a covering leads to growth of amplitude of a signal of an acoustic emission, a frictional force increase monotonously up to 10 N. At further increase in a frictional force a fast attrition of a coating occurs.

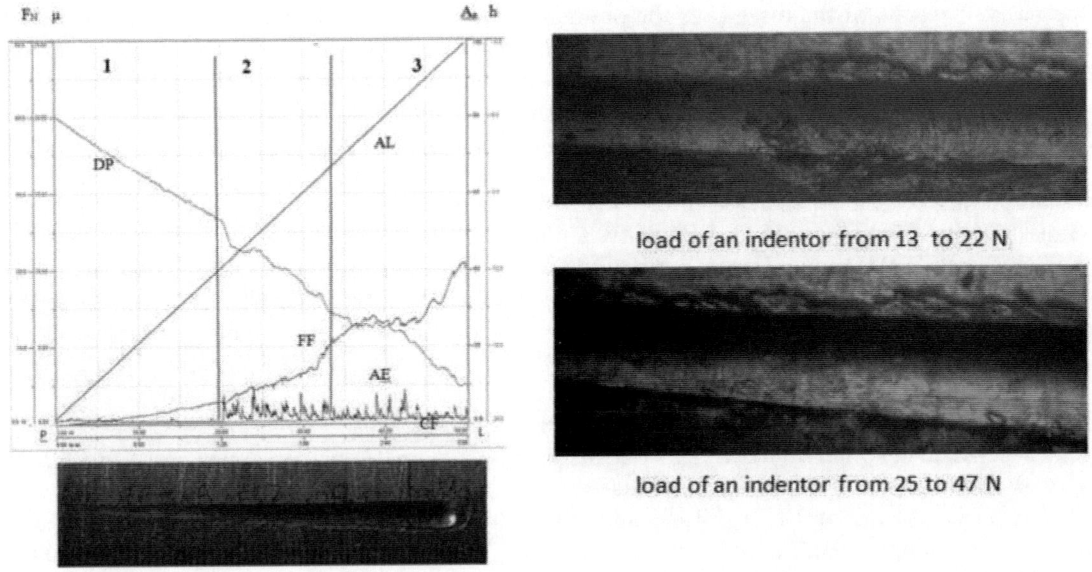

Figure 3. Panoramic shot of the scratch (x200) and curve changes of signals of different sensors at scratching the composition "TiN – steel" at the increasing loading: AL – the size of normal loading (P); DP – a depth of penetration of an indentor (h); FF – value of a frictional force (F_N); AE – value of the acoustic emission (A_i); CF – value of a friction coefficient (μ) for TiN received by the mode $I = 160$ A, $P = 0.04$ Pa.

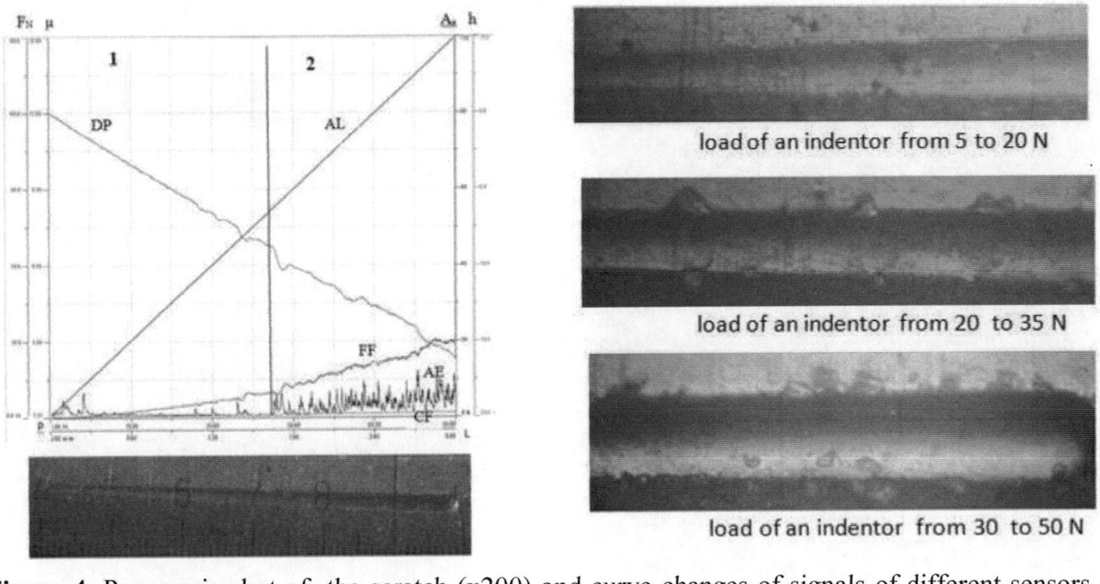

Figure 4. Panoramic shot of the scratch (x200) and curve changes of signals of different sensors at scratching the composition "TiN – steel" at the increasing loading: AL – the size of normal loading (P); DP – depth of penetration of an indentor (h); FF – value of a frictional force (F_N); AE – value of the acoustic emission (A_i); CF – value of a friction coefficient (μ) for TiN received by the mode $I = 120$ A, $P = 0.04$ Pas.

At the third stage penetration of an indentor deep into of substrate material is observed, at the same time amplitude of a signal of an acoustic emission decreases, and the frictional force (to 40 N) and a friction coefficient rise steeply.

On the basis of results of the researches the optimal mode of spraying of a coating of TiN is chosen: P = 0.04 Pa, I = 120 A, U = 140 V at which the nanostructured bar coating of TiN with an optimal combination of physic-mechanical, tribological and solidity features is formed.

4. Conclusion

The received nanostructured coating of TiN has been tested and implemented at many enterprises of machine-building branch for increase of efficiency and reliability of various details of machines: PA "Tochmash" (city Vladimir), SK «KMZ» (city Kovrov) – a limitative slideway of the gas centrifuge [4]; SK «KUZNETSOV» (city Samara) – the rods which are a part of the vibration isolations system of the RT-2 exercise machine for using on the space station ISS ; MV "VOSCHOD" (city Pavlovo) – nipples for the aircraft equipment; various enterprises of Kazakhstan – plunger couples for the «KAMAZ» cars, the compressor shovels of high pressure restored for DZh59L3 engines.

References

[1] Tarasenko Yu P, Krivina L A and Fel Ya A 2014 *News of Higher educational institutions. Volga region. Technical science* **3** 142–8

[2] Krivina L A, Tsaryov I N and Tarasenko Yu P 2016 *Collection of materials of reports of the international scientific and technical conference "Problems and Prospects of Development of Engine-building" Samara on June 22-24* **1** 120–1

[3] Tarasenko Yu P, Tsaryov I N and Romanov I G 2002 *News of Academy of Sciences. Physical Series* **8** 1223–5

[4] Tarasenko Yu P, Sorokin V A, Krivina L A *Russian Federation Patent for the useful model 87228*

IOP Conf. Series: Journal of Physics: Conf. Series **857** (2017) 012025 doi:10.1088/1742-6596/857/1/012025

Characterisation of anti-erosive properties of nanocomposite coatings by the methods of sclerometry

O V Kudryakov[1], V N Varavka[1] and V V Ilyasov[2]

[1] Material Science Department, Don State Technical University, 1 Gagarin sq., Rostov-on-Don, 344000, Russia
[2] Physics Department, Don State Technical University, 1 Gagarin sq., Rostov-on-Don, 344000, Russia

E-mail: kudryakov@mail.ru

Abstract. Results of research of coatings of the different metal-ceramics systems are given. Coatings were received by ion-plasma sedimentation in vacuum in the form of multilayered composite material, which had a thickness of layers within nanometric range. Selection of composite systems is determined by applied research problem – namely designing of the anti-erosive coatings durable in the condition of drop impingement impacts. For this purpose the sclerometric studies, the bench erosive tests and optimization of the obtained data were done.

The 2D-cermet nanocomposite coatings of the following systems were selected for study: Ti/C, Ti/Mo, TiN/MoN, Ti/AlSi, TiN/AlSiN, Ti/Zr(Nb), TiN/Zr(Nb)N. Coating was performed at the ion-plasma sputtering facility with unbalanced magnetron evaporation system. Modes of coatings subsidence were chosen so that, in accordance with the Movchan-Demchishin-Thornton's diagram a solid (non-defective) nanoscale layers with a nanocrystalline structure (figure 1) and compressive stresses in the coating were obtained. Total thickness of coatings – at least 3 μm (optimally 6–8 μm) with monolayers thickness ranging from 5...40 nm. Coating subsidence rate of 0.6...6.0 nm/s.

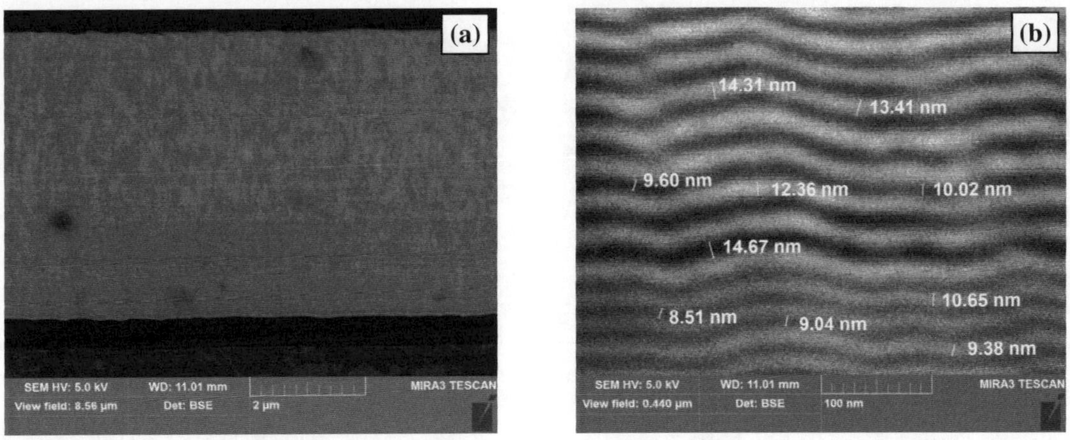

Figure 1. Microstructure of nanocomposite Ti/AlSi system coating with thickness 2.5 μm in cross section of the coating at different magnifications; SEM images.

IOP Conf. Series: Journal of Physics: Conf. Series **857** (2017) 012025 doi:10.1088/1742-6596/857/1/012025

When using the value of technological parameters of application all systems provide technologically sustainable receipt of multilayer's (2D) nanocomposite coatings except those systems containing silicon. Coverings with Si represent 3D-nanocomposites because they have an ordered (modulated) heterogeneous structure, one of phases of which is amorphous silicon-containing phase (as a binder).

Physical and mechanical properties of coatings were investigated using the sclerometric methods of indentation [1] at different samples loading schemes: discrete (microhardness), a continuous one-coordinate (nanohardness) and continuous two-cordinate (scratching with varying load). To measure microhardness according to Vickers (HV) hardness gauge “DuraScan 20" was used, which uses the method of automatic (electron-optical) measurement of the print. Measurement of nanohardness was carried out with a scanning nanohardness gauge “NanoScan-3D” in a mode of instrumental continuous indentation. To determine adhesion of the coating, its resistance to scratching, the friction coefficient of the coating and the base metal, fracture energy, and other characteristics in micro- and nanorange a scratch tester with micro- (MST) and nano- (NST) modules based on an open platform of CSM Instruments (Switzerland) – a universal sclerometric instrument for measuring a number of characteristics by scratching modes either with a constant scratch load or variable one was used. In our measurements the last mode of loading, which provides scratch indenter penetration through the entire thickness of the coating to the base metal was implemented.

Erosion tests were carried out on a unique specialized stand of the Scientific and Research University of the Moscow Power Engineering Institute "Erosion-M", which is used to study resistance of materials and coatings to the effect of high-velocity liquid drops.

When analyzing the physical and mechanical properties of the coatings obtained by different methods of sclerometry, it appeared that none of them separately can be unambiguously and reliably characterize erosion properties of the coatings. Therefore multiparametric optimization is carried out for the strength properties of coatings using integration ray diagrams method. An example of such a diagram for three systems nanocomposite coatings is shown in figure 2. The axes of radiation diagram locate relative values of nine physical and mechanical characteristics: $HV_{0.01}$, nanohardness, elasticity modulus E, friction coefficient f_1 and normal (tangential) force N_1 in coating, critical applied load and depth at formation of cracks in coating (P_1, p_1) and at full coating chipping (P_2, p_2). Since the number of parameters is large enough (nine), and the individual correlation of each of them with erosive properties is not revealed, in transition to the relative values of parameters it allows to take their boost factors (densities) as equally probable.

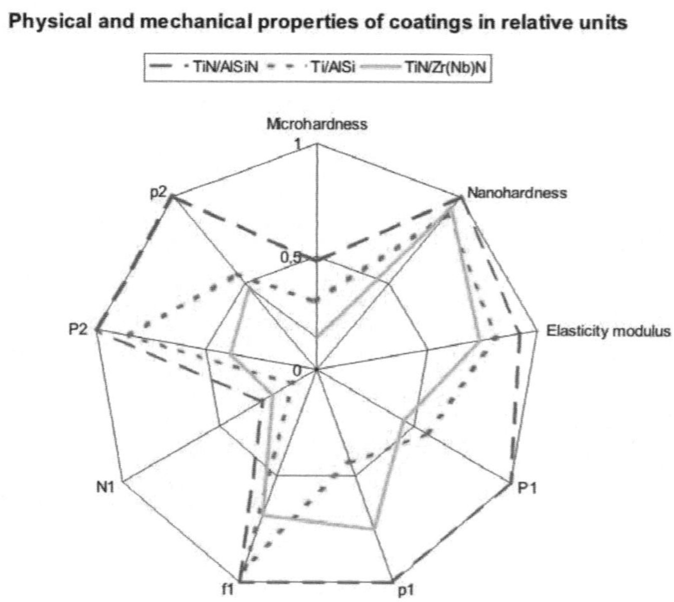

Figure 2. Radiation diagram of physical and mechanical properties of TiN/AlSiN, Ti/AlSi and TiN/Zr(Nb)N systems nanocomposite coatings.

The maximum value of features obtained in tests was taken as a unit of each parameter (all coatings of the seven systems studied) except the coefficient of friction where the minimum value is taken as relative unit. The area of the polygon on the diagram corresponding to each coating was used as an integrated statistical estimation of the relative strength of coating and was a parameter optimization: the larger the area of the polygon, the greater the strength of the coating is as a whole. The area of the polygon S was calculated as the sum of squares of its constituent n triangles by the following expression:

$$S = \frac{1}{2}\left[a_1 \cdot a_n + \sum_{i=1}^{n-1}(a_i \cdot a_{i+1})\right] \cdot \sin\frac{2\pi}{n},$$

where n – is the number of rays (properties) of the diagram; a_i – relative value of the property.

According to the results of multiparameter optimization of ion-plasma nanocomposite coatings of the studied systems can be arranged in the following series of descending strength (i.e., the *complex* of nine physical and mechanical properties): TiN/AlSiN, TiN/MoN, Ti/AlSi, Ti/C, Ti/Mo, Ti/Zr(Nb), TiN/Zr(Nb)N (figure 3(a)). The last place in the series of TiN/Zr(Nb)N nitride system is not due to the nature and properties of the coating, but technological defects of that shape coating (topography of the substrate, discontinuities, growth effects, tension).

Comparison of the erosion resistance of the studied experimental nanocomposite coating of small thickness between themselves and other materials can only be properly carried out for the duration of the erosion's incubation period value m_0, during which the wear isn't beginning yet and the coating maintains its continuity [2–4]. These results are presented in figure 3(b).

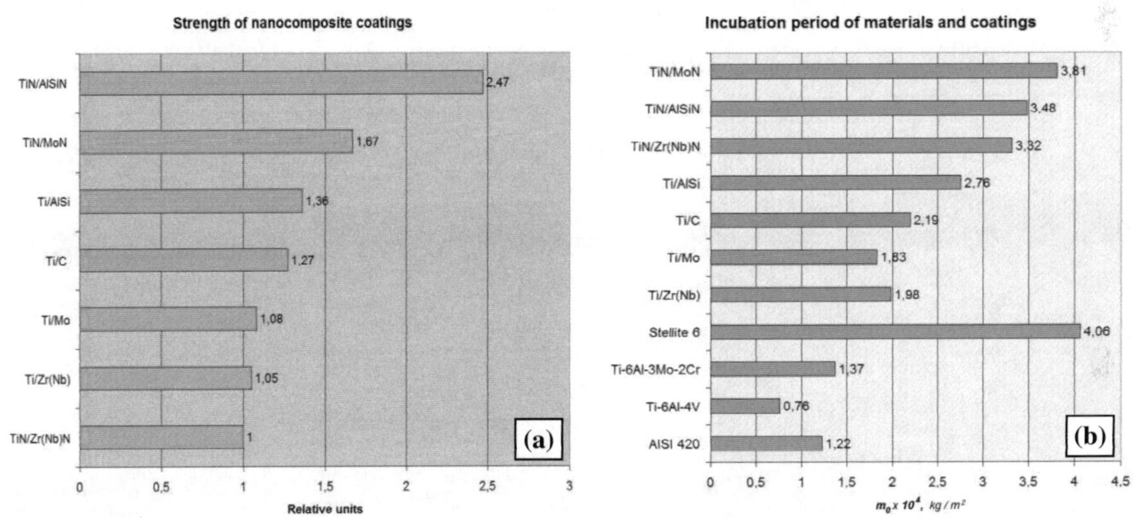

Figure 3. Chart of the integral strength of nanocomposite coatings of different systems (**a**) and chart of erosion resistance of the some materials and nanocomposite coatings (**b**).

The most important result of the research appears from the comparison of data of these strength tests and erosive properties of nanocomposite ion-plasma coatings: the gradation of coating in figure 3(a) and figure 3(b) is almost the same. Thus, there is a satisfactory correlation between the *complex* of physical and mechanical properties of nanocomposite ion-plasma coatings and their erosive properties. This provides comparative qualitative predictions of resistance for various types of coatings to droplet impingement erosion, which can measure only when the benchmark or in-situ tests performed for the complex physical and mechanical properties, as measured in the laboratory by the methods of sclerometry.

Acknowledgment
The work was carried out with financial support from Russian Foundation for Basic Research (project code 15-08-06181).

References

[1] Kudryakov O V and Varavka V N 2015 *Inorganic Materials* **51**(15) 1508–15
[2] Kudryakov O V and Varavka V N 2015 *Journal of Surface Investigation. X-ray, Synchrotron and Neutron Techniques* **9**(1) 203–14
[3] Varavka V N and Kudryakov O V 2015 *Journal of Friction and Wear* **36**(1) 71–9
[4] Varavka V N and Kudryakov O V 2015 *Journal of Friction and Wear* **36**(2) 153–62

Modeling of thermal processes in the modification of the surface by means of the cathode spot of vacuum arc on equipment with non-cooled rotating anode

V G Kuznetsov, E S Babushkina and T A Kurbanov

Institute of Problems of Mechanical Engineering, Russian Academy of Sciences, 199178, Saint Petersburg, Russia

E-mail: kvgipme@gmail.com

Abstract. Theoretically and experimentally the thermal processes in the case of modification the inner surface of the cylindrical cavity by means of the cathode spots of vacuum arc discharge are investigated. As the anode was used uncooled graphite electrode in the form of a half cut in diameter of a cylindrical rod. At the modeling of thermal processes was taken into account the process of heating the surface of the cylinder by radiation from the hot surface of the anode. In this case the cylindrical surface has been uneven wall thickness around the perimeter.

A fundamentally new field of use of the vacuum-arc discharge [1, 2], developing in recent years, demanded to study the behavior of cathode spots of a vacuum arc on the moving surfaces and, in some cases, covered with films of other material. In such circumstances the researches of cathode spots is not practically implemented. The behavior of cathode spots has been primarily addressed on fixed and clean from surface contamination of the electrodes.

An important parameter in determining the efficiency of the process of vacuum-arc treatment of metal surfaces is the temperature of the product. It should not lead to a change in material properties of the product. The uneven heating of the product must also be insufficient to change its geometrical shape by the generated mechanical stresses.

In some cases, on the contrary, the impact of cathode spots of a vacuum arc can cause change in the surface properties of the metal, for example, can be used for surface hardening [3] or for the recovery of plastically deformed areas of the surface [4]. Vacuum-arc technology can be efficiently applied for machining internal surfaces of cylindrical articles, simultaneously performing the role of the cathode and the vacuum chamber. As such products can be tubes, couplers, axle boxes of railway wagon [5] and many others.

Figure 1 shows the installation for vacuum-arc cleaning of the inner surface of the axle box of a railway wagon.

We consider the solution to the heat problem during the glowing of a vacuum-arc discharge in the electrode system (figure 2): cathode – is a cylinder of height H with an irregular perimeter by wall thickness (similar to a box); the anode – electrode in the shape of a half cylindrical rod, cutting along the diameter, of length L and radius R_a, rotating along the inner surface of the cathode.

(a) **(c)**

(b) **(d)**

Figure 1. Installation of vacuum-arc cleaning of the inner surface of axle boxes of wagon **(a)**; a rotating anode **(b)**; surface boxes before cleaning **(c)**; surface boxes after cleaning **(d)**.

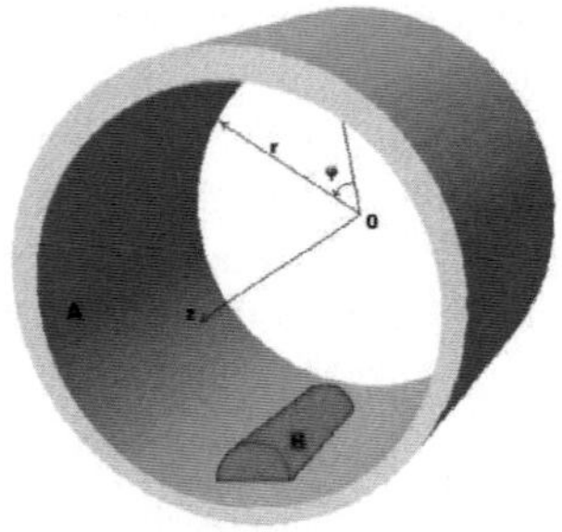

Figure 2. Scheme of the computational domain: A – cathode; B – anode.

The inner surface of the cathode was set as a cylindrical surface of radius R_1. The outer surface was set as a cylindrical surface with radius R_2, which depends on the angle φ:

$$R_2 = \begin{cases} \frac{R_4 - R_3}{\pi}\varphi + R_3, \varphi \in [0, \pi] \\ \frac{R_3 - R_4}{\pi}\varphi + 2R_4 - R_3, \varphi \in [\pi, 2\pi] \end{cases},$$

where R_3 – is the radius of the outer surface of the cylinder at the thinnest point of the wall of the cylinder; R_4 – is the radius of the outer surface of the cylinder at the thickest point of the wall of the cylinder.

The process of heat transfer in the cylinder describes a three-dimensional unsteady heat conduction equation in cylindrical coordinates in the form:

$$\rho c_p \frac{\partial T}{\partial t} = \lambda \left(\frac{1}{r} \frac{\partial}{\partial r} \left(r \frac{\partial T}{\partial r} \right) + \frac{\partial^2 T}{\partial z^2} + \frac{1}{r^2} \frac{\partial^2 T}{\partial \varphi^2} \right), \quad (1)$$

where ρ – is the density of the cylinder material; c_p – is the heat capacity of the cylinder material; λ – coefficient of thermal conductivity; T – temperature. On the outer and end surfaces of the cylinder was set the boundary conditions of free convective heat transfer with the surrounding environment in the form of:

$$\lambda \left. \frac{\partial T}{\partial r} \right|_{r = R_2(\varphi)} = \alpha_1 (T_\infty - T); \quad (2)$$

$$\lambda \left. \frac{\partial T}{\partial z} \right|_{z = 0, H} = \alpha_2 (T_\infty - T); \quad (3)$$

where α_1, α_2 – are the coefficients of heat exchange with the environment.

Inside the cylinder convective heat transfer can be neglected. Here we simulated the radiative heat transfer, which is determined by the temperature of the inner surface of the cylinder, the temperature of the rotating electrode and their instant mutual arrangement.

In each moment of time the surface of the cathode S_c and rotating anode S_a was divided into 2 parts. Surface S_c divided into S_{c1} – cathode surface under the anode and $S_{c2} = S_c - S_{c1}$. Surface S_a divided into S_{a1} – flat part of the surface of the anode and $S_{a2} = S_a - S_{a1}$.

Then, the total radiant heat flux between surfaces S_{c1} and S_{a1} can be written as:

$$Q_1 = \sigma\varepsilon \left(S_{a1}(T_a + 273)^4 - S_{c1}(T_{c1} + 273)^4 \right), \quad (4)$$

where T_{c1} – average surface temperature S_{c1}; T_a – temperature of anode; σ – is the Stefan-Boltzmann constant; ε – is the integral radiation coefficient.

Similarly, the full radiant heat flux between surfaces S_{c2} and S_{a2} can be written as:

$$Q_2 = \sigma\varepsilon \left(S_{a2}(T_a + 273)^4 - S_{c2}(T_{c2} + 273)^4 \right), \quad (5)$$

where T_{c2} – average surface temperature S_{c2}.

On a time-variable part of the inner surface asked the flow of heat from the heat source. Considering this condition, and given (4) and (5) the boundary condition on the inner surface boxes was set as:

$$\lambda \left. \frac{\partial T}{\partial r} \right|_{S_{c1}} = \frac{Q_1}{S_{c1}} + \frac{UIh_c}{S_{c1}}; \quad (6)$$

$$\lambda \left. \frac{\partial T}{\partial r} \right|_{S_{c2}} = \frac{Q_2}{S_{c2}}, \quad (7)$$

where U is the arc voltage; I – current; h_c – the ratio of cathode capacity.

The temperature change of the rotating electrode is described by the equation:

$$\rho_a V_a c_{ap} \frac{\partial T_a}{\partial t} = UIh_a - Q_1 - Q_2, \quad (8)$$

where ρ_a is the density of the material of the rotating electrode; V_a is the volume of the rotating electrode; c_{ap} – heat capacity of the material of the rotating electrode; h_a is the ratio of the anode capacity.

The solution of equations was performed numerically. The computational domain was divided differencing three-dimensional uniform grid in the directions of r, φ and z. To solve this problem (equations (1)–(7)) used the numerical method of variable directions [4]. Before the solution of the problem on the computer in the table (figure 3) are made to the source data.

IOP Conf. Series: Journal of Physics: Conf. Series **857** (2017) 012026 doi:10.1088/1742-6596/857/1/012026

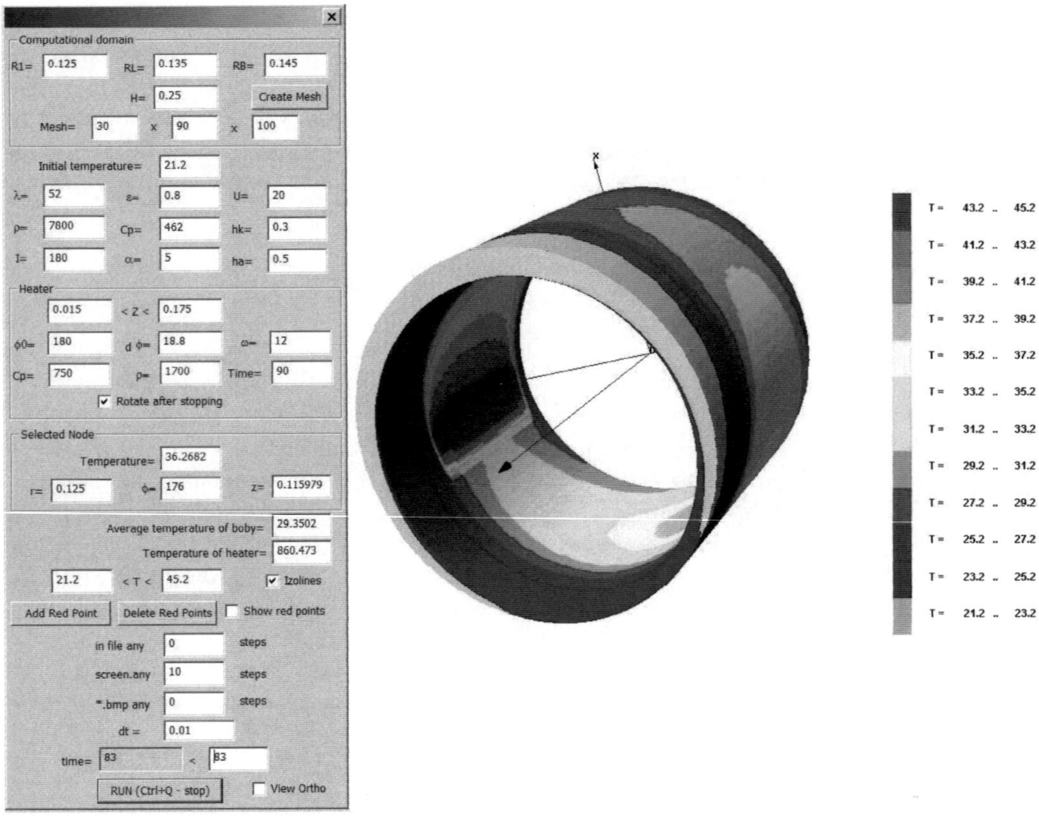

Figure 3. Table with the source data for calculation and temperature field on the surface of cylindrical product.

When the program is run in real time the temperature field on all surfaces of a cylindrical product in the form of contour lines with different colors is displayed. On color can be judged the temperature. For any number of points with given coordinates is selected on the surface of the cylinder (inside and outside), temperature data is written in a separate file. For one of the points of the temperature value is displayed in the table (figure 3). There you can observe the temperature change of the rotating anode. Figure 4 presents the calculated and experimental dependences of change of temperature on time in one of the selected points on the outer surface of the cylinder.

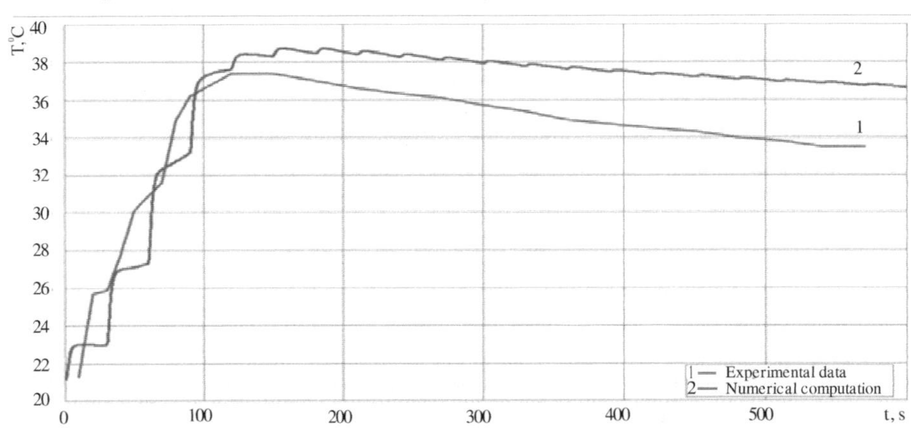

Figure 4. The calculated and experimental dependences of change of temperature on time at the point on the outer surface of the cylinder.

IOP Conf. Series: Journal of Physics: Conf. Series **857** (2017) 012026 doi:10.1088/1742-6596/857/1/012026

Experimental study of the temperature of the external surface of the cylinder during heating and cooling was carried out using an infrared optical pyrometer "PYTHON 102". As can be seen from figure 4 experimental and calculated data differ by no more than 10 %.

References

[1] Kuznetsov V G 2005 *Films and coatings* 57–62
[2] Kuznetsov V G 2009 *Vacuum technique and technology* **2** 81–4
[3] Kuznetsov V G and Ashihmin A A 2012 *Wagons and wagon farm* **2** 36–8
[4] Samarskiy A A 1971 *Introduction to the theory of difference schemes* 552

IOP Conf. Series: Journal of Physics: Conf. Series **857** (2017) 012027 doi:10.1088/1742-6596/857/1/012027

Process parameters influence on formation of oxides in coatings during pulsed laser deposition of titanium on inner surface of tubes

A A Lozovan[1], S V Prishepov[1], E P Kubatina[1], S V Frangulov[1], P N Rizakhanov[2] and S K Sigalaev[2]

[1] Moscow Aviation Institute (National Research University), 121552, Moscow, Russia
[2] SSC FSUE «Keldysh Centre», 125438, Moscow, Russia

E-mail: loz-plasma@yandex.ru

Abstract. This study concerns droplet phase influence and process parameters influence (laser radiation power, oblique deposition, residual oxygen pressure, etc.) on morphology and composition of the coating and formation of oxides in the coating during pulsed laser deposition of Titanium on inner surface of tubes. It is shown that droplets in the coating in the presence of residual oxygen significantly affect the composition of Ti-coatings.

Pulsed laser deposition (PLD) is widely used for the deposition of films of various materials and functional purposes [1, 2]. This is due to a number of advantages of PLD over other methods, such as the possibility of spraying both single- and multicomponent films, stoichiometric transfer into the film of the composition of the target material, simplicity of the scheme, deposition at lower substrate temperatures, etc. [1].

However, the PLD method has a significant disadvantage – when the target is ablated (with the laser radiation power density significant for practical applications), droplets and solid microparticles along with ions and atoms are generated to the deposited flow, which substantially degrade the properties of the coatings. To obtain high-quality thin films, it is necessary to avoid precipitation of such particles on them. Many methods of filtering these particles have been developed for deposition on open surfaces. However, it is very difficult to do this while spraying inside small cavities, for example long, small diameter pipes [3–5] (with a rare exception, when drops precipitate specifically to solve specific problems, as was done in [3, 6] during coating deposition on the internal walls of chambers).

This can be achieved by significantly reducing the intensity of laser radiation to values not much exceeding the evaporation threshold (above $25 \cdot 10^6$ W/cm^2) [7], but this results in an extremely low productivity of the PLD method. In [8], the drop phase was reduced by the additional evaporation with second laser, which complicates the scheme.

When depositing nano-sized Ti films, residual oxygen in the sputtering chamber has a significant effect on the process of their formation. Recently, it's influence on the properties and composition of nano-films has been investigated in the interests of a wide range of applications ranging from solving plasmonics problems [9] to increasing the physico-mechanical properties of metals and alloys [10], carried out at a residual oxygen partial pressure of $5 \cdot 10^{-9}$ and $2 \cdot 10^{-5}$ Pa, respectively.

In this paper, we continued the investigation of the PLD coating process on the walls of long (~1 m) small diameter pipes (~10 mm). Sputtering occurred in the quasi-closed volume of the pipe, the pressure of the residual gases in which was 10^{-3} Pa. Due to the complexity of controlling the processes inside the pipe, the experiments were carried out under conditions simulating the deposition process in the pipe, but in an open space.

A laser stand was used for the experiments [11]. Ablation was carried out with the help of a Gaussian beam laser LTI-215 (M) on yttrium-aluminum garnet activated by neodymium, with pulsed pumping and modulated Q, at room temperature of the substrate. The parameters of the laser radiation are as follows: wavelength λ = 1064 nm with a maximum pulse energy of I = 180 mJ, pulse duration τ = 10 ns, repetition rate f = 30 Hz, the radiation divergence γ = 5 mrad. Laser radiation was focused on the target by a lens with a focal length F = 100 mm. The size of the laser mark on the target was 0.8 mm, then the power density of the laser radiation on the target was $W = 5\cdot10^9$ W/cm^2. Thus, in order to find out more clearly the effect of droplets on the presence of residual oxygen in films during investigation of deposition regimes with high productivity, an ablation regime with a knowingly large generation of droplets in the deposited stream was used.

As a target, the rod end was used with a diameter of 10 mm, cut at an angles of 60 and 70° from the axis (thus, the axis of the plasma plume had the same angle from the normal to the substrate). As a substrate were used plates 1 mm thick. Target material: Ti alloy (Al ~ 7 %, Si ~ 0.5 %), and substrates – polished steel SUS 304. Irradiation time – 30 and 5 min. The distance from the spot on the target (the axis of the rod) to the substrate was 10 mm. Target's and substrate's surfaces were chemically cleaned and placed in a working chamber, which was evacuated by turbomolecular and rotary pumps.

Analysis of the chemical composition of the applied coatings was carried out using the Rutherford backscattering method (RBS) He^+ with energy of $E \approx 1$ MeV on the ion-beam complex "Sokol-3" of IPTM RAS. Morphology and elemental composition studies were also performed with the FEI Quanta 600 FEG scanning electron microscope with the EDAX Trident energy dispersive system, as well as the KLA Tencor MicroXAM-100 optical profilometer.

Figure 1 shows the RBS spectra measured at points outside the droplets on Ti films deposited on SUS 304 steel at different times. Analysis of both spectra showed the amount of oxygen in the film sufficient to form oxides of TiO_2 and $TiO_{1.5}$.

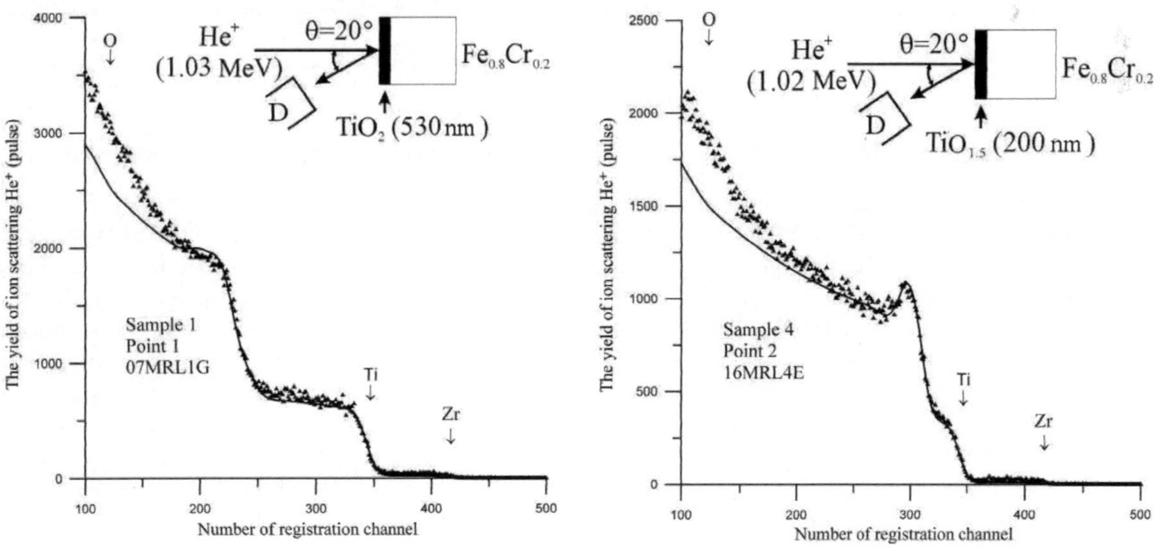

(a) **(b)**

Figure 1. RBS spectra of Ti-film on SUS 304 steel, deposition time: **(a)** – 30 min, **(b)** – 5 min. Impact angle – 60°.

However, at a pressure of 10^{-3} Pa, the amount of residual oxygen on the wall and in the volume of the pipe is small and, therefore, to ensure that the amounts of titanium and oxygen in the film correspond to these oxides compositions, the titanium flux onto the substrate (without droplets) seems to be very low despite the high intensity of laser radiation. Such situation possibly takes place due to the fact that most of the ablated material is in droplets (see figure 2). In figure 3 large droplets (more than 30 μm) are visible, which spread over the surface due to the oblique angle deposition. Droplets, among other things, significantly worsen the morphology of the film (see figure 4). Thus, we see that the presence of droplets has a complex effect on the composition and properties of the films and on the areas outside the droplets.

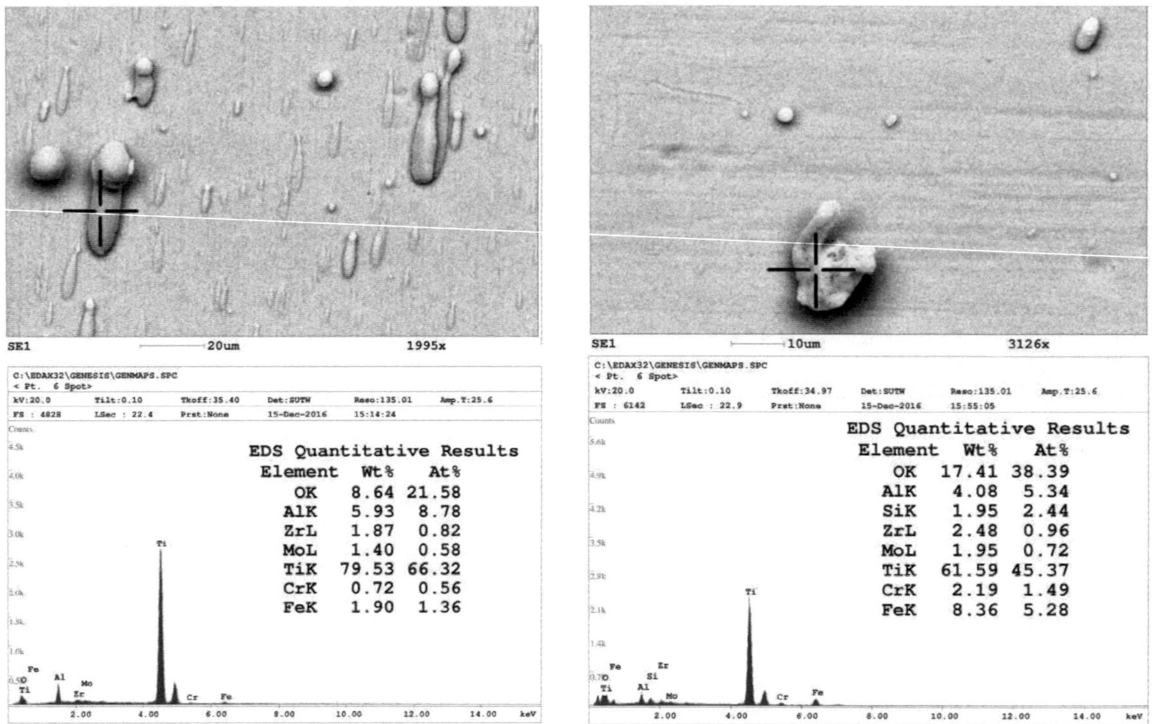

(a) **(b)**

Figure 2. The general form and composition of the films shown in figure 1: **(a)** and **(b)**, respectively.

In our opinion, several factors affect the composition of the film in areas outside the drops. First, as mentioned above, this is a small fraction of the atomic particles in the flow relative to the total ablated titanium mass. This, in turn, significantly increases the relative oxygen content of the film. The smaller relative content of oxygen in the drops is explained by the large mass of droplets. Oxygen can get into the drops only during the flight of a drop in the liquid state, and also as a result of adsorption and diffusion into the drop after its solidification on the substrate. The latter is confirmed by figure 4, which shows a drop with a very developed surface and the highest atomic concentration of oxygen.
Secondly, a lower oxygen value in the film at 5 min irradiation is associated with higher ablation efficiency in the initial period, which then decreases due to the formation of a crater on the target [12]. Thirdly, the inclined deposition of particles on the substrate influences. As is known, in this case there is an effect of shading, which leads to the growth of inclined columns on the surface.

The RBS analysis also showed a significant depletion of Fe (up to 50 %) of the substrate layer under the coating, which may be due to the mass transfer of titanium into the substrate material by diffusion along the grain boundaries and a decrease in the relative titanium content in the film. This diffusion is quite effective in columnar structures. In addition, certain additional oxidation of the film surface can occur after the air is introduced into the chamber.

To reduce the observed effect, it is necessary to take traditional measures to reduce amount of droplets in the deposition flow, and also to reduce the partial pressure of residual oxygen by improving the ultimate vacuum in the working chamber and substantially warming up the pipe or cleaning its inner wall with hollow cathode glow discharge plasma [11].

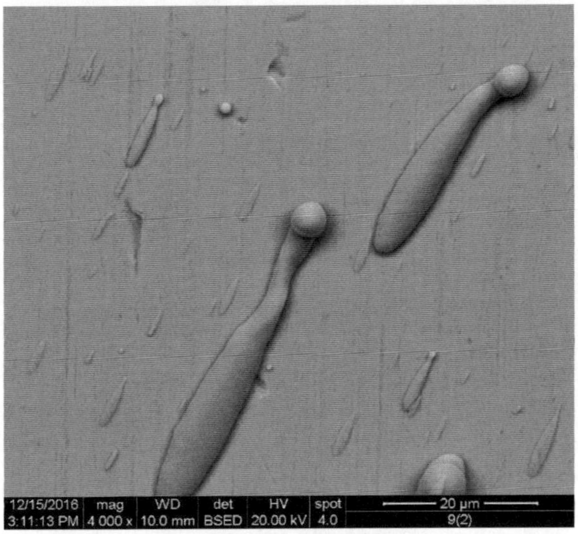

Figure 3. Spreading of large drops.

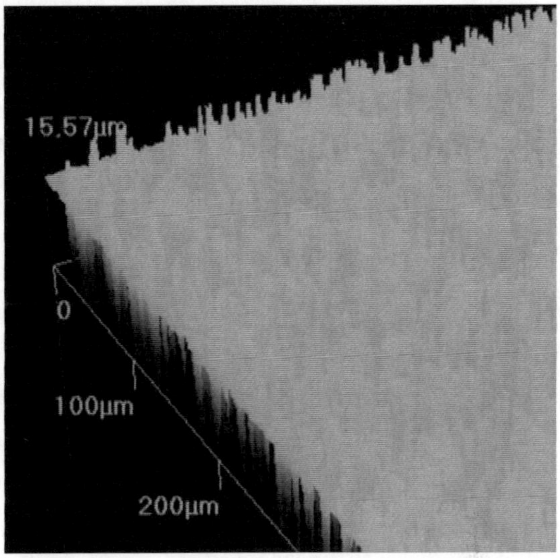

Figure 4. The section of the profilogram of the film shown in figure 1(a).

It should be noted that up to 12 % of nitrogen was detected in some drops. Apparently, it was captured by a drop when flying away from the target. In the future, it is planned to continue studying the interaction of droplets with residual gases, including during ablating the target inside the tube.

Acknowledgments
The authors thank Egorov V K for the production of RBS spectra.

References

[1] Cheung J T 1994 *Pulsed laser deposition of thin films* ed. Chrisey D B and Huber G K (New York: Wiley) 1–19
[2] Lorenz M and Rao M S R 2014 *J. Phys. D: Appl. Phys.* **47** 030301–04
[3] Lozovan A A 2003 *Vacuum* **70** 477–81
[4] Gawlitza P, Jacob K, Sebald T et al. 2004 *Appl. Phys. A* **79** 1043–6
[5] Lozovan A A, Alexandrova S S, Mishnev M A and Prishepov S V 2014 *J Alloys Compounds* **1**(1) 387–90
[6] Beyer S, Klemm D, Bobeth M et al. 2005 *Adv. Eng. Mater.* **7**(1–2) 54–8
[7] Hermann J, Thomann A L and Boulmer-Leborgne C 1995 *Appl. Phys. Lett.* **71**(19) 2752–4
[8] Lozovan A A and Alexandrova S S 2014 *J. Surf. Investig. X-ray, Synchrotron and Neutron Techniques* **8**(3) 607–11
[9] Braic L, Vasilantonakis N, Mihai AP et al. 2017 *arXiv:1703.09467*
[10] Suzuki K, Endo T, Fukushima T et al. 2013 *Mater. Transactions* **54**(7) 1140–4
[11] Lozovan A A, Frangulov S V and Chulkov D V 2011 *Laser Beams: Theory, Properties and Applications* ed. Thys M and Desmet E (New York : Nova Science Publishers, Inc) 331–49
[12] Borisov O V, Xianglei Mao and Richard E R 2000 *Spectrochim. Acta Part B* **55** 1693–704

IOP Conf. Series: Journal of Physics: Conf. Series **857** (2017) 012028 doi:10.1088/1742-6596/857/1/012028

The study of microstructure of wear-resistant coatings applied for protection from abrasive wear of horizontal and tilt drilling drill bits

I Yu Markova, E S Zakharova, A L Maslov, N I Polushin, A I Laptev and M S Ovchinnikova

National University of Science and Technology «MISiS», Moscow, Russia

E-mail: anatmaslov@mail.ru

Abstract. Drill bits of the cutting type over the period of their existence have undergone significant changes – from the use of carbide cutters to diamond composite PDC elements, in which the diamond layer is applied to a hardmetal substrate. Using such elements, it was possible to significantly increase the service life of the drill bits, however, during work, there is a significant abrasive deterioration of the bit body, which does not fully realize the advantages of PDC elements. Therefore, to protect the body from wear use special wear-resistant coatings. This work is devoted to research of microstructural coatings, namely coatings brands WokaDur NiA, HR-6750, HR-6750 with sublayer Rock Dur 47 on various steel substrates which applied by the gas-thermal spraying in Ltd "Oerlikon Metko Rus". They were examined with the use of scanning electron microscopy, X-ray phase analysis and a Vickers micro-hardness tester. It was established that the microhardness of the coating matrix is 590–660 HV, and the microhardness of tungsten carbide particles reinforcing the coating, is 2145–2455 HV.

1. Introduction

During operation, the drill bit body experiences considerable abrasive wear in its contact zone with the cuttings of the drilled rock, which leads to a weakening of the fastening and premature removal of the PCD-elements from the bit body. Due to the incomplete development of the PCD-elements is significantly reduced resource of drill bit [1].

There are various methods for applying wear-resistant coatings, such methods include various variations of low-temperature CVD and PVD deposition or electrochemical and gas-thermal spraying that meet the following requirements for the application of a wear-resistant coating on the bit body [2–6].

Firstly, the coating must have a high hardness (higher than the hardness of the drill bit body, preferably above the hardness of the drilled rock) and corrosion resistance.

Secondly, there should be a close value of the thermal linear expansion coefficient (TLEC) of both applied coating and the drill bit body materials. If the condition is not met, then at high temperatures the coating will peel off.

Thirdly, the application temperature should not exceed the melting point of the drill bit body. The most commonly used wear-resistant coatings are applied by gas-plasma method such as hand-gas-plasma welding. Therefore, the purpose of this work is a microstructural study of wear-resistant coatings containing tungsten carbide particles obtained by manual gas flame deposition in order to increase the resistance of the PDC-bit body to abrasive wear [7, 8].

2. Materials and equipment

As substrates used samples were made from steels of marks St3, St45 and KHVG, which can be used as a drill bit body, and the coatings have been chosen as the following wear-resistant layer:

– HR-6750 with 2.6 mm thickness, directly on the surface of steels;
– HR-6750 with 2.6 mm thickness on a pre-applied Rock Dur 47 sub-layer with 0.2 mm thickness;
– WokaDur NiA with 2.6 mm thickness directly on the surface of the used steels.

Complex study of the microstructure of wear-resistant coatings was carried out on a scanning microscope JSM-6480LV, located at the Center for Collective Use in NUST «MISiS».

X-ray diffraction (XRD) measurements were performed at NUST "MISiS" using DRON-3M (Co K_α radiation). The micro-hardness of matrix and carbide particles was determined by Knoop/Vickers Hardness Tester the Tukon 1102, the load on the indenter was 200 g, the exposure time under load – 5 s.

3. Results and discussion

Figure 1 shows the typical microstructure of the obtained coatings deposited on different substrates.

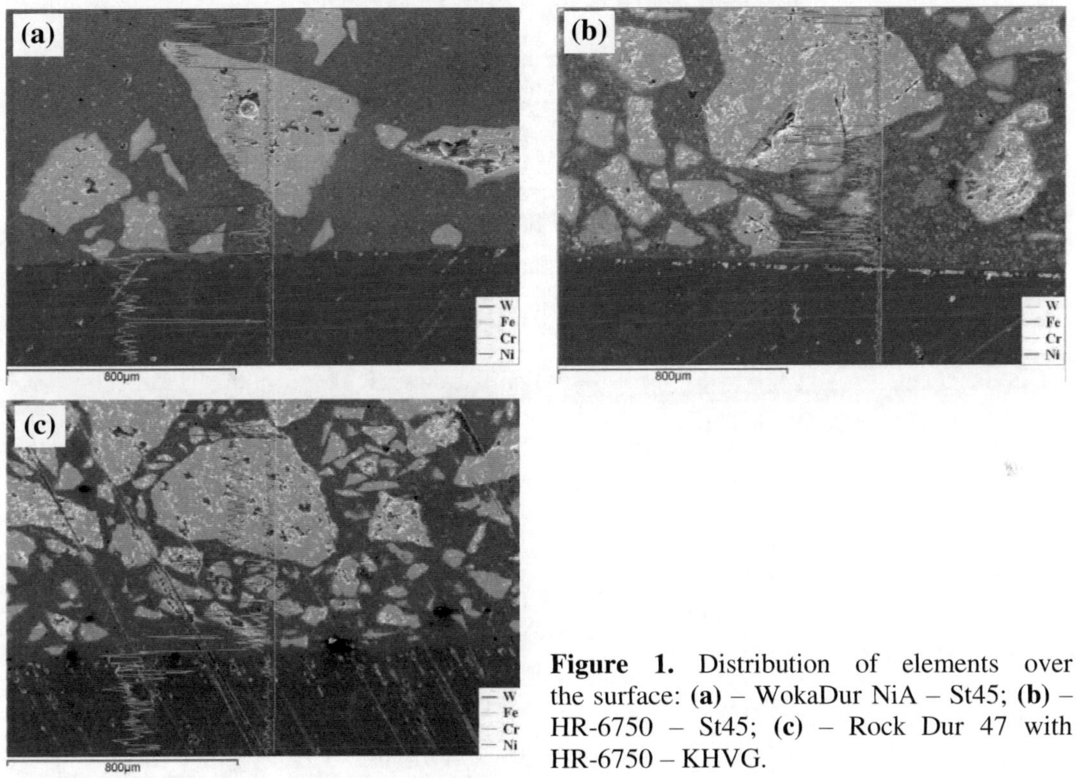

Figure 1. Distribution of elements over the surface: **(a)** – WokaDur NiA – St45; **(b)** – HR-6750 – St45; **(c)** – Rock Dur 47 with HR-6750 – KHVG.

A microscopic examination of the structure of the coatings is showed differences in all samples. Primarily due to the size of the carbide particles. Coating brand HR-6750 consists of particles with highly developed structure (figure 1(b)). The basis of all coatings is a nickel matrix with additive elements (Cr, B, Si, C) and carbide particles. Elemental analysis of the matrices revealed the presence of the following elements and their quantitative ratio, which are summarized in table 1.

The study of samples with a sublayer of Rock Dur 47 is showed partial dissolution in the process of applying HR-6750. As an example, figure 2 shows the distribution of elements on the surface of the thin section of the transition zone (coating-substrate) of the sample St45.

As a result of measurements of samples with different substrates and coatings, it was found that the microhardness of the matrix was 590–660 HV, and tungsten carbide particles 2145–2455 HV (table 2). The microhardness of the KHVG steel substrate is twice or more than the microhardness of the

IOP Conf. Series: Journal of Physics: Conf. Series **857** (2017) 012028 doi:10.1088/1742-6596/857/1/012028

substrates of St3 and St45 steels, which is associated with partial hardening of the substrates from this steel when the samples are cooled after the application of wear-resistant coatings.

Table 1. Elemental analysis of the matrix coatings

Coating	Weight, % wt.	
	Ni	Cr
HR-6750	81.09	18.91
HR-6750 with sublayer Rock Dur 47	85.56	14.44
WokaDur NiA	91.03	8.97

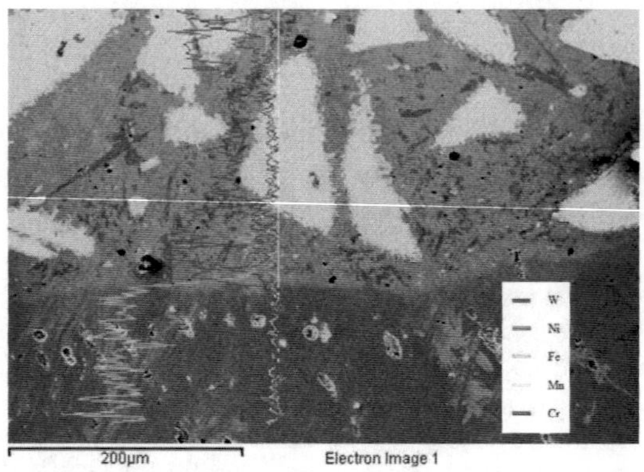

Figure 2. Distribution of elements over the surface.

Table 2. Average values of microhardness of HV matrices, particles and substrates.

	Microhardness, HV		
	St3 – WokaDur NiA	St45 – WokaDur NiA	KHVG – WokaDur NiA
Coating	590	588	620
Strengthening particles	2452	2351	2442
Steel housing	213	275	580
	St3 – HR-6750	St45 – HR-6750	KHVG – HR-6750
Coating	599	759	653
Strengthening particles	2145	2165	2282
Steel housing	206	242	1011
	St3 – HR-6750 with sublayer Rock Dur 47	St45 – HR-6750 with sublayer Rock Dur 47	KHVG – HR-6750 with sublayer Rock Dur 47
Coating	595	614	662
Strengthening particles	2303	2233	2185
Steel housing	210	241	591

According to the results of the XRD (figure 3) it was found that in samples with coating HR 6750 and sublayer Rock Dur 47 there are phases of WC, W_2C, Cr, Cr_7Ni_3. The ratio of the phases WC and W_2C in these samples is different, the sample HR 6750 dominated by the phase of the WC compared to

W_2C, the sample underlayer Rock Dur 47 inverse relationship. In samples coated WokaDur NiA discovered the phases of WC, W_2C, CrNi.

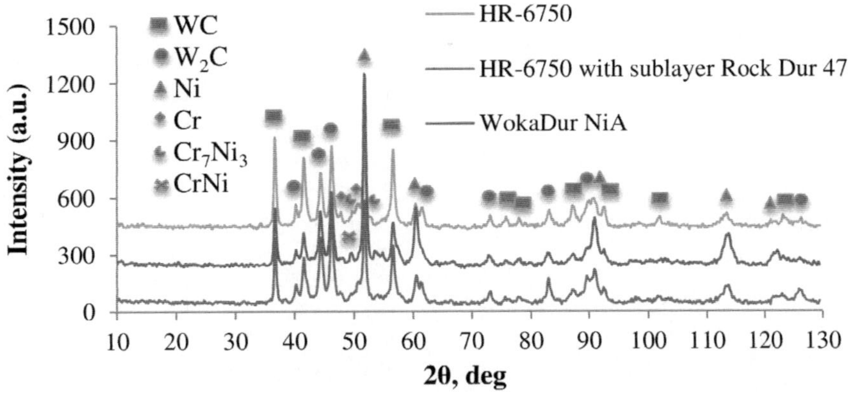

Figure 3. X-ray pattern of samples with coating deposited on a substrate St45.

4. Conclusions

In the process of conducting microscopic studies of wear-resistant coatings used to protect the cases of diamond blade PDC-bits from abrasive wear during drilling, it has been established:

– the microhardness of the matrices of the coatings obtained is 590-660 HV, and the microhardness of the tungsten carbide particles is 2145-2455 HV;

– a micro-X-ray spectral study of the Rock Dur 47 sublayer revealed its partial dissolution in the process of surfacing a wear-resistant coating HR-6750.

Based on the conducted studies, we can recommend the use of WokaDur NiA as a wear-resistant coating, which is explained by the highest hardness of carbide inclusions in comparison with other coatings. As a substitute for WokaDur NiA coating, it may be recommended to coat the HR-6750 with a sublayer of the Rock Dur 47. Among the substrates, we can recommend structural carbon steel grade St45. Its mechanical properties satisfy the requirements imposed on the drill bit housing, and its quenching does not take place under the conditions of obtaining a wear-resistant coating.

Acknowledgment

This work was carried out within the framework of the federal target project of the program "Research and development in priority areas of development of the scientific and technological complex of Russia for 2014-2020" on the topic "Development of technology for manufacturing high-performance horizontal and tilt drilling drill bits for the oil and gas industry" agreement No. 14.581.21.0012 (unique identifier of the agreement RFMEFI58115X0012) with financial support of applied scientific research by the Ministry of Education and Science of the Russian Federation.

References

[1] Nekrasov I N, Bogomolov R M, Ischuk A G, Gavrilenko M V, Morozov L V, Mukhametshin M M and Krylov S M *Patent of the Russian Federation № 2389857* 10.11.2009 **31**

[2] Polushin N I, Ovchinnikova M S and Maslov A L 2014 *Advanced Materials Research* **1040** 199–1

[3] Polushin N I, Ovchinnikova M S, Maslov A L and Kuchina I Yu 2015 *Izv. Vyzov. Series "Chemistry and chemical technology"* **58**(7) 55–7

[4] Polushin N I, Bogatirev A V, Laptev A I and Sorokin M N 2016 *Izv Vyzov "Powder metallyrgy and functional coatings"* **1** 60–6

[5] Karasev M V, Stonnikin V S and Petrov G K *Patent 1835865 USSR* 10.04.1996 **10**

[6] Bogomolov R M, Ischuk A G, Gavrilenko M V, Neupokoev V G, Morozov L V, Krylov S M and Mokrousov V P *Patent of the Russian Federation № 2275440* 27.04.2006 **12**

IOP Conf. Series: Journal of Physics: Conf. Series **857** (2017) 012028 doi:10.1088/1742-6596/857/1/012028

[7] Langford J V and Delvice R *Patent of the Russian Federation № 2167262* 10.01.2001
[8] Besson A A 2002 *Oil and Gas Review* **7** 4–31

IOP Conf. Series: Journal of Physics: Conf. Series **857** (2017) 012029 doi:10.1088/1742-6596/857/1/012029

Tribological tests of wear-resistant coatings used in the production of drill bits of horizontal and inclined drilling

A L Maslov, I Yu Markova, E S Zakharova, N I Polushin and A I Laptev

National University of Science and Technology «MISiS», Moscow, Russia

E-mail: anatmaslov@mail.ru

Abstract. It is known that modern drilling bit body undergoes significant abrasive wear in the contact area with the solid and the retracted cuttings. For protection of the body rationally use wear-resistant coating, which is welded directly to the body of bit. Before mass use of the developed coverings they need to be investigated by various methods that it was possible to characterize coatings and on the basis of the obtained data to perform optimization of both composition of coatings and technology. Such methods include microstructural studies tribological tests, crack resistance and others. This work is devoted to the tribological tests of imported brand of coatings WokaDur NiA and and domestic brand of coating HR-6750 (both brands manufactured by Ltd "Oerlikon Metco Rus"), used to protect the bit body from abrasive wear.

1. Introduction

One of the important moments in the development of modern drill bits with PDC-elements is to protect the body from abrasive wear in the contact area with drillings cuttings and sludge. Thus reducing the abrasive wear at the place of fixing PDC-element to the body, it is possible to significantly increase the service life of the bit [1].

Protection of the bit body may be achieved by various methods. For example, it is possible to apply wear resistant coatings which contain nano- or microparticles. Due to the first coating can significantly reduce the friction coefficient between contacting cuttings and sludge from zone of drilling and body of the bit. As such particles the most rationally using nanodiamonds or aluminum oxide nanoparticles. Electroplated deposited coatings have high hardness and wear resistance [2, 3] and it is promising avenue for increased bit life. Coatings which contain microparticles (usually from several tens to 1000 μm), provide protection of the bit body due to presence of very hard particles in the matrix (for example, tungsten carbide). During the work the coating with high-hardness particles doesn't wear. In this case, as with other coatings which contain the nanoparticles, there is a wear of the matrix with the subsequent departure of particles from it. Therefore, the development of various wear-resistant coatings which will work with abrasive slurry, it is advisable to carry out complex study of the matrix [4].

Method of gas-flame surfacing was got widely spreading on deposition of wear-resistant coatings on body of the bit. This is due to the fact that the resulting coating has a high hardness, corrosion resistance, speed of deposition, while the heating of the bit occurs locally, which does not reduce the characteristics of the bit body [5].

IOP Conf. Series: Journal of Physics: Conf. Series **857** (2017) 012029 doi:10.1088/1742-6596/857/1/012029

2. Materials, equipment and method of processing results

The samples were made of steel St3 and used as substrates, and as wear-resistant layers were selected as follow coatings:

– HR-6750, thickness: 2.6 mm;

– WokaDur NiA, thickness: 2.6 mm.

Tribological tests of samples were performed on automated machine friction (tribometer) CSM Instruments's TRIBOMETER (Switzerland) [6], № 44739-10 in GRSI using a rotary motion in "rod-drive" in accordance with the "Methodology for performance measurements of the friction coefficient *f* and *l* wear on tribometer CSM Instruments's "TRIBOMETER" (Switzerland). MVI KTI/10 (FR.1.28.2010.07504). The parameters were used in tribological tests are shown in table 1. Testing scheme of samples is shown in figure 1.

Table 1. Parameters of tribological test.

Parameter	Value
Track radius R	5, 6.5 mm
Applicable load F	1 N
Linear speed V	10 cm/s
Counterbody	Ball, 3 mm diameter
Material of counterbody	SiC (R = 5 mm)/Al_2O_3 (R = 6.5 mm)
Mileage L	22 700 cycles
Environment	Air

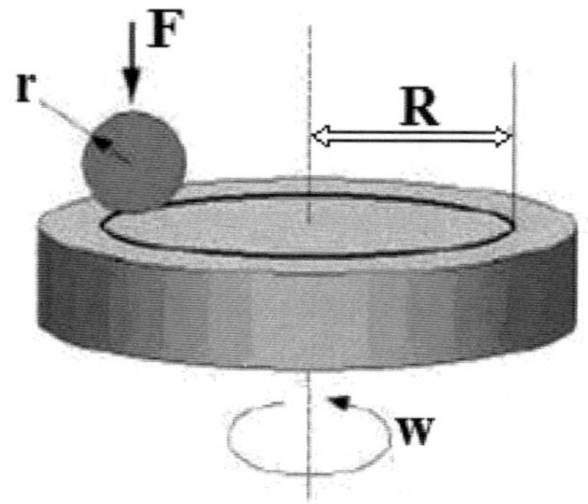

Figure 1. Testing scheme on automated machine friction TRIBOMETER: R – wear ring radius; r – radius of the ball (counterbody); F – normal load; w – rotation speed of sample.

Tests consistent with international standards (ASTM G99-05 and DIN50324). Experimental dependences of the friction coefficient on the path smoothed by 7 points.

Tribological tests were carried out on samples of the device, located in SHS Center in NUST «MISiS».

The complex study of the microstructure of wear-resistant coatings was carried out on a scanning microscope JSM-6480LV, located at the Center for Collective Using in NUST «MISiS».

3. Results of the tribological tests

Results of tribological tests of samples are shown in table 2. The test results show that the friction coefficient of WokaDur NiA coating smaller than HR-6750 coating.

In both cases established no wear of samples (less than 10^7 mm^3/N/m), which speaks of their high abrasion resistance to abrasion (example path is shown in figure 2). Less wear of counterbody from aluminum oxide due to the fact that there is "salting" the track on which moves counterbody.

Results of tests were shown that friction coefficients of HR-6750 coating were 0.57 and 0.68 for counterbodies made from silicon carbide and aluminum oxide, and WokaDur NiA coating were 0.57 and 0.66 respectively.

Table 2. The tribological characteristics of the studied sample.

Sample	Counterbody	Wear, $mm^3/N/m$		Coefficient of friction			
		Sample	Counterbody	Initial	Maximal	Average	Final
HR-6750	SiC	No ($<10^{-7}$)	$1.8\cdot10^{-6}$	0.29	0.73	0.68	0.64
	Al_2O_3		$0.3\cdot10^{-6}$	0.21	0.79	0.71	0.69
WokaDur	SiC		$3.7\cdot10^{-6}$	0.42	0.64	0.57	0.54
	Al_2O_3		$0.7\cdot10^{-6}$	0.23	0.85	0.66	0.60

Figure 2. The wear path of counterbody from silicon carbide by coating on brand of HR-6750.

The time dependences are of similar character. At the stage of the running is observed increase the coefficient of friction, which is associated with the peculiarities of formation of the contact surface in the presence of solid particles in the coating. At a later stage, the friction coefficient is stabilized, coupling work is quite stable character (figure 3).

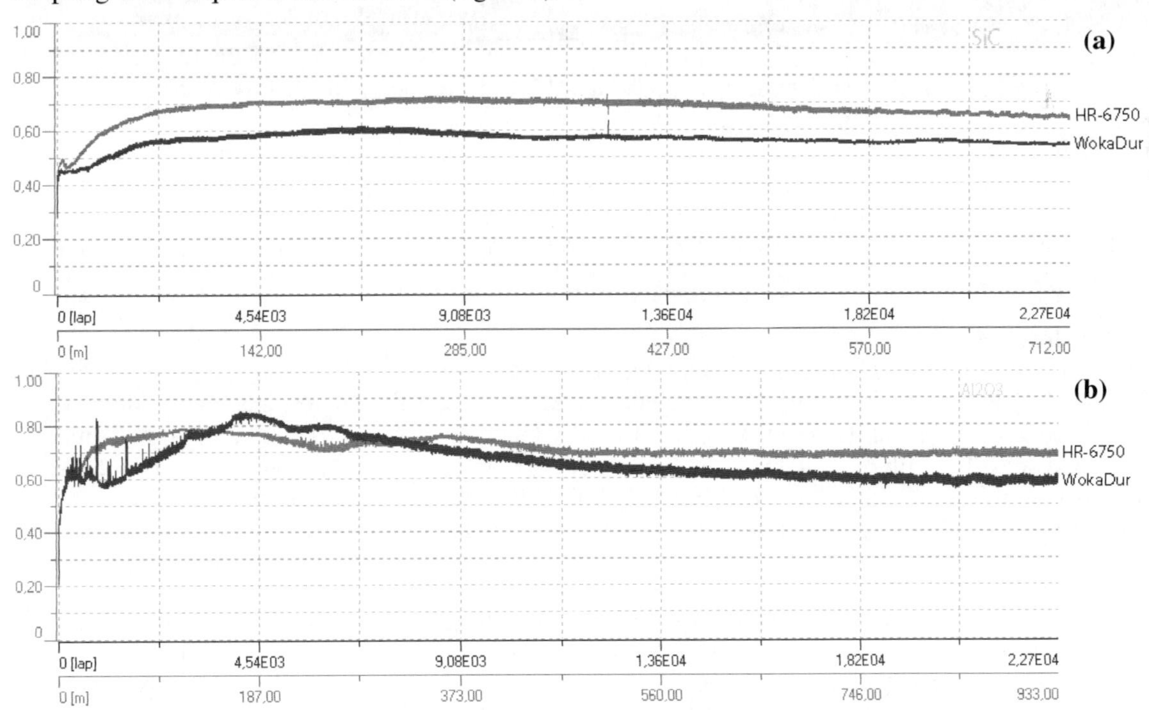

Figure 3. The time dependences of the friction coefficient for samples with coatings of HR 6750 and WokaDur NiA with counterbodies from silicon carbide **(a)** and aluminum oxide **(b)**.

Increasing wear of counterbodies in the case with WokaDur NiA coating can be explained by the fact that the ligament and the particles have microhardness greater than ligament and particles of HR-6750 coating (table 3), and as well as different structures.

Table 3. Average values of microhardness HV matrix, particles and substrates.

	St3–HR-6750	St3–WokaDur NiA
Coating	599	590
Strengthening particles	2145	2452
Steel hull	206	213

In WokaDur NiA coating is dominated by large particles of tungsten carbide size of 200–500 μm and the particles are much closer than in HR-6750 coating with tungsten carbide particles of size preferably less than 200 mm (figure 4). Closer arrangement of the particles to each other, as well as their larger size leads to a higher load at which the chipping particles from the matrix, and accordingly a lower coefficient of friction with increased wear.

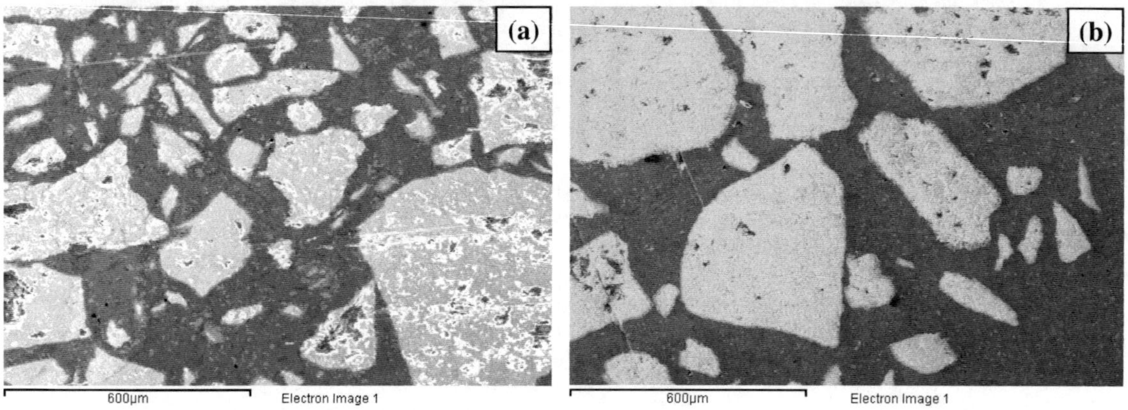

Figure 4. The microstructure of coatings HR-6750 **(a)** and WokaDur NiA **(b)**.

4. Conclusions

In the course of tribological tests of two brands of wear-resistant coating used to protect shells diamond blade PDS-bits of abrasive wear during the drilling process, it was found that the best values of the friction coefficient is observed to cover brand of WokaDur NiA, contain large particles of tungsten carbide.

Acknowledgment

This work was carried out within the framework of the federal target project of the program "Research and development in priority areas of development of the scientific and technological complex of Russia for 2014-2020" on the topic "Development of technology for manufacturing high-performance horizontal and tilt drilling drill bits for the oil and gas industry" agreement No. 14.581.21.0012 (unique identifier of the agreement RFMEFI58115X0012) with financial support of applied scientific research by the Ministry of Education and Science of the Russian Federation.

References

[1] Nekrasov I N, Bogomolov R M, Ischuk A G, Gavrilenko M V, Morozov L V, Muhametshin M M and Krilov S M 2008 *Patent of Russian Federation № 2389857* 10.11.2009 **31**
[2] Polushin N I, Ovchinnikova M S and Maslov A L 2014 *Advanced material research* **1040** 199–1
[3] Polushin N I, Ovchinnikova M S, Maslov A L and Kuchina I Yu 2015 *Izv Vyzov. Series "Chemistry and chemical technology"* **58** 55–7

[4] Polushin N I, Bogatirev A V, Laptev A I and Sorokin M N 2016 *Izv Vyzov "Powder metallyrgy and functional coatings"* **1** 60–6
[5] Langford J V and Delvice R *Patent of Russian Federation № 2167262* 10.01.2001
[6] Petrjik M I and Levashov E A 2007 *Crystallography* **52** 1002–10

The application of instrumented indentation test for thin coatings mechanical properties evaluation

V M Matyunin, A Yu Marchenkov, A N Demidov, M A Karimbekov and N A Stasenko

Metals technologies department, Moscow Power Engineering Institute, 111250, Moscow, Russia

E-mail: MatiuninVM@mpei.ru

Abstract. The high efficiency of instrumented indentation application for thin coatings mechanical properties evaluation is performed. New methods of coatings hardness and crack resistance estimation using the instrumented indentation curves analysis are proposed. It is shown, that the comparison of several coatings hardness values should be made with the equal strained volumes beneath the indenter.

The indentation is known as one of the most efficient method for thin films and coatings mechanical properties evaluation. Usually the indentation test of thin coatings includes the diamond pyramidal indenter penetration under the specified load F with the subsequent microhardness calculation. However, in case of hardness evaluation of several different coatings by indentation under the specified load the strained volumes beneath the indenter may be significantly different. Indentation depth as well as the strained volume beneath the indenter is the smallest for the hardest coating. The difference in strained volumes causes the unequal size factor impact on hardness values. This impact consists in the increase of hardness values of the material with the strained volume decrease [1].

Thus, the influence of size effect should be taken into account for correct comparison of hardness values of different coatings. Therefore, coatings hardness should be measured not under the specified load. Indentation load should be chosen so as to obtain the equal values of strained volumes beneath the indenter for each tested coating. As it was established earlier [2], the strained volume beneath the indenter depends on indent volume, so the obtainment of the equal strained volumes during the indentation of several coatings consists in obtainment of equal indent volumes. Indent volume for Vickers pyramidal indenter is known to be explicitly defined by indent diagonal d or depth h. In view of this, it is necessary to gain the specified value of d or h for each tested coating to obtain the equal indent volumes and, as a result, to provide the equal size factor impact for all tested coatings.

If all the same to compare several coatings hardness values under the specified load, the measured hardness value of the hardest coatings becomes overestimated. According to the results of authors' experiments, hardness values of the hardest coatings in case of indentation under the specified load proved to be 5…30 % higher than hardness values in case of indentation with equal indent volumes.

The results of microhardness evaluation of two TiN-based coatings using both of methods (indentation under the specified load and with equal indent volumes) are shown in table 1. The coatings were produced by physical vapor deposition method (PVD) with various mode parameters.

For the substrate impact exception on tests results the relation between coating thickness and indent depth was provided not less than 10 during all the tests.

Table 1. The results of two TiN-based coatings microhardness HV evaluation using both of methods.

Coating	Test under the same specified load F			Indentation with equal indent volumes		
	F, kg	d, μm	HV, kg/mm^2	F, kg	d, μm	HV, kg/mm^2
I	0.01	4.25	1041	0.01	4.25	1041
II	0.01	2.22	3785	0.03	4.25	3079

As it appears from the table 1, the hardness of coating II evaluated by the test under the specified load $F = 0.01$ kg is significantly (23 %) higher, then the hardness value evaluated by the proposed method. This difference may even be increased upon transition to the nanolevel of indentation ($h \leq 0.2$ μm in accordance with ISO 14577-1:2002). In view of this, it is necessary to measure the hardness not under the specified load, but with the equal indent depths or diagonals for all compared samples to estimate the efficiency of coating hardening technologies. However, the proposed hardness evaluation method is too laborious, because it usually requires the preliminary experimental selection of indentation load for obtaining the specified value of indent diagonal. The implementation of this method becomes more simply in case of instrumented indentation application including the "load F–indent depth h" curve registration and calculation of Martens hardness HM in accordance with ISO 14577-1:2002.

It should also be noticed that in accordance with ISO 14577-1:2002 Martens hardness HM evaluation is correct for $h \geq 6$ μm. In case of small indent depth ($h < 6$ μm) the Oliver-Pharr method [3] is advisable to use. According to this method the indentation hardness H_{IT} is calculated using the following equations for Vickers and Berkovich pyramidal indenters respectively:

$$H_{IT} = \frac{F_{max}}{24.5h_c^2}; \tag{1}$$

$$H_{IT} = \frac{F_{max}}{23.96h_c^2}, \tag{2}$$

where $h_c = h_{max} - \varepsilon(h - h_t)$; h_{max} – the indent depth at the ultimate load F_{max}; h_t – the intersection point of h-axis and the line tangent to unloading curve in F_{max}; ε – the coefficient depending on indenter shape (ε = 0.73 for Vickers and Berkovich pyramidal indenters).

Among the other advantages of instrumented indentation the following should be marked: possibility of test process automation, conversion of indentation curve into the tension diagram (in the area of uniform strain) and possibility of additional materials physical and mechanical properties evaluation. The procedure of the instrumented indentation test with pyramidal indenter is regulated by standard ISO 14577-1:2002, which was mentioned above. Moreover, in Russian Federation the state standard GOST R 56232-2014 regulating the instrumented indentation test using the ball indenter was also developed and enacted in 2016. Both documents describe the methods of hardness and elastic modulus evaluation by indentation curves analysis. In addition, methods of indentation curve conversion into the tension diagram in the uniform strain area are presented in GOST R 56232-2014.

Besides the hardness and elastic modulus the crack resistance should be included in the list of the most important mechanical characteristics of coatings. There are several methods of the critical stress intensity factor evaluation by indentation with pyramidal indenter, for example [4], which can be applied in case of radial cracks formation. However, in case of other types of cracks formation, for example, ring cracks, these methods cannot be applied. In this regard we suppose a method of evaluation of specific elastic-plastic strain energy required for first crack formation using the "F–h" indentation curve analysis data. It was established that the first notch on the loading curve during brittle coating indentation test corresponds to the moment of first crack formation. It was proved by indentation tests in combination with the acoustic emission method and observation of coating surface around the indent using the microscope. The instrumented indentation diagram for the TiN-based coating is shown in figure 1. The first notch appearing at the moment of first crack formation as well as the other notches responsible for subsequent ring cracks formation around the indent can be seen at the indentation diagram.

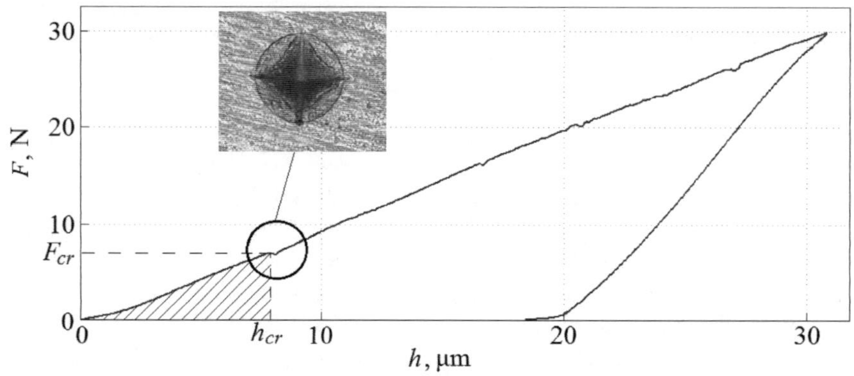

Figure 1. Instrumented indentation diagram for the TiN-based coating.

The elastic-plastic strain energy required for first crack formation W_{cr} is calculated approximately using the equation:

$$W_{cr} \approx \frac{F_{cr} \cdot h_{cr}}{2}. \qquad (3)$$

Here F_{cr} and h_{cr} are the load and the depth, corresponding to the moment of first crack formation. The specific elastic-plastic strain energy required for first crack formation ω_{cr} is calculated as the ratio of W_{cr} to the indent volume V_{cr}:

$$\omega_{cr} = \frac{W_{cr}}{V_{cr}} \approx 0.0612 \cdot \left(\frac{F_{cr}}{h_{cr}^2} \right). \qquad (4)$$

The values of ω_{cr} for several TiN-based coatings calculated using (4) are shown in table 2.

Table 2. The results of TiN-based coatings crack resistance estimation by indentation.

Sample	F_{cr}, N	h_{cr}, μm	V_{cr}, mm^3	W_{cr}, J	ω_{cr}, J/mm^3
1	6.70	7.4	$3309.5 \cdot 10^{-9}$	$2.48 \cdot 10^{-5}$	7.49
2	0.83	1.7	$40.1 \cdot 10^{-9}$	$7.06 \cdot 10^{-7}$	17.61
3	1.47	1.8	$47.6 \cdot 10^{-9}$	$1.32 \cdot 10^{-6}$	27.73
4	1.86	3.2	$267.6 \cdot 10^{-9}$	$2.98 \cdot 10^{-6}$	11.14
5	1.27	4.0	$522.7 \cdot 10^{-9}$	$2.54 \cdot 10^{-5}$	4.86

According to the results shown in table 2 the values of ω_{cr} for tested coatings are varied significantly. Thus ω_{cr} can be used as the parameter for coatings crack resistance estimation in case of appearance of any cracks types: high value of ω_{cr} indicates the high crack resistance.

Acknowledgment
The work was carried out in Moscow Power Engineering Institute (Moscow, Russia) with financial support from Russian Science Foundation (project code 15-19-00166).

References

[1] Matyunin V M 2012 *Zavodskaya laboratoriya. Diagnostika materialov* **2** 64–8

[2] Matyunin V M et al. 2015 A size effect in the indentation of materials at the micro- and nanoscale strained volumes *Piezoelectrics and Nanomaterials: Fundamentals, Developments and Applications* ed. I A Parinov (New York: Nova publishers) 133–68
[3] Oliver W C and Pharr G M 2004 *J. Mater. Res. Soc.* **1** 3–20
[4] Evans A G and Charles E A 1976 *J. Amer. Ceram. Soc.* **7-8** 371–2

Thermal processes during reactive sputtering of hot titanium target

E A Minzhulina, V I Shapovalov, V V Smirnov, A V Zav'aylov and V S Levitskiy

Saint Petersburg Electrotechnical University "LETI", 197376, Saint Petersburg, Russia

E-mail: vishapovalov@mail.ru

Abstract. The substrate heating processes during the DC magnetron sputtering of a hot titanium target were investigated. The sensitive element of a thermocouple was used as a substrate. It was found, that the heating curves of the substrate have the inflection point when the magnetron with a hot titanium target is on. The position of the inflection point is directly proportional to the argon pressure and inversely proportional to the discharge current density. The temperature of the target in the stationary mode increases when the pressure decreases. The time constant of the substrate heating process is directly proportional to the pressure and inversely proportional to the discharge current density. Power radiated on the substrate, when the magnetron with the hot target is on, is proportional to the discharge current density. For a cold target, this dependence is similar, but has a significantly smaller derivative.

1. Introduction

The substrate heating process during the film deposition using magnetron still draws an attention [1–7]. In these studies, it was found that, when the magnetron with a direct target cooling is on, the temperature of the substrate can be 50–200 higher. The magnetrons with a so-called hot target have being studied recently [8–13]. Such magnetron has an additional thermal impact on the substrate due to the target material evaporation and radiation [11].

During the substrate heating, heat supply and heat rejection with corresponding rates of $(dT/dt)^{+}$ and $(dT/dt)^{-}$ compete in such a manner that kinetic equation is written as [7]

$$\frac{dT}{dt} = \left(\frac{dT}{dt}\right)^{+} + \left(\frac{dT}{dt}\right)^{-}. \quad (1)$$

Heat rejection can occur due to radiation, heat transfer through gas and through the structural elements of the substrate holder. In [6], the kinetic equation (1) is expressed as temperature dependence

$$\frac{dT}{dt} = a - bT^{4} - cT, \quad (2)$$

where a, b and c are the quantities proportional to the power radiated on the substrate, radiating capacity of the substrate, gas heat conductivity and the structural elements of the substrate holder, respectively.

2. Experimental details

In this work, the substrate heating processes during the magnetron sputtering of a hot titanium target were investigated. A flat DC magnetron equipped with a 160 mm in diameter titanium target mounted in a vacuum chamber of $7.8 \cdot 10^{-2}$ m^3 was used for the experiment. The residual pressure did not exceed 10^{-2} mTorr. Argon was used as a sputtering gas at a pressure of p = 3–7 mTorr and discharge current density of j = 20–40 mA/cm^2. A thermocouple sensor thoroughly described in [7] was used for the research of thermal processes occurring on the substrates. The variation of the sensitive element temperature was measured in each experiment at fixed values of p and j, when the magnetron was on. The experimental accuracy appeared to be within ±5 %. To eliminate systematic errors and simplify the calculations the initial temperature is taken to be 0.

3. Results and discussion

For example, figure 1(a) demonstrates experimental dependences obtained at different argon pressures. Decrease of pressure leads to increase of target stationary temperature at any current density, as it is seen in figure 1(a). This obvious result is caused by the character of the current voltage characteristic of the magnetron [14].

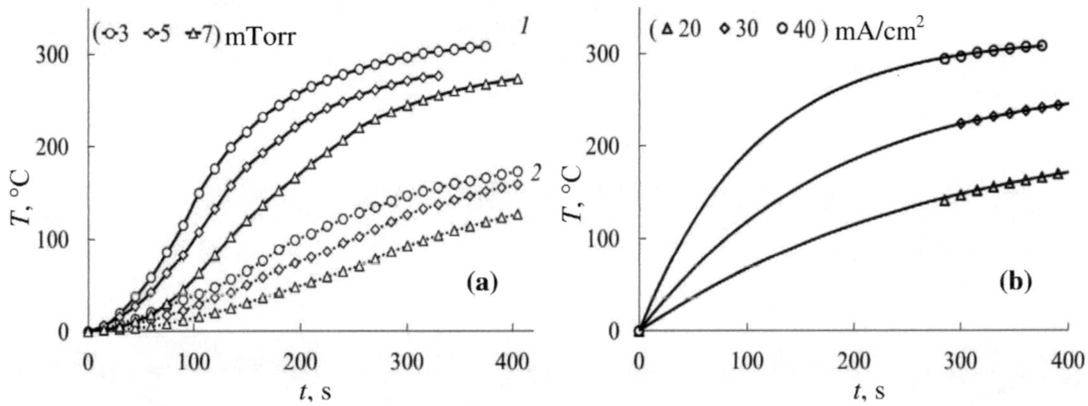

Figure 1. Sensitive element heating: **(a)** – discharge current of 40 mA/cm^2 (*1*) and 20 mA/cm^2 (*2*); **(b)** – stationary target mode and argon pressure of 7 mTorr.

Main feature of the $T(t)$ curves in figure 1 is the inflection points. The t_0 position of the points depends on the argon pressure and the discharge current density. Table 1 contains t_0 values obtained by solving $d^2T/dt^2 = 0$ equation for each experimental case. As it is seen from table 1, the t_0 value is proportional to the argon pressure and inversely proportional to the discharge current density.

Table 1. Position of inflection point t_0, s.

p, mTorr	j, mA/cm^2		
	20	30	40
3	170	135	90
5	210	175	115
7	255	205	160

The theoretical analysis showed that the inflection point appearance is connected with the relatively slow heating of the target after turning on the magnetron. The target gets into the stationary mode after 120–240 s, as it is seen in figure 1(a). The main influence on the temperature of the substrate will come from the target that works in the stationary mode since sputtering of the films is carried out for a longer period of time (up to 1800 s and more). In figure 1(b) the corresponding regions of the experimental curves from figure 1(a) are shown. Points in figure 1(b) with adequacy of 0.999 describe the expression

$$T(t) \approx T_{\infty}(p, j)[1-\exp(-\alpha(p, j)t)], \quad (3)$$

where $T_{\infty}(p, j)$ is the ultimate temperature and $\alpha(p, j)$ is the value defining the heating rate.

Table 2. Parameters of the expressions (3) and (4).

p, mTorr	j, mA/cm^2	T_{∞}, °C	α, s^{-1}	$\tau = 1/\alpha$, s	a, °C/s	c, 1/s	q, W/cm^2
	20	220	– 0.0036	280	1.84	– 0.0036	0.6
3	30	270	– 0.0057	175	3.17	– 0.0056	1.1
	40	320	– 0.0095	105	5.62	– 0.0092	1.9
	20	215	– 0.0033	305	1.67	– 0.0030	0.6
5	30	270	– 0.0053	190	3.02	– 0.0050	1.0
	40	300	– 0.0083	120	4.77	– 0.0080	1.6
	20	210	– 0.0031	320	1.56	– 0.0031	0.5
7	30	255	– 0.0048	210	2.56	– 0.0047	0.9
	40	290	– 0.0069	145	3.36	– 0.0058	1.1

The analytical model (3) shows that the substrate also has a stationary mode, and it gets into it with $\tau = 1/\alpha$ time constant. Values of T_{∞}, α and τ calculated based on experimental data are shown in table 2. As it is seen from table 2, the time constant of the heating process (τ) is proportional to the pressure and inversely proportional to the discharge current density.

We consider further the heating kinetics dependence on the temperature. A numerical differentiation of the experimental data was performed. One of calculation results in a $dT/dt = f(T)$ form is shown by dots in figure 2.

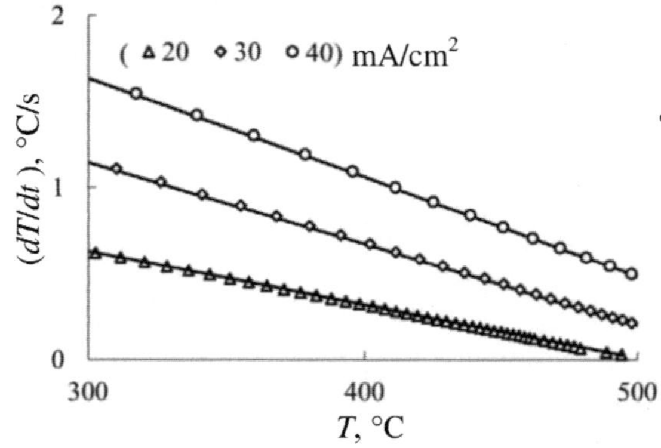

Figure 2. Substrate temperature impact on heating rate at argon pressure of 7 mTorr. Dots: experiment; solid lines: equation (4).

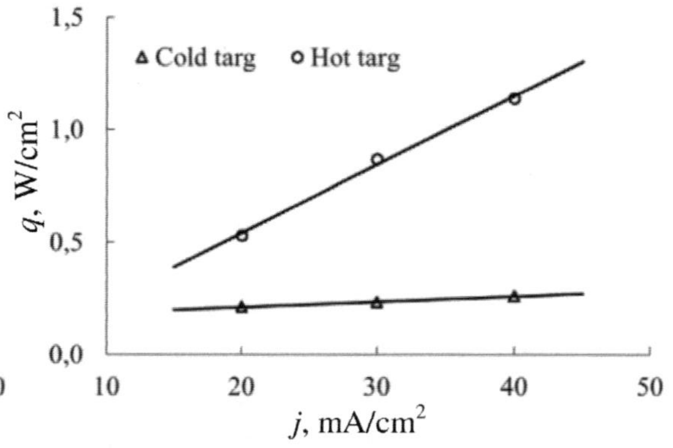

Figure 3. Power density radiated on substrate at argon pressure of 7 mTorr. Dots: experiment; solid lines: approximation.

For an analytical description of these dependences, the equation (2) was used:

$$\frac{T(t)}{dt} \approx a - cT. \quad (4)$$

Hence the radiation of the sensitive element can be neglected for $T < 400$ °C, the equation (2) loses the second term ($b = 0$). Calculated values of a and c for solid lines in figure 2 are presented in table 2.

The power density radiated on the sensitive element is equal to:

$$q = \frac{mca}{S}, \quad (5)$$

where m, c and S are the mass, the specific capacity and the element area, respectively. The results of the calculation by equation (5) can also be seen in table 2 and figure 3. The power radiated on the substrate, when the magnetron with the hot and cold target is on, is proportional to the current density as it is seen from the figure 3. But in the first case it increases faster.

4. Conclusions

The results of the research performed in this paper using a thermocouple sensor allowed us to find that:

- heating curves of the substrate have the inflection point when the magnetron with the hot titanium target is on;
- the position of the point is directly proportional to the argon pressure and inversely proportional to the discharge current density;
- the temperature of the target in the stationary mode increases with the decrease of the pressure;
- the time constant of the substrate heating is directly proportional to the pressure and inversely proportional to the discharge current density.
- power radiated on the substrate, when the magnetron with the hot target is on, is proportional to the discharge current density.

Acknowledgments

The authors thank Russian Science Foundation for the support on the study (grant 15-19-00076).

References

[1] Britun N, Palmucci M, Konstantinidis S et al. 2015 *J. Appl. Phys.* **117** 163302
[2] Wen L, Kumar M, Sahu B B et al. 2015 *Sur. Coat. Technol.* **284** 85–9
[3] Sahu B B, Han J G, Hori M et al. 2015 *J. Appl. Phys.* **117** 023301
[4] Kim H R, Wen L, Jin S B et al. 2013 *Jap J. Appl. Phys.* **52** 11NB01
[5] Wiese R, Kersten H and Wiese G 2015 *EPJ Techn. Instrum.* **2** 1–10
[6] Bornholdt S, Itagaki N and Kuwahara K 2013 *Plasm. Sourc. Sci. Techn.* **22** 025019
[7] Shapovalov V I, Komlev A E, Bondarenko A S et al. *Phys. Lett. A.* **380** 882–5
[8] Cormier P A, Thomann A L, Dolique V et al. 2013 *Thin Solid Films* **545** 44–9
[9] Lapshin A E, Levitskii V S, Shapovalov V I et al. 2016 *Glass Phys. Chem.* **42** 359–62
[10] Bondarenko A S, Kolomiytsev A A and Shapovalov V I 2016 *J. Phys.: Confer. Series* **729** 012006
[11] Shapovalov V I, Karzin V V and Bondarenko A S 2017 *Phys. Lett. A.* **381** 472–5
[12] Bleykher G A, Krivobokov V P and Yuryeva A V 2015 *Techn. Phys.* **60** 1790–5
[13] Bleykher G A, Borduleva A O, Krivobokov V P et al. 2016 *Vacuum* **132** 62–9
[14] Komlev A E, Shapovalov V I and Shutova N S 2012 *Techn. Phys.* **57** 1030–3

IOP Conf. Series: Journal of Physics: Conf. Series **857** (2017) 012032 doi:10.1088/1742-6596/857/1/012032

Formation of biocompatible surface layers depending on the sputtering distance

E O Nasakina, A V Seregin, A S Baikin, M A Kaplan, S V Konushkin, K V Sergiyenko, E D Kovaleva, A A Kolmakova, A V Leonov, M A Sevost'yanov, A G Kolmakov and S V Simakov

Institution of Russian Academy of Sciences A. A. Baikov Institute of Metallurgy and Material Science, Russian Academy of Sciences, 119334, Moscow, Russia

E-mail: nacakina@mail.ru

Abstract. Nano- and micro-dimensional surface layers of silver and tantalum on flat and wire NiTi substrates by the method of magnetron sputtering in vacuum were produced. The structure and composition of the samples were determined using SEM and Auger spectroscopy. With an increase in the sputtering distance, the thickness of the surface layers decreases, and the thickness of the transition layer and the dependence of the thickness change as a whole depend on the nature of the sputtered substance.

1. Introduction

Methods of physical vapor deposition make it possible to quickly and efficiently produce thin films of a diverse nature in a short time on virtually any substrate. Among other things this technology can be successfully used in the formation of medical composite materials that need to have complex properties, combining only the required characteristics of classical materials – for example, for the production of noninvasive stent implants designed to restore the patency of hollow parts of the cardiovascular, excretory, digestive and respiratory systems [1]: biomechanical compatibility and shape memory effect of titanium nickelide [2] with high resistance to dissolution in aggressive media, radiopaque and nontoxicity of tantalum [3–4] and silver (also exhibiting antitumor and antibacterial action) [5]. At the same time, the parameters of the resulting composites depend on a number of process conditions that can be conveniently varied over a wide range. And the purpose of this work was to determine the influence of the sputtering distance on the parameters of the layers being formed.

2. Materials and methods

Layered composites were produced by magnetron sputtering in argon at residual and working pressures of ~$4 \cdot 10^{-4}$ and ~0.4 Pa, respectively. The substrates were ion etched: surface cleaning, activation, and polishing through argon ion bombardment. Under any conditions, the substrate surface temperature did not reach 150 °C.

The surface layers were produced in the following conditions: direct current magnetron, I ~ 865 mA, in the case of silver layers U ~ 830 V, in the case of tantalum layers U ~ 400 V; sputtering time t = 30 min, the distance from the target to the substrate was 40–160 mm. To examine surface morphology and obtain composition–depth profiles, we used a TESCAN VEGA II SBU scanning electron microscope equipped with an INCA Energy energy-dispersive spectrometer system, and a

JEOL JAMP_9500F Auger electron spectrometer in combination with ion etching through argon ion bombardment at 30°.

3. Results and discussion

Externally, the surface layer repeats the substrate morphology. At small distances surface micro-defects appear as point depressions (figure 1), recalling effect of high bias voltage and ion implantation [6–7], which is correlated with a more intense flow of the sprayed material reaching the surface of the substrate, in comparison with larger distances.

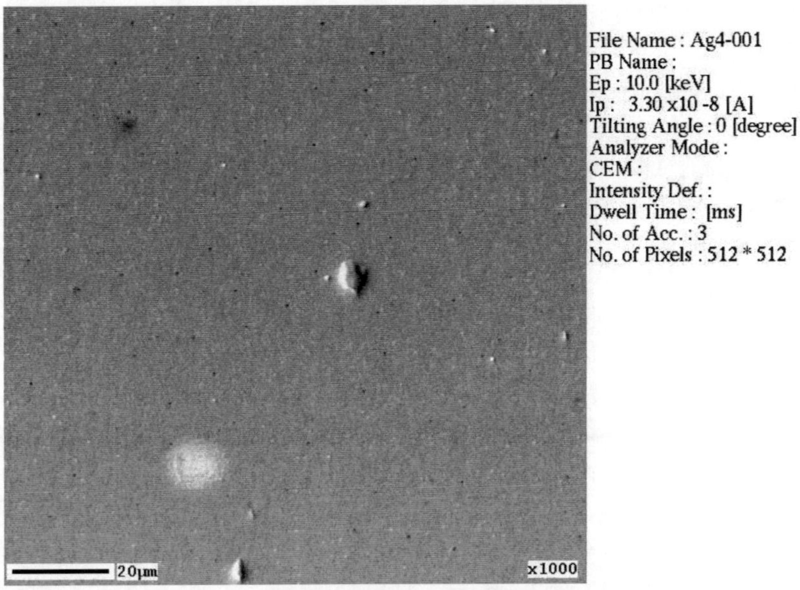

Figure 1. Morphology of the silver surface layer at a distance of 40 mm.

Researches of regularity of structure change of surface layers have shown, in general, identical results (figure 2): to 20 nm in depth from a surface the layer is enriched with oxygen, surface layer only of tantalum or silver, a transition layer (where the oxygen content is also increased and elements of the deposited layer and the substrate are contained) and the volume of the substrate itself lie deeply.

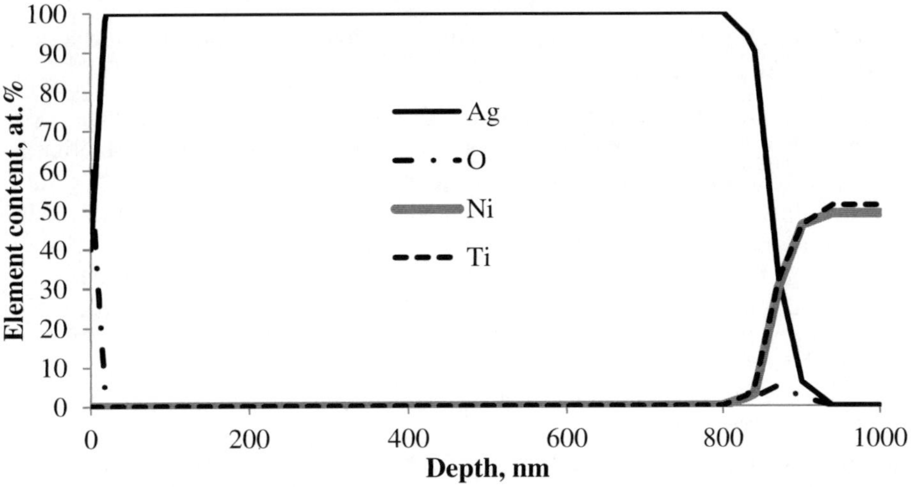

Figure 2. Composition of the composite with a silver surface layer on a titanium nickelide substrate obtained at a distance of 150 mm.

IOP Conf. Series: Journal of Physics: Conf. Series **857** (2017) 012032 doi:10.1088/1742-6596/857/1/012032

On the one hand, with distance increasing at other equal conditions the thickness of the surface layer naturally decreases (figure 3) because larger volume of the sputtered substance is scattered away from the substrate; on the other hand, the thickness of the transition layer increases, which can be explained by a more intense flow of the sputtered substance at a shorter distance, uniformly but faster filling the surface and less diffusing into the substrate; and the total thickness of the layers eventually reaches a certain plateau, practically unchanged when the distance is more than 80–90 mm. Since the presence of a substantial transition layer is a presumable reason for the good adhesion of a new surface to the substrate [3, 4, 6], the surface layer must be adjusted to the mechanical properties of the substrate, and also considering the microdefects of the surface at small distances, the distances from the target to the substrate in the range of 100–150 mm are more optimum.

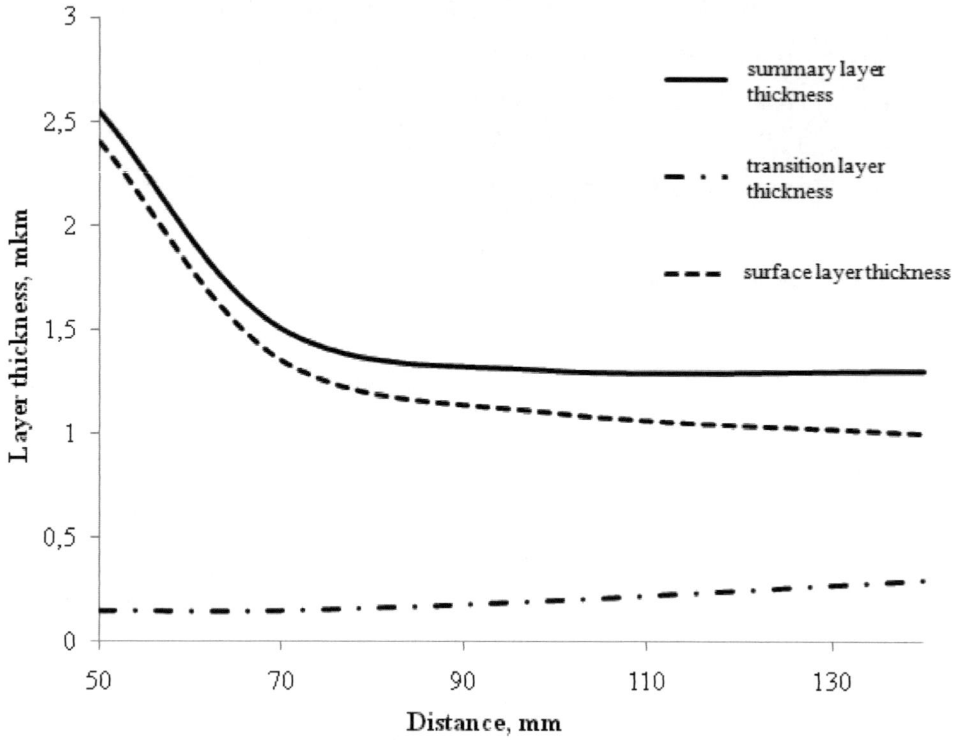

Figure 3. Change in the tantalum surface layers thickness as a function of the distance between the target and the substrate.

In contrast to tantalum, where the appearance of the surface layer thickness curve completely corresponds to the calculated models [7], in the case of a silver layer two plateaus are observed (figure 4): almost constant thicknesses at small sputtering distances can be attributed to saturation of the surface at high intensities of flux falling on the substrate. Also, unlike the previous case, both the surface layer and the transition layer are thinned with increasing distance, because a smaller volume of substance reached its goal, but the intensity of the flow did not affect the formation of the transition layer.

Formation of layers on the side that is opposite to the sputtered flow was noted. In this case, the structure and patterns of these layers changes are analogous to the straight side, but every 10–15 times thinner (figure 5).

4. Conclusions

Nanoscale and microscale layered composites with surface layers of silver and tantalum on flat and wire NiTi substrates were produced by vacuum magnetron sputtering. The presence of the transition layer that contains both substrate and target elements and presumably provides high adhesion of the surface layer to the substrate has been demonstrated.

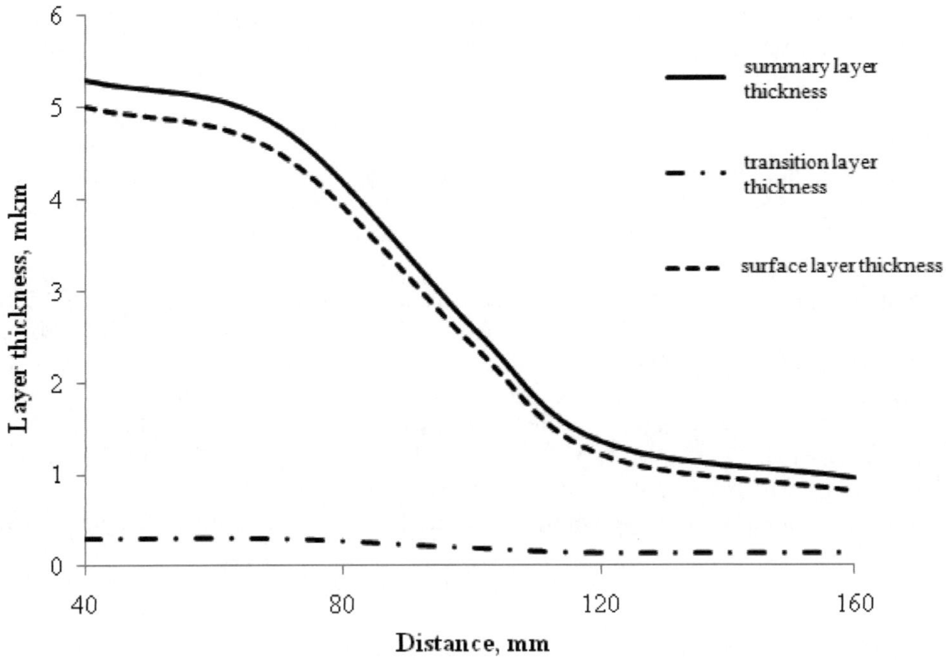

Figure 4. Change in the silver surface layers thickness as a function of the distance between the target and the substrate.

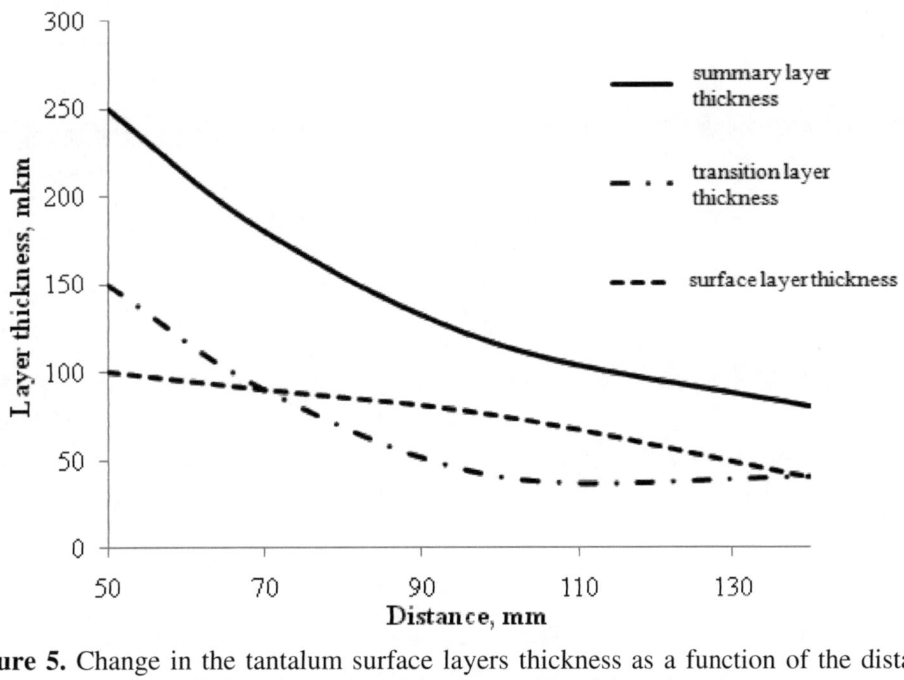

Figure 5. Change in the tantalum surface layers thickness as a function of the distance between the target and the substrate on the reverse side of the substrate.

The morphology of a new surface repeats the substrate state regardless of the sputtering conditions. Our results demonstrate a nonlinear increase in the thickness of the growing surface layers with decreasing sputtering distance under otherwise equal conditions. But the thickness of the transition layer and the dependence of the thickness change as a whole depend on the nature of the sputtered substance.

It has been shown that, at distances of 40–160 mm, insignificant deposition on the substrate side that is opposite to the sputtered flow is observed, with the thickness of formed layers also depending on the distance between the target and the substrate.

Acknowledgment
The authors wish to thank Dyomin K Yu for his help in sample analysis. This work was supported by RFBR program number 15-33-70006 "mol_a_mos" and 14-29-10208 "ofi_m".

References

[1] Zabolotniy V T, Kolmakov A G, Sevost'yanov M A and Nasakina E O 2013 *Integral* **4**(72) 42–5
[2] Gunther V E, Hodorenko V N, Yasenchuk Yu F et al. 2006 *Titanium nickelide. Medical material of new generation* (Tomsk: MITs Publishing house) 296
[3] Nasakina E O, Baikin A S, Sevost'yanov M A, Kolmakov A G, Zabolotnyi V T and Solntsev K A 2014 *Theoretical Foundations of Chemical Engineering* **48**(4) 477–86
[4] Nasakina E O, Sevost'yanov M A, Mikhailova A B, Gol'dberg M A, Demin K Yu, Kolmakov A G and Zabolotnyi V T 2015 *Inorganic Materials* **51**(4) 400–4
[5] Nasakina E O, Baikin A S, Sergiyenko K V et al. 2017 *Inorganic materials: Applied research* **8**(1) 112–7
[6] Nasakina E O, Sevostyanov M A, Dyomin K Yu, Mikhaylova A B, Goldberg M A, Kolmakov A G and Zabolotnyy V T 2015 *Inorganic Materials: Applied Research* **6**(4) 293–7
[7] Kuz'michev A I 2008 *Magnetron Spray Systems. Introduction to the physics and technology of magnetron sputtering* (Kiev: Avers) 244

IOP Conf. Series: Journal of Physics: Conf. Series **857** (2017) 012033 doi:10.1088/1742-6596/857/1/012033

Influence of thin glass films coherence on the Schottky and Rebinder effects in sensors

S N Novikov, N E Korobova, S P Timoshenkov and E P Goryunova

National Research University of Electronic Technology "MIET", 124498, Moscow, Zelenograd, Russia

E-mail: korobova3@mail.ru

Abstract. On the basis of the experimental study common property of three vitreous systems (SiO_2, alumina borosilicate glass fiber and $GeSe_2$), which are two- or multilevel group of atoms presence capable of changing their electronic configuration in these systems under the influence of external influences has been established. The change in the structure of film glassy systems is accompanied by changes in the work electron function (Schottky effect). It was found that the strength of alumina borosilicate glass fibers was due to coherent domains that form a kind of surface film "cover". A sharp decrease in the strength of glass fibers from 400 to ~250 kg/mm^2 (Rebinder effect) was established in a water-containing medium and heat treatment.

Physical glass nature is constantly at the center of scientific research attention, not only because of its practical importance, but also because of the fundamental problem importance of the vitreous state properties. Analysis of a large number of theoretical and experimental works [1–2] devoted to the structure and properties of Se, $GeSe_2$, SiO_2, $SiSe_2$, AsSe, As_2O_3, Sb_2O_3, TiO_2, GeO_2, P_2O_5, $BeCl_2$, BeF_2, $ZnCl_2$ allowed to create "concept of polymer-polymorphoid glass and glass-forming liquid". The basis of this concept is the provision in the glass-forming liquid existence at least two structural units – polymorphoids: low-temperature (LTPM) and high-temperature (HTPM), capable of transitioning from one polymorphic modification to another (LTPM ↔ HTPM) under certain conditions [3]. It can be considered and established that structure glasses is a system of two- (or multilevel) particles capable of transferring from one energy level to another under certain circumstances. It was known [4] that in such systems energy is released in the form of collective spontaneous emission or "super radiation Dicke". The cause of this radiation is the correlation between spatially separated emitters, interacting with each other through the radiation field. Thus, the main feature of the vitreous state can be considered the presence of multilevel molecular groups capable of performing a mutual transition (polymorphic transformation) under certain conditions (temperature, pressure), accompanied by a change in their electronic configuration.

The experimental results during the study of various glassy systems, such as (1) study of changes in the electron work function (EWF) of Si(100) plates in different oxidation states, (2) as well as the study alumina borosilicate glass fibers properties have been obtained during their heat treatment. The EWF values for silicon plate having a thick oxide film are close to the corresponding value for silicate glass, therefore, when discussing the processes occurring in glass fibers, the EWF data obtained for thick SiO_2 films were used. Oxidation of Si(100) plates was carried out thermally by the method adopted in the electronics industry [5]. The samples were heat treated in a chamber oven (~ 1 h). The study results in the form of the dependencies $\varphi = f(t°)$ are shown in figure 1.

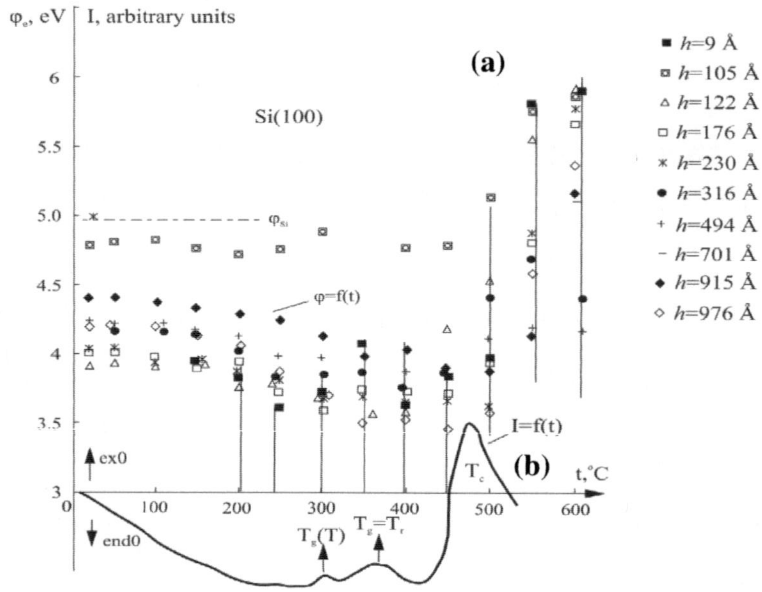

Figure 1. Dependence of the electron work function φ of Si(100) plates vs. heat treatment temperature **(a)**, *h* is the oxide film thickness on Si (100); intensity (*I*) dependence of differential-scanning calorimeter (DSC) vs. annealing temperature of $GeSe_2$ **(b)**.

Mechanical properties of industrial alumina borosilicate glass fibers samples (SiO_2 – 53.8; Al_2O_3 – 13.4; CaO – 13.4; B_2O_3 – 10.4; MgO – 4.5; other – 4.5 %) with average tensile strength $\sigma \approx 250$ kg/mm^2, average diameter of ~6.5 μm have been determined by standard techniques under atmospheric conditions at the length of all test samples 10 mm. The loading rate was constant and equal to 0.5 g/s, Density values (ρ_{gf}) were measured in addition to the mechanical properties of glass fibers. The measurement accuracy for ρ_{gf} was ~1–3 %. The results are shown in figure 2.

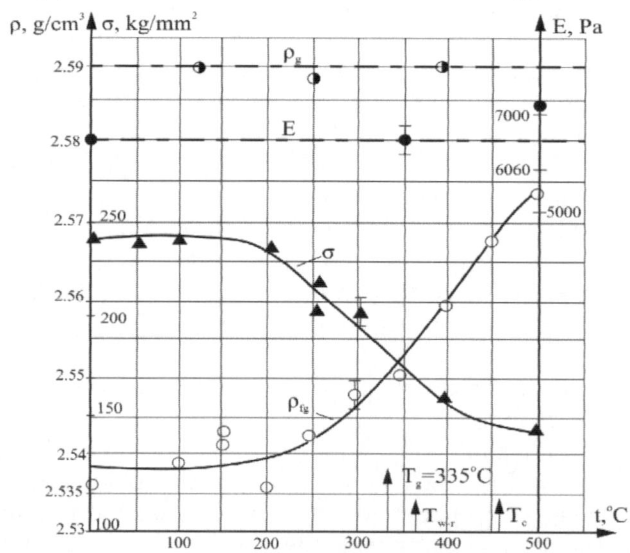

Figure 2. Dependence of tensile strength σ of alumina borosilicate fiber, elastic modulus *E*, glass fiber density ρ_{gf} and glass density ρ_g vs. annealing temperature *t*°.

The results of differential scanning calorimetry (DSC) obtained in the study of the structural relaxation of vitreous $GeSe_2$ in isothermal annealing have been used in this study to assess the energy effect that takes place at the restructuring of glass elements during their heat treatment. These data are shown in figure 1(b). If the change in the glass fiber strength is traced directly after its manufacture and in time, then the glass fiber strength during tension σ has two levels: a) high value closes to the theoretical (~400 kg/mm^2) for so-called "untouched" fibers, i.e. measured immediately after drawing from a glass vessel with special precautions in the absence of contact with other glass fibers; b) working (industrial) value σ after fabrication (~250 kg/mm^2), which is usually used for design development. Large difference σ between these levels is associated with non-reinforcing effect of the water-containing medium, which renders an adsorption ("wedging") effect on brittle bodies' strength (Rebinder effect). The mechanism of this phenomenon was investigated in [6]. It was shown, that sharp strength drop from 360 to ~200 kg/mm^2 when literally the first portions of water (relative humidity ρ/ρ_s from 0.001 to 0.03) occurs when precision weighting method has been used for measuring water adsorption on glass fibers (figure 3).

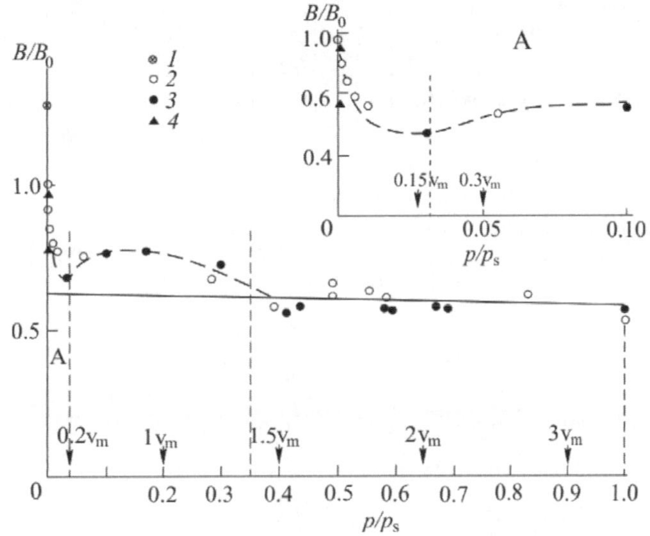

Figure 3. Dependence of the relative tensile strength B/B_o vs. relative air humidity ρ/ρ_s (υ_m – value of H_2O molecules monolayer) [6].

It is interesting to note (figure 1(a)) that the EWF value for oxidized Si(100) plates decreases sharply (~1 eV) when measured in atmospheric moist air medium (Schottky effect). The physical mechanism of this effect on Si(100) has also been investigated by authors [7] and explained by the presence in the structure of the sorbed H_2O coherent clusters having a large dipole moment. Sharp strength decrease effect in various materials has been explained in the scientific works of P A Rebinder [8] due to "a decrease in free surface energy, i.e. formation work of new surfaces". Thus, it is natural to assume that this kind of process must be accompanied by noted above a change in the electronic properties of the material. So, it can be concluded that: a) the main glass feature is the presence of two or multi-level groups of atoms in the system capable of changing their electronic configuration under the influence of external influences. The change in the structure of vitreous systems is accompanied by change in the electron work function (Schottky effect); b) during transition from liquid state to the brittle solid state at high cooling rate, which is realized in modern technology for the glass fibers production, structure of the liquid state was fixed, i.e. low-density fraction predominates, so the "untouched" glass fiber has a tensile strength σ close to the theoretical (~ 400 kg/mm^2).

In the structure of such fiber, the main role is played by coherent domains, which form a kind of surface polymer "cover". The stay of glass fiber in the water-containing medium leads to a sharp decrease in its strength to ~250 kg/mm^2 (Rebinder effect); c) fiber heat treatment to ~500 °C is accompanied by a further decrease in strength (up to ~150 kg/mm^2) and increase in density (by 3.7 %). Main property changes were occurred in the range from 250 to 400 °C (to the crystallization temperature ~460 °C). Electron work function growth (by ~2.0 eV) has been observed in the same range of annealing temperatures, which indicates the destruction of coherent domains in the glass structure.

Acknowledgements
This work was financially supported by the Russian Science Foundation with unique identifier for applied research is 16-19-00177.

References

[1] Novikov S, Timoshenkov S, Korobova N and Goryunova E 2016 *Journal Physical Chemistry A* **4** 856–60
[2] Minaev V S 2004 *Semiconducting Chalcogenide Glass, Ser. Semiconductors and Semimetals* (Elsevier Academic Press) 139–79
[3] Minaev V, Timoshenkov S, Vasiliev V, Aleksandrovich V, Kalugin V and Korobova N 2016 *Journal of Optoelectronics and Advanced Materials* **1–2** 10–23
[4] Andreev A V, Emelyanov V I and Ilinsky Yu A 1980 *Successes of Physical Sciences* **4** 653–94
[5] Krasnikov G Ya and Zaitsev N A 1999 *Physical and technological basis of VLSI quality assurance* (Micron-Print) 166
[6] Novikov S N 1999 *Russian Journal of Physical Chemistry* **8** 1452–59
[7] Novikov S N and Timoshenkov S P 2010 *Russian Journal of Physical Chemistry* **7** 1394–7
[8] Rebinder P A 1978 *Surface phenomena in disperse systems. Physico-chemical mechanics* (Nauka) 368

IOP Conf. Series: Journal of Physics: Conf. Series **857** (2017) 012034 doi:10.1088/1742-6596/857/1/012034

Wear-resistance of nanostructured coatings based on diamond-like carbon and compounds of titanium with carbon

S A Plotnikov[1], A B Vladimirov[1], A B Rinkevich[1], A P Rubshtein[1], J Zhang[2], V A Zavalishin[1] and N A Sokolkina[3]

[1] M. N. Miheev Institute of Metal Physics, Ural Branch of Russian Academy of Sciences, 620990, Ekaterinburg, Russia
[2] Lanzhou Institute of Chemical Physics, Chinese Academy of Sciences, 730000, Lanzhou, China
[3] Ural Federal University, 620002, Ekaterinburg, Russia

E-mail: rubshtein@imp.uran.ru

Abstract. Multilayer coatings $[(TiC_x/Ti/a\text{-}C)+ta\text{-}C]_n$ with different composition of composite (TiC_x/Ti/a-C) layers are studied. The dependences of abrasive wear resistance and H^3/E^2 ratio (H – hardness, E – elastic modulus) of multilayer coatings on the carbon content in the composite layer are determined. The phase composition of the (TiC_x/Ti/a-C) layer, the ratio of volume fractions of the phases and the volume fraction of interface component, interlayer adhesion and adhesion to the substrate of the multilayer coating have a synergistic effect on the wear resistance of the testing coatings.

1. Introduction

Recent decades, the quality of wear-resistant coatings has been improved significantly through multicomponent nanocomposite structure. Amorphous matrix, grain size and dopants affect their physical, mechanical and tribological properties. For example, the introduction of Al, Ti, Cr improve the strength properties of diamond-like carbon coatings [1]. Creating gradient structures without sharp boundaries and multilayer structures with a large number of interfaces to prevent the cracks propagation is the way for optimizing the mechanical properties of coatings [2–3]. Multilayer structures can be created by uniformly embedding layers with different components between amorphous diamond-like carbon layers [2]. The mechanical and tribological properties of the multilayer coatings should depend on the composition and structure of the embedded layers. The development of protective coatings, including multilayer coatings, depends largely on the ability to predict their properties using the experimentally measured characteristics. For a long time, hardness is considered the most important property for wear-resistant materials. Nevertheless, the parameters combining hardness (H) and elastic modulus (E), such as H/E and H^3/E^2, are no less important characteristics of coatings' wear resistance [4–5]. The critical load for the beginning of the plastic yielding is scaled with H^3/E^2 ratio. For a given load, elastic contact will be preferred, if H^3/E^2 of coating increases.

In present work there were fabricated and investigated the multilayer coatings $[(TiC_x/Ti/a\text{-}C) + ta\text{-}C]_n$, where the composite layers (TiC_x/Ti/a-C), consisting of titanium (Ti) or carbon (a-C) matrix with titanium carbide TiC_x grains, are embedded between amorphous diamond-like carbon (ta-C) layers.

2. Materials and methods

Multilayer coatings were deposited on polished tool steel and stainless steel substrates by the PVD technique using UVNIPA-1-001 machine. The substrates were cleaned with distilled water and ethanol in an ultrasonic bath and by ion etching in vacuum chamber. Ta-C layers were deposited by arc pulse sputtering of graphite target, while the composite layers (TiC_x/Ti/a-C) were deposited by simultaneous sputtering of titanium and graphite targets [6]. The titanium target was sputtered using an arc technique at a direct current of the arc source, whereas a graphite target was sputtered using a pulsed arc technique at various arc pulse frequencies (f = 1, 3, 5, 10, 15, 20 and 25 Hz). The thickness of multilayer coatings and the composition of composite layers were determined using scanning electron microscope QUANTA 200 with EDAX analyser. The hardness and elastic modulus of the multilayer coatings were evaluated using a nanoindentation method (NanoTest 600, Micro Materials Ltd, UK) with a Berkovich diamond indenter. The loading was carried out with the automatic control mode of indenter depth penetration at loads from 1 to 256 mN. There were made five nanoindentations at each load.

The scratch tests were evaluated according to the following scheme. A diamond ball of 50 μm in diameter slid over the coating surface at a rate of 1 μm/s. The ball passed a distance of 20 μm without load (stage I, figure 1), then a distance of 100 μm with a constant loading rate of 5 mN/s up to load P = 500 mN (stage II, figure 1), and a distance of 80 μm at P = 500 mN (stage III, figure 1). Figure 2 shows the SEM images of scratches for various multilayer coatings. Abrasive wear resistance of multilayer coatings were tested in a jet of corundum particles of 120–150 μm in size at a speed of 20 m/s and an attack angle of 90^0. The relative wear resistance was determined from the following relation: $(I/I_0)^{-1} = (\Delta m/m_c + 1)$. I, I_0 are the wear rate of the multilayer coating and the substrate material, respectively, Δm – the difference between the wear of the test substrate and the wear of substrate with completely worn multilayer coatings at the same dose of abrasive particles, m_c – weight of worn multilayer coating.

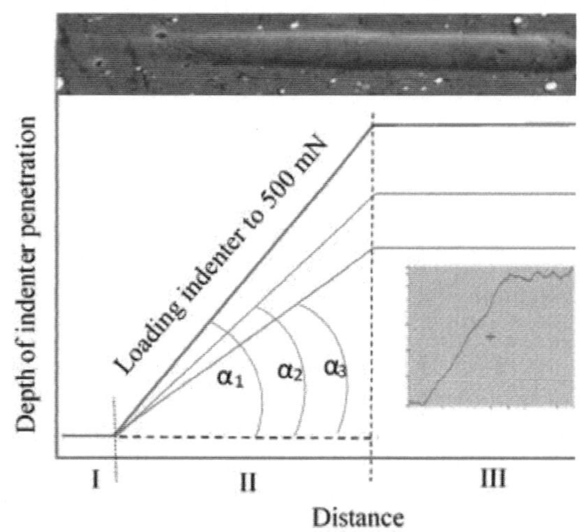

Figure 1. Scheme of the scratch test: α – angle of slope of the straight line in the loading region (stage 2) for different coatings.

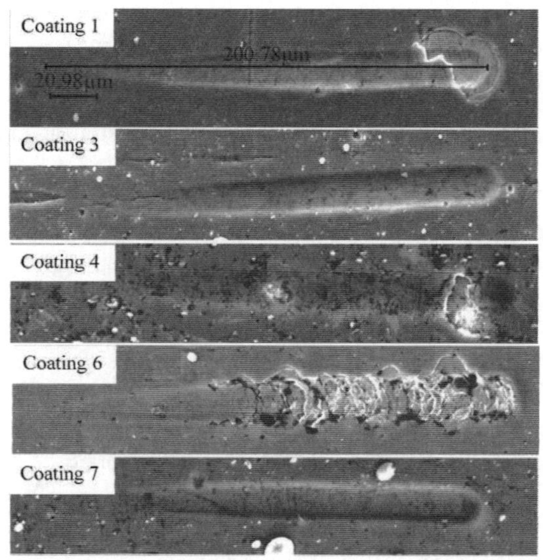

Figure 2. Images of scratches, created during scratch test (scanning electron microscopy).

3. Results and discussion

The deposition conditions and characteristics of the multilayer coatings are presented in table 1. Figure 3 demonstrates the $f = H^3/E^2(C_C)$ and $f = (I/I_0)^{-1}(C_C)$ dependences. The increase of carbon content in the composite layer to ~25 wt. % leads to an increase of the H^3/E^2 ratio. Further increase of C_C is accompanied by a decrease and then by an increase of the H^3/E^2 ratio. In the ranges of carbon content $6 < C_C < 15$ wt. % and at $C_C > 30$ wt. % a direct relationship between relative wear resistance of multilayer coating $(I/I_0)^{-1}$ and the H^3/E^2ratio is observed: the wear resistance increases with increasing H^3/E^2 ratio. Multilayer coatings 1 and 6 have the lower relative wear resistance in compare with the coatings 1 and 7 (table 1, figure 3). After the scratch test, multilayer coatings 1 and 6 partially are peeled off at the edges of the scratch. Numerous cracks are marked on the scratch of coating 6. The scratches of coating 3 and 7 have not visible peeling. Only small crosscut cracks formed inside the scratch of coating 7.

Table 1. Deposition conditions and characteristics of multilayer coatings.

Coating	f, Hz[a]	C_C, wt. %[b]	d, nm[c]	H, GPa	E, GPa	H^3/E^2
1	1	5.8	1500	17.9	266	0.08
2	3	11.2	1500	18.6	243	0.11
3	5	14.6	1700	20.8	264	0.13
4	10	18.4	2500	21.1	249	0.15
5	15	25.3	2000	22.1	255	0.17
6	20	34.2	1500	18.1	221	0.12
7	25	51.0	2000	22.0	267	0.15

[a] Frequency of graphite target sputtering during deposition of (TiC_x/Ti/a-C) layer.
[b] Carbon content in (TiC_x/Ti/a-C) layer.
[c] Thickness of multilayer coating.

In the range of C_C from 18 to 25 wt. %, the discrepancy between $(I/I_0)^{-1}$ and H^3/E^2 is observed. In this range, the embedded composite layers have the maximum volume fraction of TiC_x grains and the minimum volume fraction of amorphous matrix [6–7]. Moreover, in this range the phase composition of composite layer changed. The composite layer consists of titanium and titanium carbide grains (TiC_x/Ti) at C_C < 20 wt. %, and an amorphous carbon matrix with grains of titanium carbide (TiC_x/a-C) at C_C > 20 wt. %. The structural features of composite layer can affect both the wear of the embedded layer and the interlayer adhesion. After the scratch test, coating 4 (C_C = 18 wt. %) has a no uniformity surface with a partial delamination of separate layers (figure 2). The theoretically calculated volume fraction of the interface component in the (TiC_x/Ti/a-C) composite coatings has a minimum at C_C = 20 wt. % and two maxima near the carbon content at which the multilayer coatings demonstrate the greatest relative wear resistance.

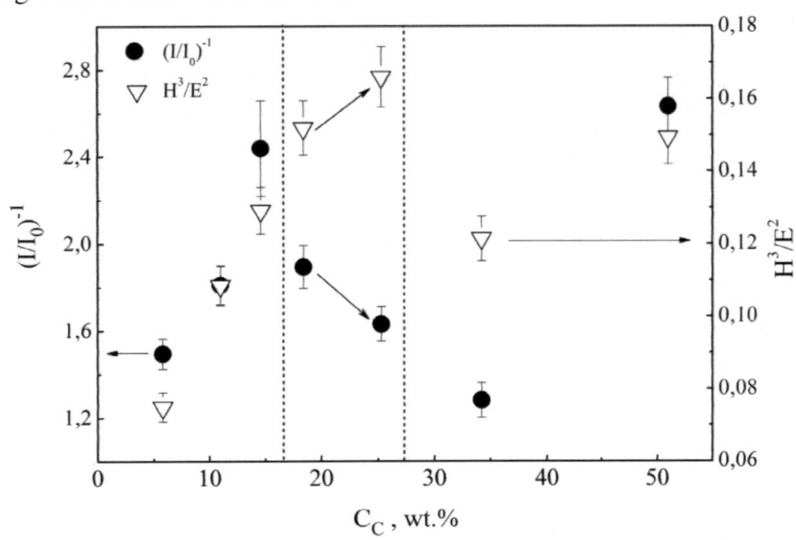

Figure 3. Dependences of the relative wear resistance of the multilayer coatings (black circles) and the H^3/E^2 ratio (open triangles) on the carbon content (C_C) in composite layer of multilayer coating.

IOP Conf. Series: Journal of Physics: Conf. Series **857** (2017) 012034 doi:10.1088/1742-6596/857/1/012034

4. Conclusion

The phase composition of (TiC_x/Ti/a-C) layer, the volume fraction of titanium carbide grains, the volume fraction of interface component, the interlayer adhesion and the adhesion of the multilayer coatings to the substrate have a synergistic effect on wear resistance of nanostructured coatings formed by layer-by-layer deposition of the ta-C and (TiC_x/Ti/a-C) layers. The results have shown that two coatings demonstrated the most wear-resistance. The one is formed by layers of ta-C and (TiC_x/Ti) with a carbon content of ~15 wt. %. The other is formed by layers of ta-C and (TiC_x/a-C) with a carbon content of ~50 wt. %.

Acknowledgments

The research was carried out within the state assignment of FASO of Russia (theme "Spin" No. 01201463330), supported in part by RFBR (project No. 17-58-53076_GFEN_a).

References

[1] Dai W, Gao X, Liu J and Wang Q 2016 *Diam. Relat. Mat.* **70** 98–104

[2] Bai W Q, Li L L, Xie Y J, Liu D G, Wang X L, Jin G and Tu J P 2016 *Surf. Coat. Technol.* **305** 11–22

[3] Cai J B, Wang X L, Bai W Q, Zhao X Y, Wang T Q and Tu J P 2013 *Appl. Surf. Sci.* **279** 450–7

[4] Leyland A and Matthews A 2004 *Surf. Coat. Technol.* **177–178** 317–24

[5] Beake B D, Vishnyakov V M and Harris A J 2017 *Surf. Coat. Technol.* **309** 671–9

[6] Rubshtein A P, Vladimirov A B, Korkh Y V, Ponosov Y S and Plotnikov S A 2017 *Surf. Coat. Technol.* **309** 680–6

[7] Vladimirov A B, Plotnikov S A, Trakhtenberg I Sh, Rubshtein A P and Volkova E G 2015 *Protection of Metals and Physical Chemistry of Surfaces* **51** 230–3

Structure and functional properties of TiNiZr surface layers obtained by high-velocity oxygen fuel spraying

P O Rusinov, Zh M Blednova and O I Borovets

Department of dynamics and strength of machines, Kuban State Technological University, 350072, Krasnodar, Russia

E-mail: ruspiter5@mail.ru

Abstract. The authors studied a complex method of surface modification of steels for materials with shape memory effect (SME) Ti-Ni-Zr with a high-velocity oxygen-fuel spraying (HVOF) of mechanically activated (MA) powder in a protective medium. We assessed the functional properties and X-ray diffraction studies, which showed that the formation of surface layers according to the developed technology ensures the manifestation of the shape memory effect.

1. Introduction

The intermetallide based on nickel alumina NiAl, which has a high-temperature shape memory effect (martensitic transformation temperatures for NiAl SME alloys can reach 1000 K) is widely known and differs by the formation of several martensite variations with different structures [1]. The alloy with shape memory effect of TiNiZr is less well known. Partial replacement of titanium with zirconium increases the temperatures of martensitic transformations by more than 400 K. Thus, SME alloy Ti-Ni-Zr has a high-temperature shape memory effect [2], which has higher strength properties than NiAl alloy and can be used in various industries. Functional alloys with shape memory effect have unique applications [3], but their high cost limits their use in engineering. Efficiency and economic feasibility of using shape-memory alloys in various engineering industries can be achieved both by doping with the third component and by forming multifunctional composite surface layers of multicomponent materials with shape memory effect [4–7]. One of such composition is "steel – a layer with the shape memory effect on the basis of Ti-Ni-Zr", which is formed by the high-velocity oxygen-fuel spraying of the mechanically activated powder Ni47Ti26Zr27. This composition was studied in this article.

2. Mechanical activation

The surface layers were formed by high-velocity oxygen-fuel spraying of mechanically activated powder made of materials with shape memory effect on the basis of Ni47Ti26Zr27 on the upgraded GLC-720 aggregate. The coatings were applied in argon onto steel 321. A ball mill GEFEST-2 (AGO-2U) with the following parameters was used for mechanical activation and grinding of the powder Ni47Ti26Zr27: drum rotation frequency 1200 min^{-1}, rotational speed of the carrier 900 min^{-1}, diameter of steel balls 6 mm, operating time 10–30 min. We used mechanically activated powder Ni47Ti26Zr27 for the application of coatings. The size of the fractions was 1–10 μm, the powder was mainly in the B2 phase (~70 %).

The aim of this work is to study the structure and functional properties of the composition "steel - layer with the shape memory effect on the basis of Ti-Ni-Zr" obtained by high-velocity oxygen-fuel spraying of a mechanically activated powder in a protective medium.

3. Investigation of structure

Macro- and microanalysis of Ti-Ni-Zr alloy surface layers obtained by the developed technology showed that the layer structure is sufficiently dense. The boundary between the coating and the substrate has no visible defects (figure 1(a)). The Ti-Ni-Zr layer has a thickness of 1 mm. When particles of a powder pass through an oxygen-fuel jet, they heat up, melt, and, upon impact against the substrate, solidify in the form of deformed disks with a diameter of 20–35 μm, and thickness of 5–20 μm. The grain size in the coating obtained by high-velocity oxygen-fuel spraying in a protective medium (argon medium) is 80 to 100 nm (figure 1(b)–(d)).

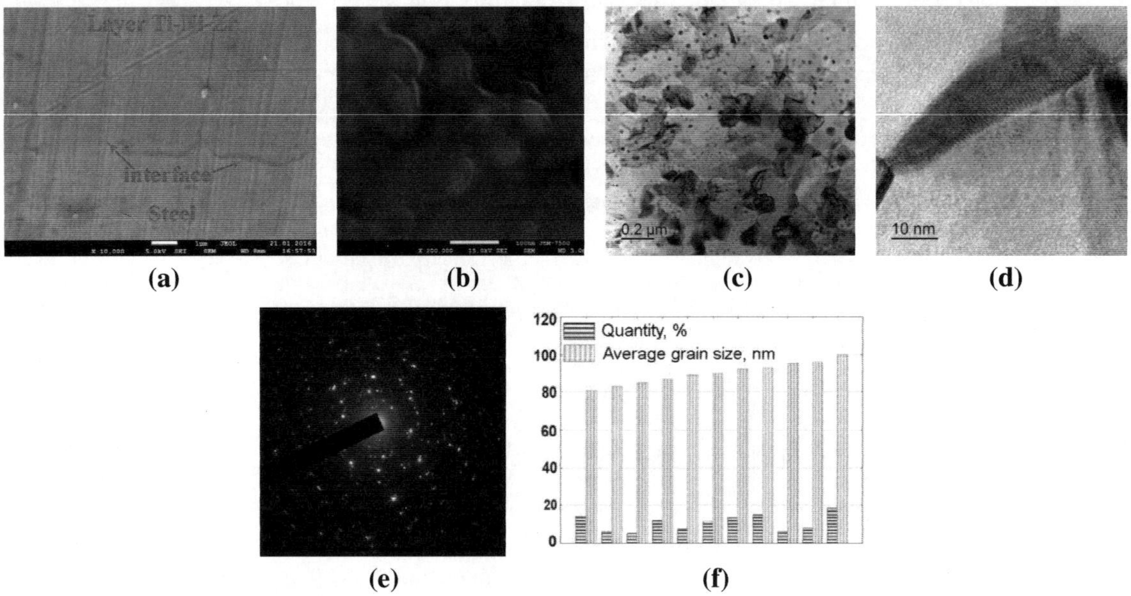

Figure 1. The microstructure of the Ti-Ni-Zr coating obtained by HVOF in a protective medium ×10 000 **(a)**; austenite structure ×200 000 **(b)**, ×120 000 **(c)**; tweed electron-microscopic contrast ×150 000 **(d)**; micro X-ray diffraction pattern of the coating **(e)**; dependence of the average grain size on its percentage **(f)**.

The dependence of the average grain size on its percentage is presented in figure 1(e). The microhardness of the Ti-Ni-Zr layer varies within H_{μ} = 9.5–12.7 GPa. The X-ray diffraction analysis showed that at room temperature the initial phase state of the Ti-Ni-Zr layer, after the HVOF of the mechanically activated powder in the protective meduim (argon), is B19 martensite phases with monoclinic lattice, austenite B2 phases with a cubic lattice, intermetallic phases Ni_3Ti, Ti_2Ni, NiZr, $NiZr_2$, with a cubic and hexagonal lattice, and a small amount of titanium oxide (TiO), less than 2 % (figure 2).

4. Investigation of functional properties

HVOF of SME coatings is a multi-stage process, at each stage of which parameters of functional-mechanical properties are improved. One of these stages is thermal operations. The necessity of heat treatment of Ti-Ni-Zr surface layers after HVOF in a protective medium, as well as the choice of temperature-time intervals for heat treatment, is dictated by the following considerations: to achieve the effect of structure stabilization with possible relaxation of residual internal stresses after deposition. It is known that with the growth of annealing temperatures for SME alloys, the processes of stress relaxation and defect elimination are activated.

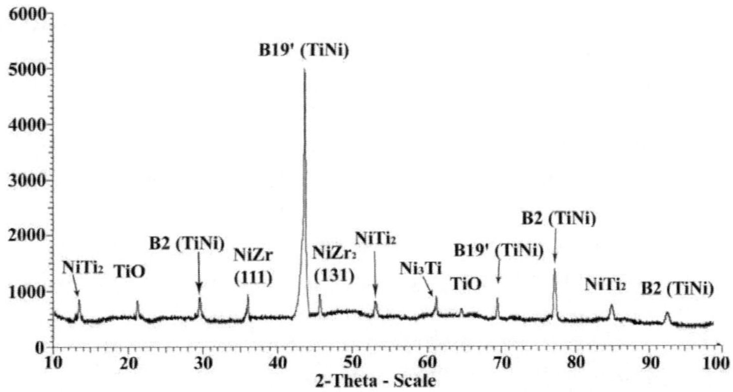

Figure 2. X-ray phase analysis of Ti-Ni-Zr alloy, after HVOF in argon protective medium.

HVOF of mechanically activated powders with SME leads to high density of crystal lattice defects and to internal stresses (first-order stresses or residual stresses), while their magnitude and character are determined by the nature of the materials and the spraying conditions. Internal stresses, preventing the mobility of twinning boundaries, can lead to decrease in the functional and mechanical properties of the resulting layers, as well as to cracking or peeling of coatings from the substrate. In accordance with the relaxation mechanism of internal residual stresses during annealing, their partial or complete relaxation is a by-product of homogenization or recrystallization annealing. In this case, annealing is used as a means of improving the shape memory characteristics, since with the relaxation of internal stresses, martensitic twins can be mobile within free subboundaries. Homogenization (diffusion) annealing eliminates the chemical heterogeneity of the deposited layers. Homogenization annealing leads to complete dissolution of the excess phases, approximating the phase composition of the alloys to the thermodynamically balanced state and improving the properties of SME alloys. Thermal treatment also intensifies diffusion processes, which occur at the boundary "steel base – transition layer". These processes increase adhesion strength of coatings to the substrate.

In the Ti-Ni-Zr layer, after the thermal treatment and surface-plastic deformation, we observe the same sequence of martensitic transformations as after the quenching B2 ↔ B19'. This is due to the fact that the austenite structure is formed in the annealing process at 873 K, and it does not differ significantly from the structure after the alloy quenching. Figure 3 shows the calorimetric curves of Ti-Ni-Zr. It can be seen that the martensitic transformation in the Ti-Ni-Zr alloy occurs from the austenite phase to the martensitic one B19: B2 ↔ B19'.

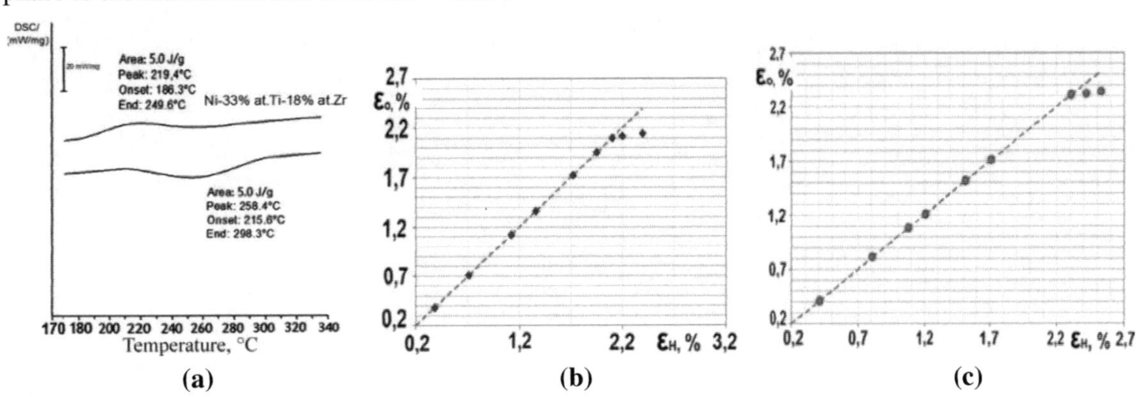

Figure 3. Calorimetric curves of surface-modified TiNiZr layers, after a full annealing cycle + surface-plastic deformation **(a)**; dependence graphs of ε_O on ε_H for TiNiZr after HVOF and surface-plastic deformation, annealing at: **(b)** – 600 °C; **(c)** – 800 °C.

In the Ti-Ni-Zr layer, after the thermal treatment and surface-plastic deformation, the following sequence of martensitic transformations is observed: B2 ↔ B19'. Figure 3(a) shows the calorimetric curves of Ti-Ni-Zr, after a complete annealing treatment + surface-plastic deformation. Temperatures of the martensitic transformations of layers made of SME materials are: M_f = 459 K, M_s = 522 K, A_s = 488 K, A_f = 571K. Figure 3(b), (c) shows the dependence graphs of the reversible strain magnitude ε_O on the induced deformation ε_H for TiNiZr after HVOF, heat treatment and surface-plastic deformation.

After HVOF and surface-plastic deformation, the largest amount of reversible deformation for the TiNiZr alloy is 2.1–2.3 % (figure 3(b), (c)), achieved in the annealing interval 600–800 °C.

5. Conclusions

After the HVOF of the mechanically activated TiNiZr powder, we obtain a nanostructured coating with a grain size of 80–100 nm. X-ray diffraction studies have shown that the formation of surface layers according to the developed technology ensures the manifestation of the shape memory effect. We also studied the functional properties of TiNiZr layer, which showed a reversible deformation of 2.1–2.3%.

Acknowledgements

This work was supported by a grant from the Russian Science Foundation No. 15-19-00202 and a grant from the President MK-5445.2016.8.

References

[1] Kim H Y and Miyazaki S 2004 *Scripta Materialia* **50** 237–41
[2] Meisner L L, Sivokha V P and Grishkov V N 1995 *Proceedings of high schools, physics* **38** 37–9
[3] Duering T W et al. 1990 *Engineering Aspects of Shape Memory Alloys* 499
[4] Rusinov P O, Blednova Zh M and Shishkalov V V 2015 *Matec Web of Conferences* **33** 1–6
[5] Blednova Zh M, Rusinov P O, Stepanenko M A and Tarbin A N 2016 *Materials Science Forum* **844** 7–12
[6] Rusinov P O, Blednova Zh M, Balaev E U and Dmitrenko D V 2016 *Procedia Structural Integrity* **2** 1506–13
[7] Rusinov P O and Blednova Zh M 2016 *Key Engineering Materials* **723** 497–502

IOP Conf. Series: Journal of Physics: Conf. Series **857** (2017) 012036 doi:10.1088/1742-6596/857/1/012036

Formation and investigation of thin MoS_2 films for electronics

A S Samartsev and Kyaw Zin Phyo

Department of electronic technologies in mechanical engineering, Bauman Moscow State Technical University (BMSTU), 105005, Moscow, Russia

E-mail: StromtrooperLLL@rambler.ru

Abstract. The article is devoted to obtaining and studying of films of molybdenum disulfide for applications in electronics. The paper presents the results of processing images obtained using optical and atomic force microscopy of films deposited by magnetron sputtering. The structure of the substrate has shown significant influence on structure of film of MoS_2.

1. Introduction

Recently, a large number of researches have been published works about thin films molybdenum disulphide. Researchers are attracted to this material with its prospective application and useful properties for devices in the field of electronics and optoelectronics [1].

Films of molybdenum disulfide are represented a great number of monomolecular layers, which are connected in the intermolecular by Van-der-Waals forces. The sulfur and molybdenum atoms in layers are hexagonally packed crystal structure with three layers of S-Mo-S [2, 3]. Each molybdenum atom is in the center of trigonal prismatic coordination and is linked by covalent polar bonds to six sulfur atoms. Chemical formula: MoS_2. When the form of one monolayer of disulfide is meant, this formula is supplemented. Designation of the single layer of molybdenum disulfide: ML-MoS_2.

MoS_2 is a member of the transition metal dichalcogenides (TMDs) family. Atomic single-layer molybdenum disulphide (MoS_2) has a direct band gap of about 1.8 eV. MoS_2 can be served as an important addition to the place of graphene which has zero band gap, for the creation of new semiconductor elements, such as thin-film transistors [4], phototransistors [5], photodetectors [6]. The charge carrier mobility in a bulk MoS_2 is relatively high – from 200 to 500 $cm^2 \cdot V^{-1} \cdot s^{-1}$[4]. Prospects for creating these devices determine interest of scientists to develop high-quality (without defect) and continuous of thin films ML-MoS_2 on a large scale and in a commercially advantageous way.

Variety of synthesis methods such as chemical vapor deposition (CVD) and physical deposition in a vacuum (PVD) are widely used for the production of TMDs, which possible to obtain grains size of ML-MoS_2 films with a diameter of 10–100 μm [7, 8]. Another method widely used for obtaining TMDs is known mechanical exfoliation. PVD methods have been produced TMD films of materials such as MoS_2 [9], WS_2 [10], WSe_2 [11]. In this paper we will examine the production of MoS_2 thin films by the method of magnetron sputtering, which could be subsequently used as a raw material for the production of ML-MoS_2 for electronics.

2. Effect of deposition factors on the structure of molybdenum disulfide

It is known, when deposited by method magnetron sputtering, the structure and properties of textured films were influenced by factors: substrate temperature [12], working argon gas pressure [13], bias voltage [14], bombarding ion energy [15], etc. Depending on these factors, different structure could be obtained: polycrystalline, single-crystal. High crystalline quality is an important requirement to obtain the desired texture of the films for a wide range of applications. It is tested the effect of different factors on the formation and properties of films WS_2 and WSe_2 obtained by magnetron sputtering method [16]. The authors of previous work show, that the influence of temperature on the formation and structure formation of thin films molybdenum disulphide films on a silicon and graphite substrate by the CVD method [17]. In the another work [18] molybdenum disulphide film was obtained by the vacuum sputtering method, moreover, layered structures 5–20 nm thick with a texture (002) formed in the initial phase of deposition, then formed columnar structures in the later phases. Currently, it was established [19] when sputtering in a vacuum by ion-plasma methods, the influence effect of substrate temperature on the structure and its properties of obtained films MoS_2. At the different temperatures, the coating films MoS_2 formed hexagonal structures: quasi-amorphous (T < 383K), textured and polycrystalline with preferential orientation along the axes of the crystallites [100] (at T = 473–573 K and T = 773–923 K) and [110] (at T = 673–773 K) perpendicular to the surface of the substrate. crystallites [100] (at T = 473–573 K and T = 773–923 K) and [110] (at T = 673–773 K), perpendicular to the surface of the substrate.

The authors of [20] discovered that the films began to grow with the (001) orientation, where the planes of the monolayers are preferentially parallel to the surface of the substrate. The diffraction analysis in these studies shows that the textured films (002) are obtained under conditions of high substrate temperature and a low deposition rate.

In these work [21] it was analyzed the structural and electrical properties of thin film tungsten disulfide with different temperatures and substrates. Increasing the substrate temperature can be raised mobility of the deposited atoms, that can be achieved by partial vapor ionization of the material.

In our experiment, the temperature and the type of substrate were chosen as the variable factors, as the most essential for the formation of the film.

3. Experimental

Molybdenum disulphide films were deposited by magnetron sputtering on silicon substrate (100) and glass ceramics at constant current in argon atom at two different temperatures at 200 and 300 °C. For this work for deposition we used vacuum processing unit of the Department MT-11, Bauman Moscow state technical University (BMSTU), equipped with independent ion source and balanced magnetron sputtering systems with target (MoS_2) diameter of 78 mm.

Before the deposition was performed preparation of the substrate surfaces, consisting of two stages: stage one – the substrates were cleaned in an ultrasonic bath in sequence: in alkaline solution and in ethanol (two minutes in each liquid); the stage two is the treatment of the substrate surface by the flow of argon ions from an independent ion source in a single vacuum, immediately before sputtering of the target material. The choice of these types of substrates is mainly because of they are the most commonly used types of substrates. The substrates also have a different structure: a monocrystalline structure of the silicon substrate and an amorphous structure of the glass ceramics.

MoS_2 films were deposited at the same parameters on the on pre-cleaned Si substrates № 1, 2 and glass-ceramic № 3, 4. Parameters of the process of film deposition: target – MoS_2; the distance between the substrate and the target – 50 mm; working gas – argon; pressure – $9 \cdot 10^{-3}$ mbar; substrates – Si, glass ceramic; discharge voltage – 650 V; current – 0.1 A; deposition time – 3 min. Variable parameters are shown in table 1.

Table 1. Variable parameters of the process of film deposition.

Parameter	Sample			
	№ 1 Si	№ 2 Si	№ 3 glass ceramic	№ 4 glass ceramic
Substrate temperature, °C	200	300	200	300

4. Processing of the obtained results

The MoS_2 films were studied by using the multi-function microscope "Solver-Next" by means of atomic force microscopy, and the optical microscope from Carl Zeiss.

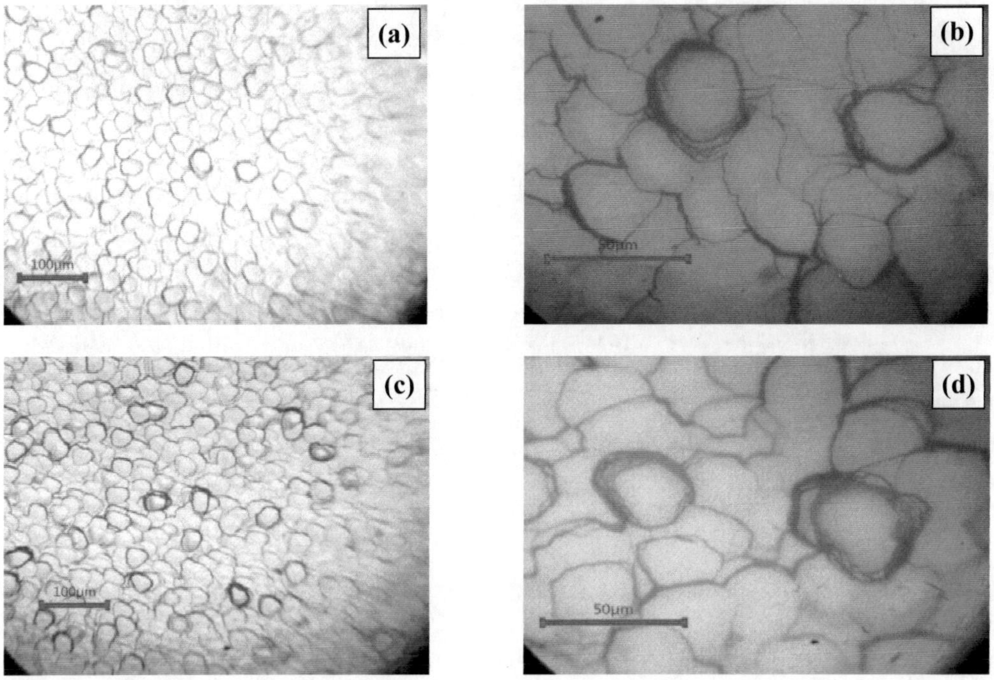

Figure 1. Samples on silicon substrate: № 1: **(a)**, **(b)**; № 2: **(c)**, **(d)**. Images obtained by an optical microscope: **(a)**, **(c)** low magnification; **(b)**, **(d)** high magnification.

In figures 1, 2 images of coatings obtained on silicon substrate are presented for comparison itself. It is clearly that the temperature has not a significant effect on the structure and surface topography. In both cases, the surface consists of many regions of grains with a diameter of about 40 μm. Figure 2 show that the surface topography – multistage.

There are regions of grain separation on the surface of film, which show the layered structure of individual grains (figure 3). According to the results of the measurements can be predetermined that the films deposited on a silicon substrates, have the same structure: coating structures are layered on top of each other gain, each of which, also uniformly consists of individual layers. Typical grain diameter is approximately 40 μm.

Deposited coating on glass ceramic substrate also not shown dependence structure on the selected temperature range. Unlike the structure on silicon substrate, the structure of these coatings is less orderly and is not oriented preferentially parallel to the plane of the substrate. Film coated on glass ceramic shown in figure 4.

5. Results and discussion

According to the experiments and obtained analysis of microscopy images, it is possible to predetermine that the selected range of temperatures: from 200 to 300 °C, the variable factor was insignificant.

At the same time the effect of the structure of the substrate was significant. We can assume that the crystal structure of a silicon substrate to have a beneficial effect on the growth of the film and, ultimately, on its structure. Also it should be noted that the method of magnetron sputtering can be applied for coating preferably orientated MoS_2.

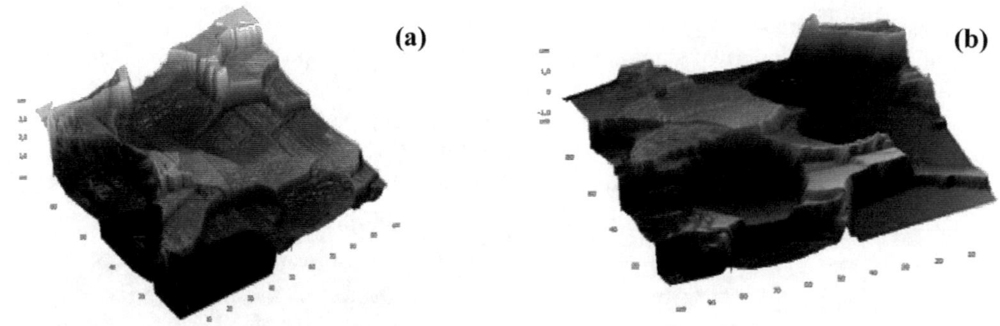

Figure 2. Samples on silicon substrate: № 1: **(a)**; № 2: **(b)**. Images obtained by atomic-force microscopy.

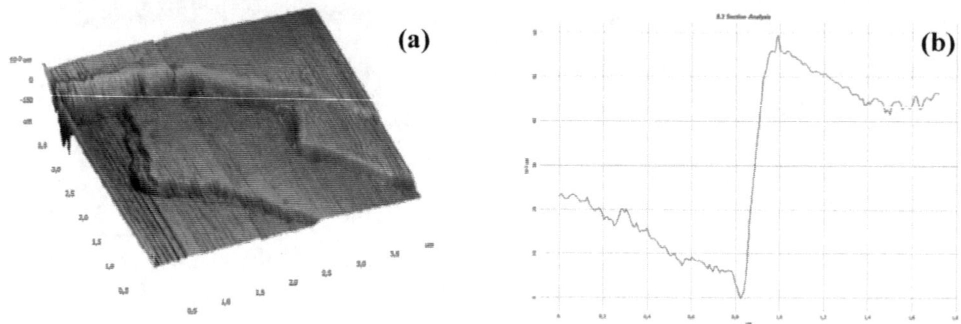

Figure 3. Layered of grains: three-dimensional image **(a)**; the profile of the step **(b)**.

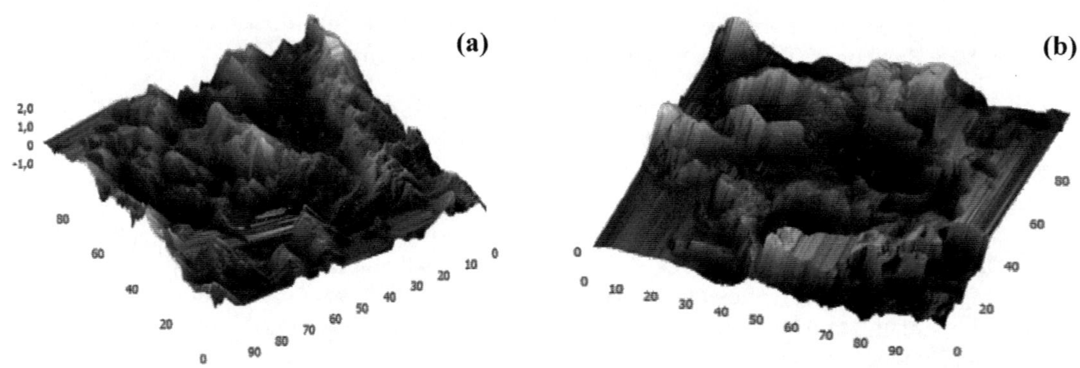

Figure 4. Samples on glass ceramic: № 3: **(a)**; № 4: **(b)**. Images obtained by atomic-force microscopy.

6. Conclusions

MoS_2 coating on silicon substrate is layered structure with grain diameter about 40 μm and is oriented properly (parallel to the plane of the substrate). Because of that, we can try to attempt to get a random collection of films from the entire surface of the substrate, or in other words from many columns at a time, resulting in a nonzero percentage of the films of ML-MoS_2, with a size of about 40 μm. This group approach for producing films of ML-MoS_2 may have sufficient yield to be manufacturing applicable and commercially justified, allowing for the mass production of nanoelectronic industry.

References

[1] Li X and Zhu H 2015 *Journal of Materiomics* **1** 33–44
[2] Lince J R and Carre D J 1987 *Phys Rev* **36** 1647
[3] Zeng H, Dai J, Yao W, Xiao D and Cui X 2012 *Nature Nanotechnology* **7** 490–3

IOP Conf. Series: Journal of Physics: Conf. Series **857** (2017) 012036 doi:10.1088/1742-6596/857/1/012036

[4] Radisavljevic B, Radenovic A, Brivio J, Giacometti V and Kis A 2011 *Nature Nanotechnology* **6** 147–50
[5] Yin Z, Li H, Jiang L, Shi Y, Sun Y, Lu G, Zhang Q, Chen X and Zhang H 2012 *ACS Nano* **6** 74–80
[6] Lopez N P, Lin Z and Pradhan N R 2014 *2D Materials* **1** 011004
[7] Lin Y C, Zhang, Huang J K, Liu K K, Lee Y H, Liang C T, Chu C W and Li L J 2012 *Nanoscale* **4** 6637 W
[8] Martin P M 2010 (William Andrew Publishing) **1** 912
[9] Tao J, Chai J, Lu X, Wong L M, Wong T I, Pan J, Xiong Q, Chi D and Wang S 2015 *Nanoscale* **7** 2497–503
[10] Ellmer K, Mientus R, Seeger S and Weiß V 2004 *Phys Stat Solidi* **201** R97–100
[11] Bozheyev F, Friedrich D, Nie M, Rengachari M and Ellmer K 2014 *Phys Status Solidi* **211** 2013–9
[12] Belyanin A and Samoilovich M I 2003 (JSC "TSNITI – TEKHNOMASH") 19–110
[13] Fortuna S V 2006 *The microstructure of coats on the basis of nitride of the titan, gained by vacuum methods* (Phd thesis) 231
[14] Messier R, Giri A P and Roy R A 1984 *Journal of Vacuum Science and Technology* **A2** 500–3
[15] Petrov I, Barna P B and Hultman L 2003 *Journal of Vacuum Science and Technology* **A21** 117–28
[16] Bozheev F E 2014 (Phd thesis) 65–90
[17] Lee W Y and Karren L 1995 *Journal of Materials Resea*rch **10** 49–53
[18] Moser J and Levy F 1993 *Materials Research Society* **8** 206–13
[19] Nozhenkov M V 2013 *Friction & Lubrication in Machines and Mechanisms* **6** 2
[20] Weiss V, Bohne W, Röhrich J, Strub E, Bloeck U, Sieber I and Ellmer K 2004 *Journal of Applied Physics* **95** 7665
[21] Ellmer K, Stock C, Diesner K and Sieber I 1997 *Journal of Crystal Growth* **182** 389–93

IOP Conf. Series: Journal of Physics: Conf. Series **857** (2017) 012037 doi:10.1088/1742-6596/857/1/012037

Investigation of zirconia tetragonal phase coatings formed by plasma electrolytic oxidation

S V Savushkina[1,4], A A Ashmarin[1], A V Apelfeld[2], A M Borisov[2], A V Vinogradov[2], M N Polyansky[1] and N L Bogdashkina[3]

[1] Keldysh Research Center, Onezhskaya str. 8, 125438 Moscow, Russia
[2] Moscow Aviation Institute (National Research University), Volokolamskoe sh. 4, 125993 Moscow, Russia
[3] Institute of Physical Chemistry and Electrochemistry of RAS, 119991 Moscow, Leninsky pr., 31-4

E-mail: sveta_049@mail.ru

Abstract. Zirconia coatings were formed on zirconium-niobium alloy (1 % Nb) by plasma electrolytic oxidation (PEO) in silicate-hypophosphite electrolyte with Y_2O_3 nanopowder additives. The structure, elemental composition and layer-by-layer phase composition of the PEO coatings were investigated using methods of scanning electron microscopy (SEM), X-ray microanalysis, X-ray diffraction (XRD), X-ray photoelectron spectroscopy (XPS). PEO coatings formed in an electrolyte with 6 g/L Y_2O_3 nanoparticles additive contain predominantly tetragonal *t*-ZrO_2 phase and have the superficial layer mainly consisting of electrolyte components and their compounds, including yttria and yttrium silicate. Moreover, stabilization of the *t*-ZrO_2 phase in the PEO coating leads to increase its hardness up to 4.6 GPa.

1. Introduction

Zirconia coatings are most often produced by thermal spraying and electron-beam deposition methods [1, 2]. The method of plasma electrolytic oxidation (PEO) also allows forming ZrO_2 coatings that have a high adhesion strength and low thermal conductivity [3–5]. It makes them promising for using as thermal barrier layers in thermal barrier coatings to prevent overheating of heat-stressed units of aerospace, etc. Zirconia PEO coatings usually contain a monoclinic *m*-ZrO_2 phase or a mixture of *m*-ZrO_2, tetragonal *t*-ZrO_2, and cubic *c*-ZrO_2 phases [5–7]. The partially stabilized tetragonal zirconia *t*-ZrO_2 with 7–8 % doping of Y_2O_3 is the most suitable for obtaining of the effectively working thermal barrier coatings [2, 8]. Y_2O_3 forms a solid solution with ZrO_2. It prevents the phase transformations during thermal cycling that often lead to the destruction of the coating. Moreover, *t*-ZrO_2 has higher hardness than *m*-ZrO_2 that increases the wear resistance of PEO coatings. Incorporation of stabilizing oxides in zirconia coatings can be achieved by carrying out the PEO process in suitable slurry electrolytes [9, 10].

In present work the layer-by-layer study of the structure, elemental and phase composition of PEO coatings with a thickness up to 120 μm formed in a silicate-hypophosphite electrolyte with additions of Y_2O_3 nanoparticles and consisting mainly of the tetragonal modification of zirconia was carried out.

2. Experimental setup and characterization techniques

PEO coatings were formed on specimens fabricated from zirconium-niobium alloy (1 % Nb). PEO process was carried out in electrolytes comprising 9 g/L sodium silicate, 5 g/L sodium hypophosphite

IOP Conf. Series: Journal of Physics: Conf. Series **857** (2017) 012037 doi:10.1088/1742-6596/857/1/012037

and 4 or 6 g/L Y_2O_3 nanopowder with particle size of ~30 nm. Slurry electrolytes were treated using a homogenizer with a power of 20 W for 3 min at ultrasonic vibration frequency of 40 kHz to stabilize the suspended state of suspension. The PEO process was carried out for 60 min under the AC electrical mode at equal values of anode and cathode currents and total current densities 20 or 30 A/dm^2.

The coatings structure was examined by scanning electron microscope (SEM) Quanta 600 with X-ray microanalysis system TRIDENT XM 4. The thickness of the coatings was measured by eddy current thickness gauge and on the cross-sections using SEM. The elemental composition of the coatings surface layer was investigated using X-ray photoelectron spectroscopy (XPS) on a Quantera device after preliminary surface cleaning by argon ions for 3 min. X-ray diffraction analysis of PEO coatings was carried out layer-by-layer in β-filtered copper radiation ($\lambda_{K\alpha1,2}$ = 1.54178 Å) on a X-ray diffractometer EMPYREAN. The hardness of the coatings was measured on the microhardness tester Micromet 5114 with a load on indenter of 0.25 N. The time from the start of the load application to its nominal value was 5 s. The exposure under the load was 10 s.

3. Results and discussion

SEM investigations of surface coating (~70 μm thickness) formed in the electrolyte with 4 g/L Y_2O_3 nanoparticles additive showed a typical structure for PEO coatings on zirconium alloys. It consists of crater-shaped areas with size up to 70 μm with the predominance of the elements of zirconia (Zr, O) and inter-crater areas with the predominance of the elements of electrolyte components (Si, P, Na, Y, O) (figure 1(a)) [5]. Some crater-shaped areas have in their centers pores with the diameter up to 10 μm preserved after extinction of microdischarges. Most of the crystallites in such areas are elongated and directed to the craters center.

When increased the concentration of nanopowder Y_2O_3 in the electrolyte up to 6 g/L the size of these areas enlarged up to 100 μm (figure 1(b)). The thickness of the coating increases to ~100 μm. The net of microcracks mainly outgoing from the centers of the craters is formed on the surface of the coating.

When the total current density was increased up 30 A/dm^2 the square of the areas enriched by the elements of electrolyte components was significantly grown, and the thickness of coating reached ~120 μm. On the surface of PEO coatings there are conglomerates of nanoparticles with size up to 0.3 μm. Some of them are incorporated into the coating and some are electrophoretically deposited on its surface and are welded to it by micro-arc discharges (figure 1(c)).

On the cross-section of the PEO coating formed in the electrolyte with 4 g/L Y_2O_3 nanoparticles additive at a current density of 20 A/dm^2 four layers can be distinguished (figure 1(d)). The barrier layer (1) is located at the boundary with the zirconium-niobium alloy. It is dense and has a thickness of ~1 μm. The main median layer (2) occupies the most of coating thickness and is characterized by a crystalline structure. The outer layer of the coating (3) occupying up to a third of its thickness is dense but has large pores and microcracks at the junctions of the large crystallites. This layer is characterized by a high content of silicon that incorporates into the coating from sodium silicate dissolved in the electrolyte. On the outer layer of the coating there is the superficial layer (4) with the thickness up to 5 μm. The thickness of this layer is inhomogeneous and it is the thinnest in the crater-shaped areas of the coating surface. X-ray microanalysis showed that the yttrium content in the median layer is approximately two times lower than in the superficial layer of coating. This indicates that part of the Y_2O_3 nanoparticles are deposited in the surface layer. As a result they were not incorporated inside the coating and not formed a solid solution with zirconia. It is found that the yttrium content grows in the top layer of the PEO coating with an increase of Y_2O_3 nanopowder concentration in the electrolyte from 4 to 6 g/L and with an increase current density from 20 to 30 A/dm^2.

Figure 1(e) shows the polished structure of the median layer at the boundary with the barrier layer of coating. It has the islands consisting of yttrium and silicon in addition to zirconium and oxygen as X-ray microanalysis shows. The size of islands is comparable to the size of craters on the coating surface. It means that Y and Si incorporate precisely through these craters from the electrolyte into the coating. The median layer is characterized by the structure of heteraxial crystallites with dimensions up to 0.3 μm (figure 1(f)). Such structure has formed during the PEO process as a result of the permanent influence of microdischarges on the already formed coating material.

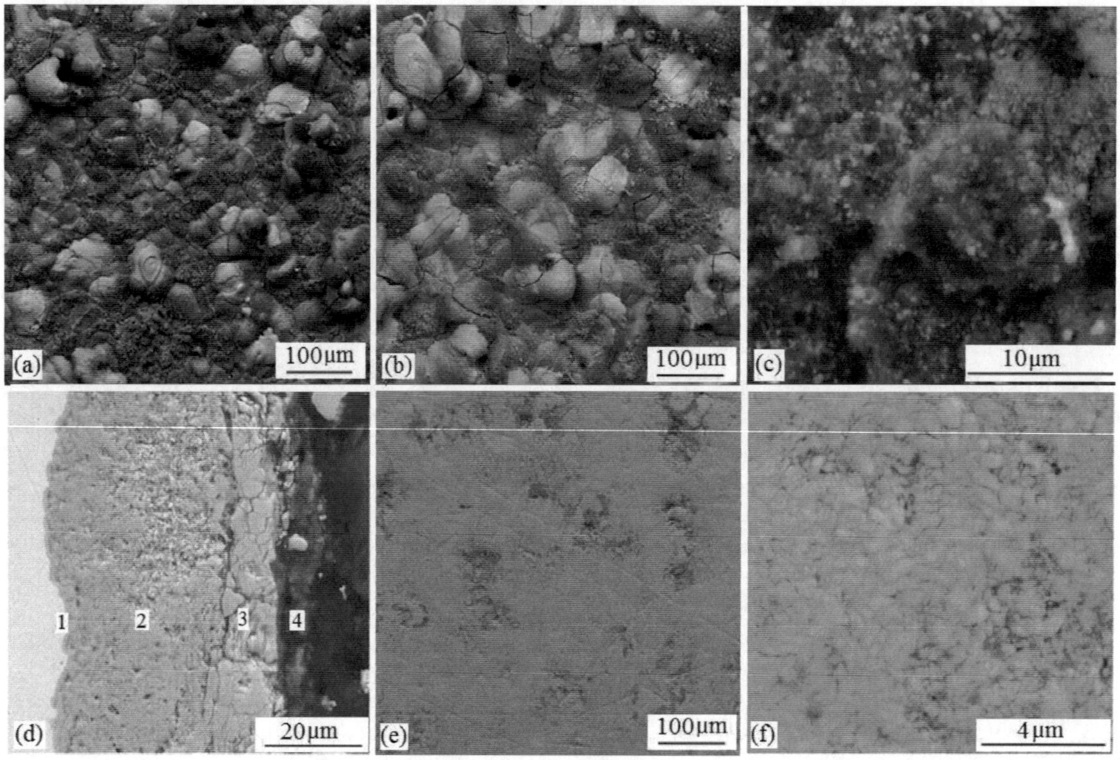

Figure 1. The backscattered electron images of PEO coatings formed at current density 20 A/dm^2 in slurry electrolyte with Y_2O_3 nanopowder additives 4 g/L **(a)** and 6 g/L **(b), (c)**; the layer structure on the cross-section of the coating formed in electrolyte with 4 g/L Y_2O_3 nanoparticles additive at current density 20 A/dm^2 **(d)** with barrier layer (1), median layer (2), outer layer (3) and superficial layer (4); the polished structure of the median layer of the coating formed at current density 20 A/dm^2 in electrolyte with 6 g/L Y_2O_3 nanoparticles additive at the boundary with the barrier layer **(e)** and its crystalline structure **(f)**.

The XPS analysis (spot area of 100 μm^2) identified the absence of zirconium in the superficial layer unlike the other layers the other layers of PEO coatings. This layer totally consists of the elements of electrolyte components: 71 at. % O; 9 at. % P; 9 at. % Si; 5 at. % Na; 6 at. % Y. The analysis of XPS peak of yttrium showed that it consists of several sub-peaks. This suggests that yttrium is presented in the PEO coatings in the form of various compounds such as Y_2O_3, yttrium silicate $Y_2Si_2O_7$, etc. The thickness of the superficial layer is less than 1 μm in the crater-shaped areas and up to 5 μm in the inter-crater areas.

The XRD analysis of the outer layers of PEO coatings formed in the slurry electrolyte with the concentration of Y_2O_3 nanoparticles 4 g/L showed t-ZrO_2 and m-ZrO_2 phases. In the electrolyte without Y_2O_3 additives the m-ZrO_2 phase usually forms. When added 6 g/L of Y_2O_3 nanopowder in the electrolyte t-ZrO_2 phase is predominate in the composition of the outer coating layer. When the current density of the PEO treatment increases up 30 A/dm^2 $Y_2Si_2O_7$ reflexes additionally appear in XRD pattern. Apparently, the microarc discharges power gets sufficient to form yttrium silicate as a result of fusing Y_2O_3 and SiO_2 formed in the process of thermolysis of the sodium silicate.

The layer-by-layer XRD analysis was carried out by successive mechanical removal the certain thickness layers from the surface of coating. The residual thickness of the coating was monitored by the eddy current thickness gauge. For quantitative phase analysis powder reference samples with different mass content of tetragonal and monoclinic ZrO_2 phases were prepared. Dependence *I*(101)*t*/*I*(111)*m* on the mass content of *m*-ZrO_2 in the samples was used to determine the content of *t*-ZrO_2 in the PEO coatings [10]. Figure 2 shows the change in the intensity of the tetragonal phase reflex *I*(101)*t* at layer-by-layer XRD analysis of the PEO coating. At depths of ~50 µm (peaks 1, 2 and 3 on figure 2) only the *t*-ZrO_2 phase was identified. In the next layer (peak 4) traces of the *m*-ZrO_2 phase were found. Its content increases up to 5 % at the distance from the surface of ~60 µm (peak 5) and up to 45 % (peak 7) at the depth of ~80 µm. Then the content of the *m*-ZrO_2 phase increases up to 70 % (peak 8) and reaches 85 % at the depth of ~100 µm (peak 9).

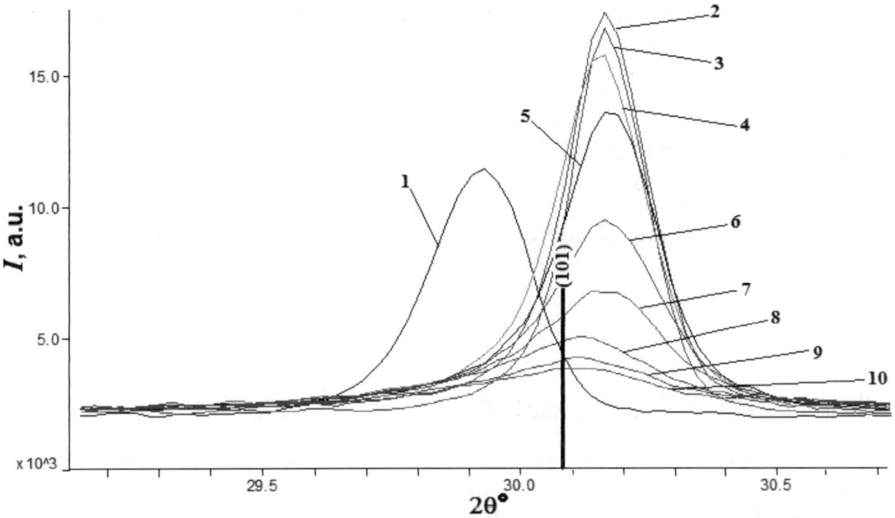

Figure 2. The intensity of the reflex *I*(101)*t* obtained by layer-by-layer XRD analysis of the PEO coating formed at current density 20 A/dm^2 in the electrolyte with 6 g/L Y_2O_3 nanoparticles additive: 1–10 – peaks corresponded the coating layers.

The presence of the *m*-ZrO_2 phase in the inner layers of PEO coating can be explained by insufficient temperature for *m*-ZrO_2 → *t*-ZrO_2 polymorphic transformation as a result of better heat removal near metal substrate. Because of low thermal conductivity of zirconia increasing of PEO coatings thickness gets worse the heat transfer from the action zones of microarc discharges in the median layer. The local temperature grows in the layer and it leads to formation of the high-temperature *t*-ZrO_2 phase, as noted in [10]. This is promoted by yttria that forms a solid solution with zirconia. The particles Y_2O_3 can enter directly from the electrolyte, as well as in the microdischarges action areas from the superficial layer enriched with yttria. At the initial stages of the PEO process this layer is practically absent, that reduces the Y_2O_3 delivery into the coating, and therefore, there are no conditions for the *t*-ZrO_2 phase formation. Enlargement of the areas enriched by the electrolyte components including yttria occurs with increasing of current density and PEO process duration.

The lattice parameters of *t*-ZrO_2 phase in the superficial layer of PEO coatings are very different from those in the underlying layers. Apparently, this is due to residual stresses as a result of excess content of yttria in the upper layers of the coating. It is indirectly confirmed by the appearance of microcracks in the outer coating layer with increasing of Y_2O_3 content in the electrolyte up to 6 g/L (figure 1(b)).

The hardness of PEO coatings was measured on cross-sections in the median layer and on the polished surface of the median layer near the barrier layer. The increase of hardness when approaching to the barrier layer of the coating was seen. The hardness of the median layer near the outer layer was 4.2 GPa and when approaching to the layer centre hardness was 4.3 GPa. The hardness reached 4.6 GPa near the barrier layer (figure 1(f)). It should be added, the hardness of PEO coatings formed in the electrolyte without Y_2O_3 nanopowder additives was 3.5 GPa, and the hardness of the Zr-Nb alloy was 1.7 GPa.

4. Conclusions

The structure, elemental composition and layer-by-layer phase composition of the PEO coatings with the thickness up to 120 μm formed on zirconium-niobium alloy (1 % Nb) in the silicate-hypophosphite electrolyte with Y_2O_3 nanopowder additives were studied. The layer-by-layer XRD analysis of PEO the coatings in the electrolyte with 6 g/L Y_2O_3 nanoparticles additive showed that more than half of their thickness contains predominantly *t*-ZrO_2 phase. When approaching to the barrier layer *m*-ZrO_2 phase begins to predominate. The lattice parameters of the zirconia tetragonal phase on the surface of the coating are very different from those in the underlying layers, that is associated with the residual stresses in the outer layer. Besides, a superficial layer mainly consisting of the electrolyte components and their compounds including Y_2O_3 and $Y_2Si_2O_7$ has been identified in the coating. Stabilization of the *t*-ZrO_2 phase in the PEO coating leads to increasing of its hardness up to 4.6 GPa.

Acknowledgments

The study was partially supported by Russian President Grant No. MK 524.2017.8.

References

[1] Slifka A J, Filla B J, Phelps J M, Bancke G and Berndt C C 1998 *J. Thermal Spray Technol.* **7** 43
[2] Zhao H, Yu F, Bennet T D and Wadley H N G 2006 *Acta Material.* **54** 5195
[3] Cheng Y, Matykina E, Arrabal R, Skeldon P and Thompson G E 2012 *Surf. Coat. Technol.* **206** 3230
[4] Savushkina S V, Polyansky M N, Borisov A M, Vinogradov A V, Lydin V B, Dankova T E and Agureev L E 2016 *J. Surf. Invest.: X-Ray, Synchrotron. Neutron. Tech.* **10** 406
[5] Apelfeld A V, Borisov A M, Krit B L, Ludin V B, Polyansky M N, Romanovsky E A, Savushkina S V, Suminov I V et al. 2015 *Surf. Coat. Technol.* **269** 279
[6] Chen Y, Nie X and Northwood D O 2010 *Surf. Coat. Technol.* **205** 1774
[7] Zou Z, Xue W, Jia X, Du J, Wang R and Weng L 2013 *Surf. Coat. Technol.* **222** 62
[8] Shin D I, Gitzhofer F and Moreau C *J.* 2007 *Mater. Sci.* **43** 5915
[9] Borisov A M, Krit B L, Lyudin V B, Morozova N V, Suminov I V and Apelfeld A V 2016 *Surf. Engineer. and Applied Electrochem.* **52** 1 50
[10] Apelfeld A V, Ashmarin A A, Borisov A M, Vinogradov A V, Savushkina S V and Shmytkova E A 2016 *Surf. Coat. Technol.* http://dx.doi.org/10.1016/j.surfcoat.2016.09.071 (in press)

IOP Conf. Series: Journal of Physics: Conf. Series **857** (2017) 012038 doi:10.1088/1742-6596/857/1/012038

Formation of optically transparent nanocomposite protective coatings on glass produced by ionic implantation and magnetron sputtering methods for space applications

V P Sergeev[1,2], M P Kalashnikov[1,2], I A Bozhko[1,2], E V Rybalko[1], O V Sergeev[1], A V Voronov[1] and M V Fedorischeva[1,2]

[1] Institute of Strength Physics and Materials Science, SB RAS, 634055, Tomsk, Russia
[2] National Research Tomsk Polytechnic University, 634050, Tomsk, Russia

E-mail: vserg@mail.tomsknet.ru, fed_mv@mail.ru

Abstract. The purpose of this research is design of a magnetron deposition technology for quartz glasses of optically transparent nanocomposite protective coatings based on Al-Si-N with preliminary modification of the substrate surface by a high-energy Ni^+ ion beam. The coating improves the mechanical properties and resistance of the glass to the impact of high-speed iron microparticles. The structural-phase state and elemental composition of the coatings and the surface layer of the substrate are investigated by electron microscopy, electron-probe analysis, and secondary ion mass spectrometry.

1. Introduction

Continuous bombardment of glass windows of spacecraft in space by the flow of high-speed micrometeoroids and space debris leads to the craters formation with the subsequent occurrence of cracks on the glass surface [1]. This phenomenon results to degradation of the optical and mechanical characteristics of the quartz glass and their failure. A possible way to solve the problem can be the deposition of protective coatings from materials with high impact resistance and melting point, which are transparent in the visible region of the spectrum [2, 3].

The possibility of using ion implantation and pulse magnetron deposition of nanocomposite coatings on the basis of the Al-Si-N system for protection of quartz glasses against the flow of high-speed solid microparticles is studied.

2. Experimental procedure

As experimental samples, industrial quartz glass plates were used, on which Al-Si-N coating with a thickness up to 10 μm with Al:Si = 3:1 atomic concentrations was formed. The deposition of the coatings was implemented using the pulse magnetron method by sputtering of the mosaic targets on the basis of aluminum with silicon inserts using the UVN-05MI "KVANT" vacuum magnetron sputtering equipment (Techimplant Ltd., Russia). A bipolar pulsed power supply with a frequency up to 100 kHz was used.

Before the coating deposition, the surface of the quartz substrate was bombarded by high-energy nickel ions using a pulsed vacuum arc ion source "DIANA-3" (Techimplant Ltd., Russia), at first the accelerating voltage was 80 kV and a fluence 10^{17} cm^{-2} and then 40 kV and $2{\cdot}10^{17} cm^{-2}$.

The structure and phase composition of the samples was investigated by the transmission electron microscope JEOL-2100F (Jeol Ltd., Japan) in Department of Nanomaterials and Nanotechnologies in Tomsk Polytechnic University. The average grain size <d> was determined using the dark-field images obtained by the TEM. To determine the phase and chemical composition of individual structural components and local micro regions in the coating the modes of microdiffraction and micro-X-ray analysis (INCA-Energy X-act, INCA-Wave, Oxford Instruments, GB) were used. The foils were prepared by the «cross-section» method using the Ion Sliser-EM-09100IS installation (Jeol Ltd., Japan). Work was performed in the "Nanotech" user center of scientific instruments of ISPMS, SB RAS.

The microhardness, the modulus of elasticity and the elastic recovery coefficient of the coatings and the quartz substrates were measured using the NanoHardnessTester (CSM Instruments, Switzerland) under a load on the indenter 20 mN.

To assess the impact-protective properties of the coatings, the experimental samples were bombarded by high-speed microparticles in the light-gas gun MPH23/8 [4] (Institute of Applied Mathematics and Mechanics affiliated to Tomsk State University). The iron microparticles of the spherical form with an average diameter of 56 μm with the velocity of 6–8 km/s were used for a bombardment.

3. Results and discussion

As a result of ion implantation in the above mode, Ni^+ ions are introduced into the surface layer of quartz plate on depth up to 150 nm (figure 1). In this case, the microhardness H_m and the elastic modulus E^* of the surface layer of quartz plate increase 1.2 time. A slightly smaller increase in the elastic recovery coefficient K_e is observed (table 1). However it has enhanced substantially an adhesion of Al-Si-N coatings with a quartz substrate.

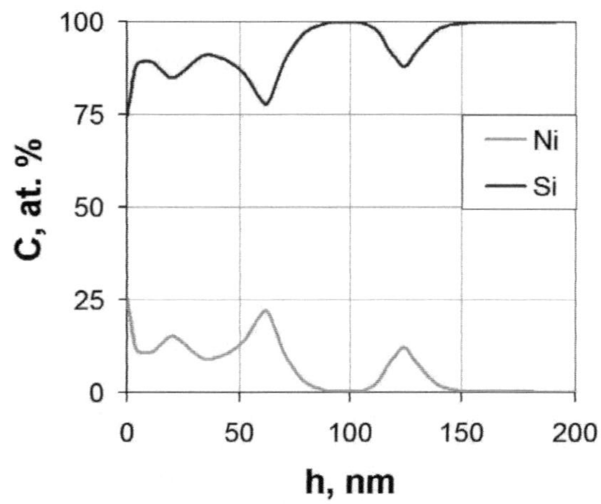

Figure 1. Concentration profile of implanted Ni^+ ions in the surface layer of SiO_2 quartz plate, obtained by the secondary ion mass spectroscopy method (SIMS).

Using the TEM method we determined (figure 2) that the Al-Si-N coating has a two-phase structure consisting of AlN nanograins (figure 2(a)) with a hcp crystalline lattice (figure 2(c), (d)) and Si_3N_4 phase in an amorphous state (figure 2(a), (b)) in the form of interlayers between nanograins of AlN similar to that observed in coatings Al-Si-N with a low Si content [5]. The average grain size of AlN was determined using the dark-field images and it was about 6 nm.

Measurements using UVIKON 943 spectrophotometer, the light-transmission factor of the quartz plate with Al-Si-N nanocomposite coating in the visible wavelength region is not less than 0.85 ± 0.02, which is close to its value for quartz without coating 0.90 ± 0.02.

Table 1 shows the values of the microhardness H_m, the elastic modulus E^*, and the elastic recovery coefficient K_e. Microhardness of ion-implanted samples with Al-Si-N coating 3.5 times and K_e 1.6 times is higher than on the initial samples without coating. The value of the elastic modulus for an ion-implanted quartz substrate with Al-Si-N coating exceeds its value for the initial quartz glass 3.1 times. After impacts on the glass samples with Al-Si-N coatings and without them by the Fe high-speed microparticls the craters are formed in the glass surface as we see in figure 3. These data were obtained using LEO EVO 50 scanning electron microscope (Carl Zeiss Group, Germany).

Table 1. An average values of the microhardness H_m, elastic modulus E^*, elastic recovery coefficient K_e

Kind of a samples	H_m, GPa	E^*, GPa	K_e	ρ, 10^6 m^{-2}
Quartz as initial without coatings	9.3 ± 0.6	78.2 ± 5.9	0.51 ± 0.04	26.9
Quartz as treated by a Ni^+ ion beam	11.1 ± 0.8	93.5 ± 9.2	0.60 ± 0.06	–
Quartz as treated by a Ni^+ ion beam with an Al-Si-N coating	32.4 ± 1.9	241 ± 25	0.82 ± 0.09	8.85

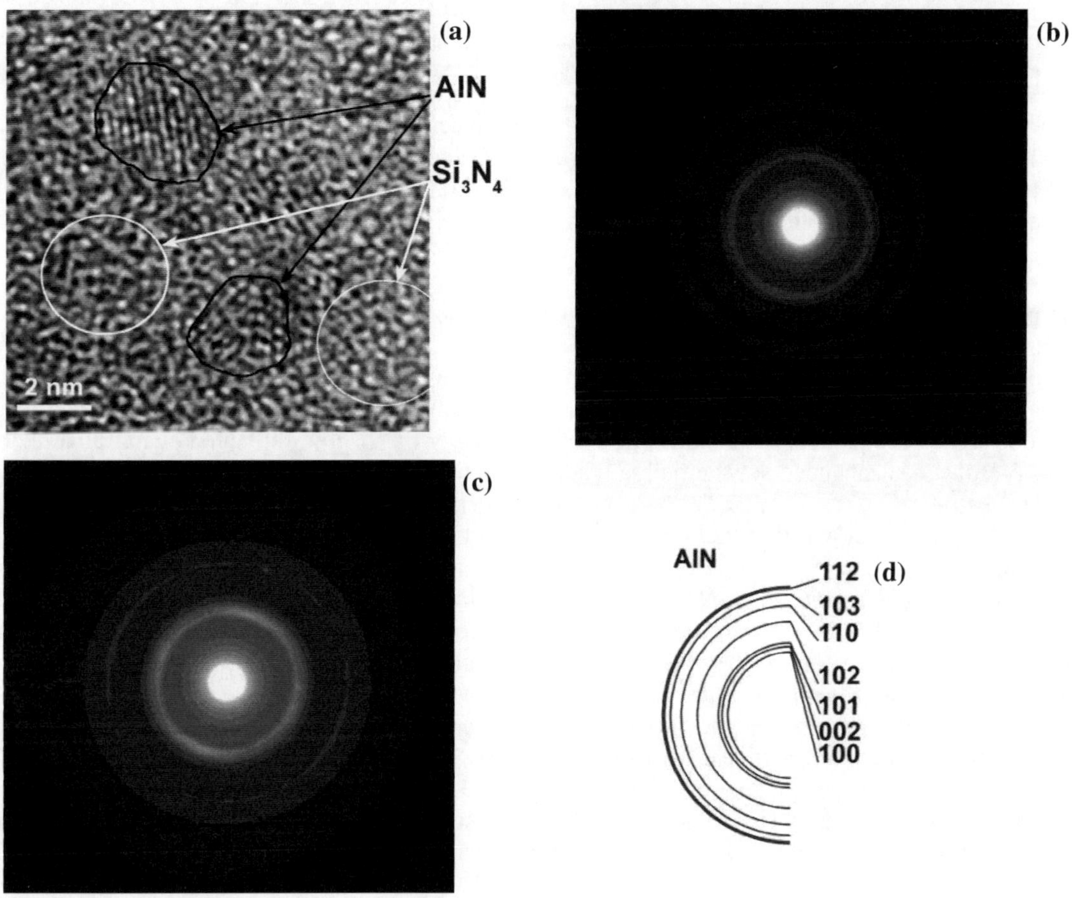

Figure 2. TEM bright-field image of cross-section of coating on the basis of Al-Si-N **(a)**, micro diffraction patterns obtained from areas of Si_3N_4 amorphous phase **(b)** and micro diffraction patterns obtained from areas of nanocrystalline AlN phase **(c)** and their indexing schemes **(d)**.

One can see (figure 3) that ion-implanted quartz glasses with coating and without them have the different surface density of the craters ρ under the same test conditions.

The surface density of craters formed on the surface of quartz glass under impacts of Fe microparticles with speed up to 6–8 km/s was decreased about 3 times by the Ni^+ ion implantation and magnetron deposition of Al-Si-N nanocomposite coatings (table 1).

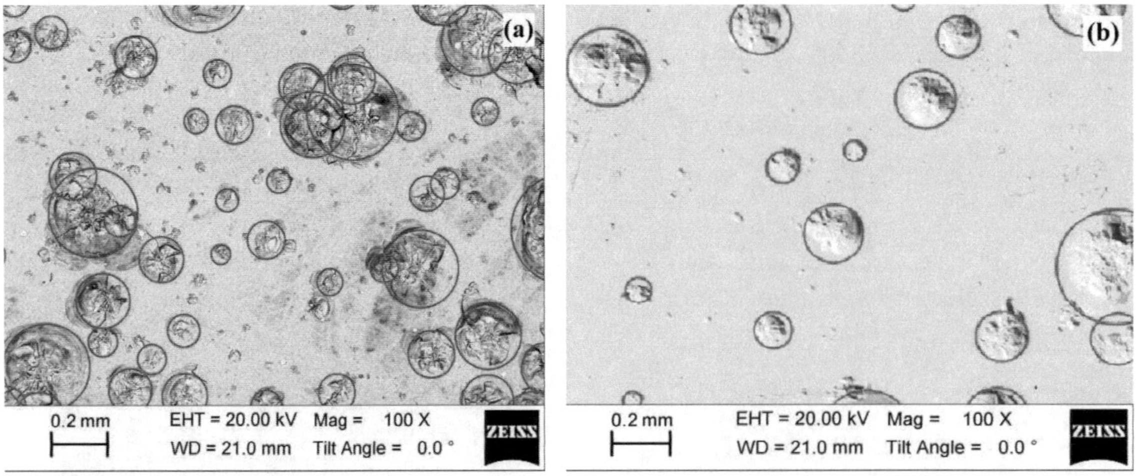

Figure 3. SEM images of the craters formed on the surface of the samples under the action of the high speed flow of the Fe microparticles: **(a)** – part of a surface of the initial quartz sample without the coating; **(b)** – part of a surface of quartz sample with preliminary Ni^+ ion implantation and with Al-Si-N coating. The craters on the SEM images are marked by circles.

4. Conclusions

The possibility of using ion implantation and pulse magnetron deposition of nanocomposite coatings on the basis of the Al-Si-N system for protection of quartz glasses against the flow of high-speed solid microparticles was studied. By SIMS method it has been established that Ni^+ ions are introduced into the surface layer of quartz glass on depth up to 150 nm. In this case, the microhardness, elastic modulus and elastic recovery coefficient of the surface layer of Ni^+ ion implanted quartz plate increase weakly. But it has enhanced substantially an adhesion of Al-Si-N coatings with the Ni^+ ion implanted quartz substrate.

By TEM it was shown that there is two-phase structure consisting of AlN nanograins with a hcp crystalline lattice and Si_3N_4 phase in an amorphous state having form of interlayers between nanograins of AlN in the Al-Si-N coating. Microhardness of ion-implanted samples with Al-Si-N coating increases 3.5 times and elastic recovery coefficient 1.6 times compare with initial samples without coating. The value of the elastic modulus for an ion-implanted sample with Al-Si-N coating exceeds its value for the initial quartz glass 3.1 times.

By SEM it was shown that the surface density of craters formed on the surface of quartz glass under impacts of Fe microparticles with speed up to 6–8 km/s decreased about 3 times by the Ni^+ ion implantation and pulse magnetron deposition of Al-Si-N nanocomposite coatings. The light-transmission factor of the quartz plate with Al-Si-N nanocomposite coating in the visible wavelength region is not less than 0.85 ± 0.02, which is close to its value for quartz without coating 0.90 ± 0.02.

Acknowledgment

Work was supported in the scope of basic scientific research of the Russian Federation state academies of sciences for 2013–2020, performed within the scope of the state task "Space Materials Science" NRTPU for 2016–2018.

References

[1] Novikov L S 2009 *Exposure of solid particles of natural and artificial origin on the spacecraft* (Moscow: Moscow State University Press) 104
[2] Sergeev V, Panin V, Psakhie S, Chernyavskii A, Svechkin V, Khristenko Yu, Kalashnikov M and Voronov V 2014 *AIP Conference Proceedings* **1623** 563–6
[3] Musil J, Javdosnak D, Cerstvý R, Haviar S, Remnev G and Uglov V 2016 *Vacuum* **133** 43–5

[4] Gerasimov A, Pashkov S and Khristenko Yu 2011 *Bulletin TSU. Mathematics and Mechanics* **16** 70–8
[5] Musil J, Šašek M, Zeman P, Čerstvý R, Heřman D, Han J G and Šatava V 2008 *Surf. Coat. Tech.* **202** 3485–93

IOP Conf. Series: Journal of Physics: Conf. Series **857** (2017) 012039 doi:10.1088/1742-6596/857/1/012039

Modelling of hot target reactive sputtering

V I Shapovalov and V V Smirnov

Saint Petersburg Electrotechnical University "LETI", 197376, Saint Petersburg, Russia

E-mail: vishapovalov@mail.ru

Abstract. The main foundations of the physicochemical models for reactive magnetron sputtering of a hot metal target in a nitrogen environment are described. The system of equations describing the model is solved for the titanium sputtering in a nitrogen environment. It is established that the width of the hysteresis loop decreases down to zero with an increase of the discharge current density.

1. Introduction

A detailed study of the cold target reactive sputtering has allowed to develop physical models of different complexity. The private physical models were proposed in the last third of the 20th century [1, 2]. In addition, a large number of studies are devoted to the development of a general model, which presents the compound formation on all surfaces of a vacuum chamber as chemisorption [3–6]. A non-isothermal physicochemical model describes the process of the cold target reactive sputtering in a more consistent way [7, 8].

2. Physicochemical model

A "hot" target means that the target is heated to a temperature at which the evaporated material flow cannot be neglected against the sputtered flow. The elements of the system are located in the Ar + N_2 mixture. Our key assumptions are described below.

In the hot target sputtering, the physicochemical processes taking place on the target surface and in the gas environment change. To describe the problem correctly, it is necessary to take into account the impact of the target temperature on these processes [9, 10].

1. The temperature of the target depends on the j_+ ion current density (or on the power density radiated on the target): $T_t = f(j_+)$.

2. The plasma-chemical reaction for the M_mN_n nitride formation on the excited target surface and surface chemical reactions on the substrate and the wall are given by an equation

$$\mathrm{M} + \frac{n}{2m}\mathrm{N}_2 \leftrightarrow \frac{1}{m}\mathrm{M}_m\mathrm{N}_n. \tag{1}$$

The reaction (1) is fed by the molecular flow of N_2 from the gas phase, which is described by the Langmuir isotherm. As a result, the gas consumption occurs on each surface having area A_i (figure 1); the consumption is given as a flow Q_i, (i = t, s, w).

3. The surface of the target is sputtered by argon ions and evaporated in accordance with the Hertz-Knudsen law, thus the densities of J_M metal flow and J_N compound from the target surface are expressed as follows (figure 1(a))

$$J_{\mathrm{Mt}}(j_+) = J_{\mathrm{Msp}}(j_+) + J_{\mathrm{Mev}}\left[T_{\mathrm{t}}(j_+)\right] \text{ and } J_{\mathrm{Nt}}(j_+) = J_{\mathrm{Nsp}}(j_+) + J_{\mathrm{Nev}}\left[T_{\mathrm{t}}(j_+)\right]. \quad (2)$$

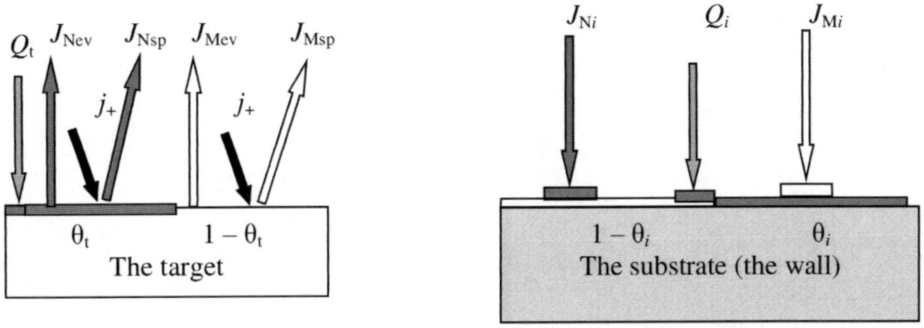

(a) **(b)**

Figure 1. The processes on the target **(a)**, the substrate (i = s) and the wall (i = w) **(b)**.

The first terms on the right of the equations (2) describe sputtered flows, the second ones – evaporated flows.

4. The film on the substrate and the wall is formed by the reaction (1) and material flow from the target (figure 1(b)). Its chemical composition depends on the mode of operation of the target.

5. The discharge current density is

$$j = (1+\gamma)j_+ + j_-, \quad (3)$$

where γ – the coefficient of ion-electron emission; j_- – the current density of thermionic emission (the Richardson-Deshman law):

$$j_-(T_{\mathrm{t}}) = AT_{\mathrm{t}}^2 \exp\left(-\frac{\varphi_{\mathrm{t}}}{kT_{\mathrm{t}}}\right), \quad (4)$$

where $A \approx 120$ A·cm^{-2}·K^{-2}; φ_t – electronic work function for the target material; k – Boltzmann's constant.

6. The j discharge current density and the Q_0 volume flow rate of nitrogen are independent variables. The main dependent variable of the process is the partial pressure of nitrogen.

The analytical description of the physicochemical model contains eight algebraic equations [7, 10]:

- equations for the stationary state of the surfaces of a target, a substrate and a wall;
- equations describing nitrogen consumption by the surfaces of a target, a substrate and a wall;
- equation of high-vacuum pumping system;
- gas balance equation.

3. Simulations

The proposed model was used to study the process of the hot titanium target sputtering in the Ar + N_2 environment. The dependence of the target temperature on the power density was obtained by numerical solution of the thermal problem. Using this dependence and the experimental current–voltage characteristics of the magnetron, the relation connecting the target temperature with the ion current density was determined:

$$T_{\mathrm{t}} \approx 1804 - 1464\mathrm{e}^{-0.00446\, j_+}. \quad (5)$$

The results of solving the system of equations are showed in figure 2. The dependence of the nitrogen partial pressure on its flow rate with the change of the current density is calculated in the range of 0 to 500 A/m^2. The study of the model showed that there is a critical temperature (power density). The process of sputtering loses its hysteresis under this temperature. The region with a negative derivative disappears, but two typical regions are kept (figure 2(b), (c)) at this and higher temperatures depending on $p = f(Q_0)$:

• the first one corresponds to the work of the magnetron under the "metal" operating mode of the target. The pressure is close to zero on it at a low nitrogen flow;

• the second one reflects the "nitride" mode. In this mode, the pressure is proportional to the flow rate of nitrogen.

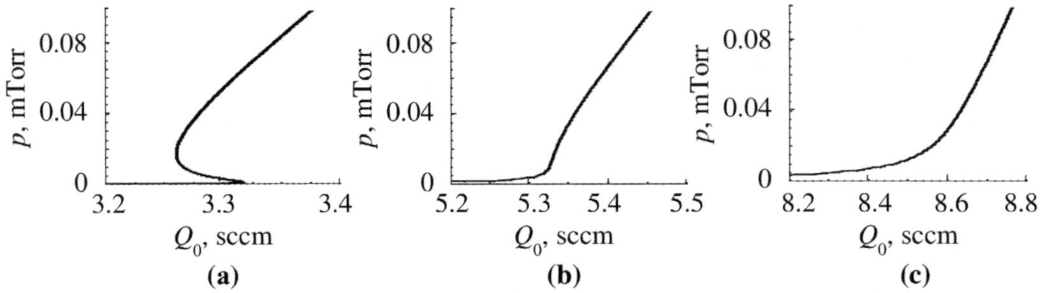

Figure 2. Dependence of the partial nitrogen pressure on the flow for a hot target at the discharge current density (A/m^2): **(a)** – 270 (1365); **(b)** – 280 (1384); **(c)** – 290 (1402). Corresponding target temperature in K calculated using the formula (5) is given in parentheses.

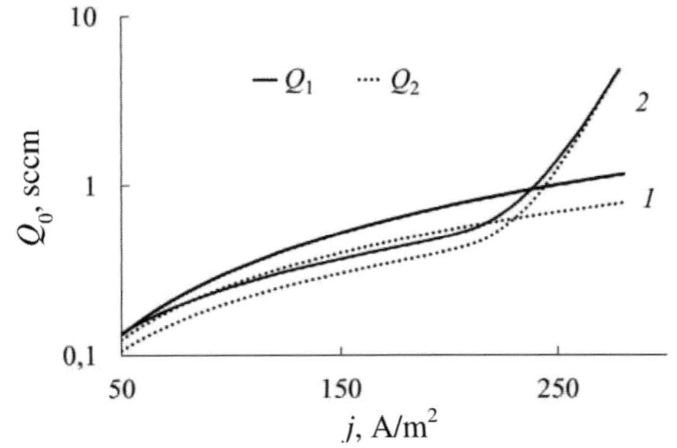

Figure 3. Current density dependences of the nitrogen input flow values when operating modes of the cold (*1*) and hot (*2*) targets change: Q_1 – a metal-to-oxide mode transition; Q_2 – an oxide-to-metal mode transition.

It is known that the hysteresis effect remains at any current density [7] during reactive sputtering of a cold target. This parameter affects the position of the points of the target transition from one mode to another and the width of the ΔQ_0 hysteresis loop. Dependences of the Q_0 nitrogen flow rate, when there are changes of operating mode of cold and hot target, on the current density are shown on the figure 3. As it can be seen from figure 3, these values for the cold target are proportional to the current density. And they are highly nonlinear for a hot target.

Figure 4(a) reflects the dependences $\Delta Q_0 = f(j)$. They are shown in figure 4(a) by the round markers for a cold target. The surface temperature of the target is a constant parameter in this problem. The surface temperature significantly affects the value of ΔQ_0. The function $\Delta Q_0 = f(j)$ differs for a hot target as it can be seen from figure 4 (triangular markers). A smooth increase of ΔQ_0 ends with an extremum and a further decrease down to zero. The critical value of the current density when the width of the hysteresis area draws in zero is equal to 458 A/m^2.

Further calculations shown in figure 4 allowed finding that evaporation is the basic physical mechanism initiating the loss of the hysteresis by the reactive sputtering process. Thermionic emission has a little effect on the process.

4. Conclusions

The physicochemical model of the hot target reactive magnetron sputtering makes it possible to analyse the process. For the titanium target sputtered in Ar + N_2 environment, two types of processes depending on the current density are possible: with and without the hysteresis. The same result was obtained for the tantalum target sputtered in Ar + O_2 environment [10].

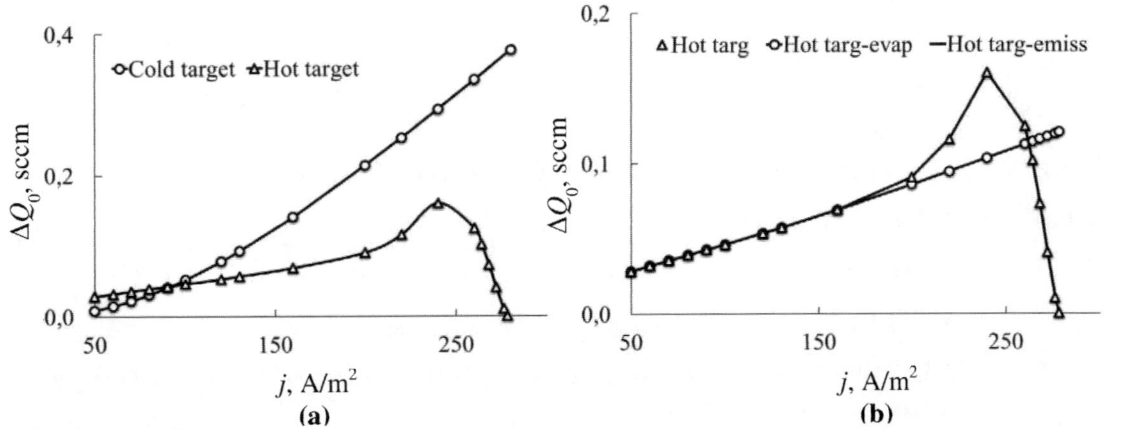

Figure 4. Dependence of the hysteresis width on the current density: **(a)** – for hot and cold targets; **(b)** – Hot targ – full model, Hot targ-emiss – neglecting thermionic emission, Hot targ-evap – neglecting evaporation.

Acknowledgments
The authors thank Russian Science Foundation for the support on the study (grant 15-19-00076).

References

[1] Shinoki F and Itoh A 1975 *J. Appl. Phys.* **6** 3381–3
[2] Lemperiere G and Poitevin J M 1984 *Thin Solid Films* **111** 339–49
[3] Berg S, Larsson T and Blom H O 1986 *J. Vac. Sci. Technol. A.* **4** 594–7
[4] Särhammar E, Strijckmans K, Nyberg T et al. 2013 *Sur. Coat. Technol.* **232** 357–61
[5] Kubart T, Nyberg T, Pflug A et al. 2010 *Sur. Coat. Technol.* **204** 3882–6
[6] Särhammar E, Nyberg T and Berg S 2015 *Sur. Coat. Technol.* **279** 39–43
[7] Barybin A A and Shapovalov V I 2007 *J. Appl. Phys.* **101** 054905
[8] Barybin A A, Zavyalov A V and Shapovalov V I 2012 *Glass Phys. Chem.* **38** 396–401
[9] Lapshin A E, Levitskii V S, Shapovalov V I et al. 2016 *Glass Phys. Chem.* **42** 359–62
[10] Shapovalov V I, Karzin V V and Bondarenko A S 2017 *Phys. Lett. A.* **381** 472–5

The effect of reactor geometry on the synthesis of graphene materials in plasma jets

M B Shavelkina, R H Amirov and T B Shatalova

Joint Institute for High Temperatures of Russian Academy of Sciences, Izhorskaya Street 13/2, 125412, Moscow, Russia
Chemical Department, Lomonosov Moscow State University, 1-3 Leninskiye Gory, 119991, Moscow, Russia

E-mail: mshavelkina@gmail.com

Abstract. The possibility of synthesis of graphene and graphane (hydrogenated graphene) using the decomposition of hydrocarbons by thermal plasma has been investigated. Investigations of the influence of the plasma-forming gas on the efficiency of synthesis and the morphology of graphene materials were carried out. The synthesis products have been characterized by the methods of scanning microscopy, Raman spectroscopy and thermal analysis. It is found that the morphology of graphene materials is affected by the geometry of the reactor. It was demonstrated that the obtained graphene materials are uniformly distributed in the volume of plastic based on cyanate ester resins under mixing.

1. Introduction

The field of films and coatings continues to develop due to new materials, new methods processing and potential applications. Automobiles, planes, ships, machinery, facades and the interior of buildings, furniture, household appliances, magazines, posters and data storage devices: The list of products covered with lacquers and paints is sheer endless. Surfaces are coated with lacquer or paint to protect them against mechanical, chemical and weather-related impacts but also to improve their aesthetic appearance. To meet the ever-growing demands on modern coatings, the paint industry continuously strives to improve their products. Therefore, over the past years nanotechnology has become more and more important in the development of coatings. Nanomaterials are used to achieve higher opacity, better interaction between coating and surface and higher durability of the coating.

At present, various composites use graphene and hydrogenated graphene (graphane), which is characterized by the presence of hydrogen in its structure. These materials have a number of unique distinctive properties that allow them to be used separately in diametrically opposite areas or together supplement each other [1].

The purpose of this work is to study the possibility of controlled synthesis of graphene and graphane using plasma jets generated by a dc plasma torch and to study the properties of the materials obtained.

2. Methods

To synthesize graphene and graphane, we used a plasma jet reactor, which was previously used for the synthesis of carbon nanotubes. A detailed description of the experimental setup was given in the studies [2]. The experiment involved a simultaneous input of gaseous carbon precursor (methane, propane-butane mixture, acetylene) with the plasma forming gas (argon, nitrogen, helium) into the plasma torch, wherein heating and decompositions occurred in the plasma jet and in the region of the arc discharge, followed by condensation of the synthesis product on the metal surface of the reaction chamber. The rate of carbon precursor, plasma gas flow and the power of the plasma torch varied independently of each other. For the experimental conditions the electric power of plasma torch was set up to 42 kW. The current value of the plasma torch was constant during the experiment and equaled to 350 A for argon and nitrogen, and 400 A for helium. The flow rate of plasma forming gas varied from 0.75 to 3.0 g/s. The experiment was carried out at three pressure values: 350, 500 and 710 Torr. The hydrocarbon flow rate varied from 0.05 to 0.37 g/s. The time duration of the experiments was 6–10 min. The plasma torch can be attached to two different reactors, without water cooled but one made of stainless-steel and other is made bronze. The first reactor was cylindrical and the second reactor was conical. A cylindrical reactor has 28 cm in diameter and 90 cm in height, and a conical reactor has 100 cm in height and an angle of expansion of 30°. Figure 1 shows the direction of the flow of the vapor-gas stream in the reactor volume.

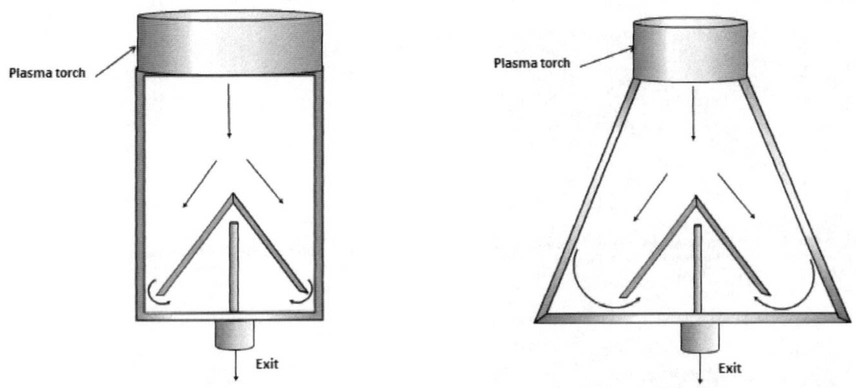

(a) **(b)**

Figure 1. The experimental setup: cylindrical reactor **(a)**, reactor in the form of a cone **(b)**.

To study the synthesis products standard methods to diagnose carbon nanostructures were used. Method of electron microscopy was used to investigate the structure of the synthesized products on a scanning electron microscope of MIRA 3 TESCAN with Schottky field emission cathode in high vacuum regime and scanning electron microscope Hitachi S5500 was used in three modes: SE (displays the surface morphology), BF-STEM (provides a transmission mode), DF-STEM (displays regions that efficiently scatter electrons). Efficiency of the synthesis, thermal stability and phase composition of carbon products were evaluated by thermogravimetry and differential scanning calorimetry on a synchronous thermal analyzer STA 409 PC Luxx (NETZSCH) with linear heating sample in air at the rate of 10 K/min at temperatures up to 1000 °C and in argon at the rate of 5 K/min at temperatures up to 1300 °C. Raman spectra were investigated by using the exciting radiation at a wavelength of 532 nm (Ntegra spectrum). The spectra were taken at three points on the surface of the prepared samples from 100 to 3000 cm^{-1}.

3. Results and discussions

Figure 2 presents the typical morphology of graphene materials obtained using the cylindrical reactor geometry. When decomposing hydrocarbons, the products of synthesis are flakes, which are usually observed in the synthesis of graphene by means of thermal plasma [3]. According to the Raman spectroscopy, there are characteristic three peaks on the spectrum. Peak D is observed in the region of 1350 cm^{-1}, peak G is at 1583 cm^{-1} and the peak of 2D is located at 2680 cm^{-1}. Figure 3 shows the Raman spectrum for a sample synthesized by the decomposition of methane in helium plasma at a pressure of 500 Torr. Analysis of the spectra of graphene flakes obtained by the decomposition of acetylene showed an increase in the intensity of the D peak, which is explained by increasing the degree of the disorder of the graphene flakes in the sample investigated due to the decrease in their sizes. The use of argon plasma and nitrogen plasma increased the dispersion of the lateral dimensions of graphene flakes. The characteristic size is in the range from 100 to 600 nm at a reactor pressure of 500 or 710 Torr. The effect of pressure changes on morphology was not found.

Using a conical reactor, it was found that the optimal synthesis parameters for the formation of graphene flakes are: pressure in the reactor of 350 and 710 Torr, flow rates of 3.0 g/s for argon, 0.75 g/s for helium and 2.0 g/s for nitrogen. The lateral dimensions of the flakes reach 1.5 μm. On the surface of the synthesized samples, the areas of swelling of the surface are observed. Figure 4 shows this effect. Research of Raman spectra showed that the intensities of peaks D and G are equalized, and for some samples the intensity of peak G decreases. According to [4], these features correspond to hydrogenated graphene (graphane).

Figure 5 shows the characteristic change in mass of the sample of graphane and the desorption rate of hydrogen in the thermogravimetric analysis. The maximum rates of hydrogen evolution are observed at characteristic temperatures [5].

Figure 2. SEM image of graphene flakes synthesized in helium plasma at 500 Torr. Precursor is methane.

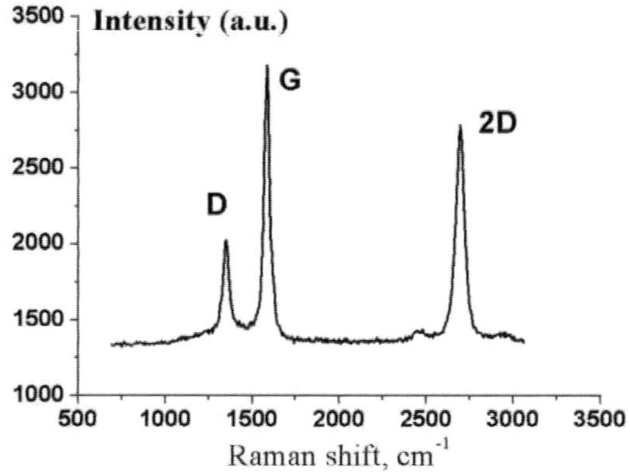

Figure 3. Raman spectrum of graphene flakes obtained in helium plasma at 500 Torr. Precursor is methane.

Figure 4. SEM image of graphane flakes synthesized in helium plasma at 710 Torr. The precursor is propane-butane.

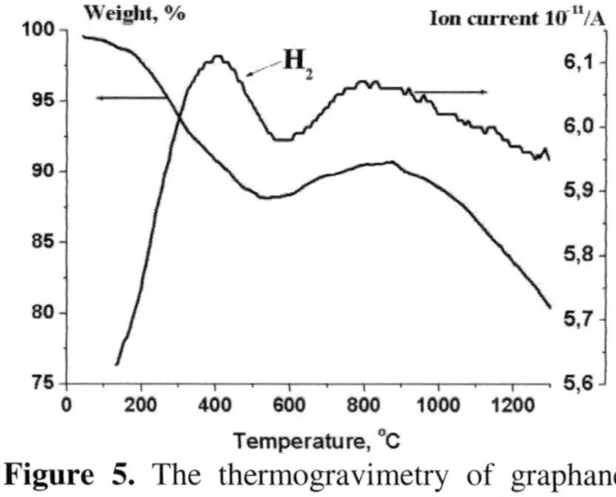

Figure 5. The thermogravimetry of graphane flakes. The precursor is propane-butane. Pressure of argon is 710 Torr.

Synthesized graphene and graphane were introduced into the cyanate ester prepolymer by simple mixing with heating to 80 °C. Cyanate ester resin (CER) is a modern and promising material with all the properties required for applications in space apparatus engineering [6]. This is thermal stability, low gas and humidity absorption, high dielectric properties, low outgassing, good resistance to radiation and radio transparency. It is known that the use of carbon nanostructures increases the thermal conductivity of plastics [7]. Studies have been carried out to determine the possibility of embedding graphene materials in a polymer mesh of plastic without affecting its dimensional stability. For this purpose, both graphene and graphene were used. It was found that the graphene, obtained in a conical reactor, is distributed faster and more evenly throughout the volume of the plastic. Figure 6 shows the distribution of carbon nanomaterial in the volume of plastic at two concentrations.

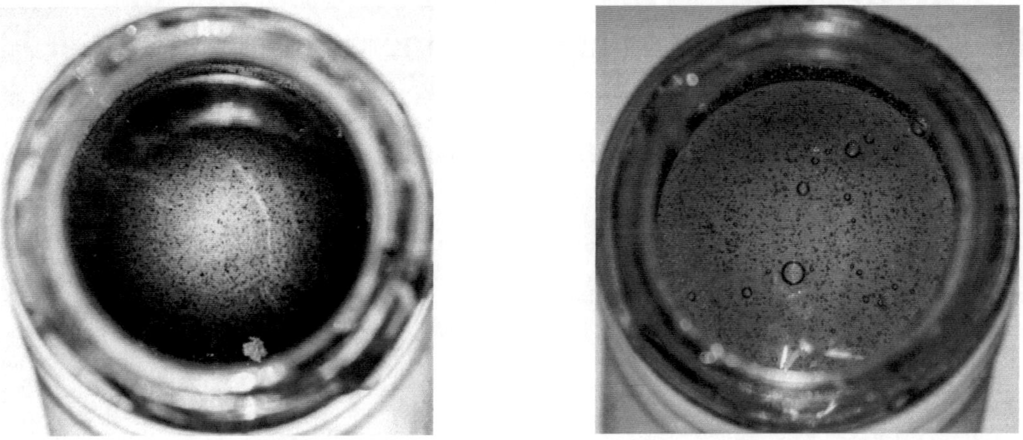

(a) **(b)**

Figure 6. Microphotographs of cyanate ester resin samples with the addition of graphane: 0.02 mass % (**a**); 0.05 mass % (**b**).

4. Conclusion

Graphene and graphene were synthesized during the decomposition of hydrocarbons in a plasma jet reactor. It has been found that the kind of the plasma-forming gas and the type of carbon precursor do not influence the morphology of the synthesis products at pressures of 350 and 710 Torr. It was found that the geometry of the reactor affects the synthesis. When a cylindrical reactor is used, graphene are usually formed, with a conical form, graphane are synthesized. The synthesized samples were flakes with a lateral size of 50 to 1500 nm. The possibility of obtaining graphene materials without the use of catalysts in the reactor volume is shown, which does not require an additional stage of purification from impurities.

Acknowledgements

The authors gratefully acknowledge Russian Foundation for Basic Research for the support by grants No. 15-08-00165, 16-08-00145 and 16-08-00081.

References

[1] Novoselov K S, Fal'ko V I, Colombo L, Gellert P R, Schwab M G and Kim K 2012 *Nature* **490** 199
[2] Amirov R, Isakaev I, Shavelkina M and Shatalova T 2014 *J. Phys.: Conf. Ser.* **550** 012023
[3] Pristavita R, Mendoza-Gonzalez N-Y, Meunier J-L and Berk D 2010 *Plasma Chem Plasma Process* **30** 267
[4] Pumera M and Wong C H 2013 *Chem Soc Rev.* **42**(14) 5987–95
[5] Ahn C C, Ye Y, Ratnakumar B V, Witham C, Bowman R C, Jr. and Fultz B 1998 *Applied Physics Letters* **73** 3378
[6] Soutis C 2005 *Materials Science and Engineering: A* **412** 1
[7] Bastiurea M, Rodeanu M S, Dima D, Murarescu M and Andrei G 2015 *Digest Journal of Nanomaterials and Biostructures* **10**(2) 521

IOP Conf. Series: Journal of Physics: Conf. Series **857** (2017) 012041 doi:10.1088/1742-6596/857/1/012041

The protective coatings for the multisectional thermoelements of the generators working at the temperatures up to 1200 K

Yu I Shtern, N V Igumnova, A A Sherchenkov, M S Rogachev and M Yu Shtern

National Research University of Electronic Technology, Russia, Moscow, Zelenograd

E-mail: m.s.rogachev88@gmail.com

Abstract. To increase efficiency structure of multisectional thermoelectric generator working up to the temperature of 1200 K was developed. Compositions were proposed, and thermoelectric materials were obtained for the fabrication of each section of thermoelement. Investigations showed that these materials have high figure of merit. Sublimation of thermoelectric materials in the working temperature of multisectional thermoelement was investigated. To eliminate sublimation of thermoelectric material, Si_3N_4 and SiO_2 protective layers were deposited by plasma-chemical method, which were effective up to the temperature of 1200 K.

1. Introduction

Low efficiency restricts wide application of thermoelectric generators – perspective alternative electric power source. One of the possible ways of increasing efficiency is connected with the increasing of temperature difference (ΔT) between the hot and cold junctions of the thermoelectric element (TE), which can be as high as 900 K. However, existing thermoelectric materials (TM) have high thermoelectric parameters only in narrow temperature range. In this case for the realization of this approach of increasing efficiency it is necessary to fabricate TE with multisectional legs. For this purpose it is required to solve complicated scientific and practical problems. In this work results of the fabrication of a number of semiconductor materials with increased thermoelectric figure of merit in the temperature range from 300 to 1200 K, optimization of geometrical sizes of each section of the TE legs, and development fabrication technology of the protective coverings for the prevention of sublimation of TM section legs at high temperatures.

2. Experimental

A number of TM perspective for the application in the temperature range from 300 to 1200 K were investigated. Taking into account temperature dependencies of thermo- and electrophysical parameters of TM, it is proper to use 4 sections for the TE legs working at the temperatures up to 1200 K (figure 1). Investigations allowed developing 8 TM for the multisectional TE.

For the low temperature sections of TE legs TM with working temperature up to 400 K were prepared by zone melting method:

– *n*-type leg, section № 1 – $Bi_2Te_{2.8}Se_{0.2}$, doped by $CdCl_2$ (0.14 wt. %);

– *p*-type leg, section № 1 – $Bi_{0.5}Sb_{1.5}Te_3$, doped by Te (2 wt. %) and TeI_4 (0.14 wt. %).

TM for the middle temperature sections of TE legs:

– *n*-type leg, section № 2 – $Bi_{2.0}Te_{2.4}Se_{0.6}$, doped by CuBr (0.18 wt. %), prepared by zone melting method, working temperature up to 650 K;

– *p*-type leg, section № 2 – $Bi_{0.4}Sb_{1.6}Te_{3.0}$, doped by $PbCl_2$ (0.12 wt. %) + Te (1.50 wt. %), prepared by zone melting method, working temperature up to 650 K;

– *n*-type leg, section № 3 – PbTe, doped by PbI_2 (0.2 wt. %) and Ni (0.3 wt. %), prepared by hot pressing, working temperature up to 900 K;

– *p*-type leg, section № 3 – $(GeTe)_{0.96}$-$(Bi_2Te_3)_{0.04}$, prepared by hot pressing, working temperature up to 900 K.

Materials for the high sections of TE legs with working temperature up to 1200 K were prepared by the method of spark plasma sintering:

– *n*-type leg, section № 4 – $Si_{0.8}Ge_{0.2}$, doped by P (2.2 wt. %);

– *p*-type leg, section № 4 – $Si_{0.8}Ge_{0.2}$, doped by B (1.8 wt. %).

Fabricated TM has high values of thermoelectric figure of merit (figure 2, designations: 1 – $Bi_{0.5}Sb_{1.5}Te_3$; 2 – $Bi_2Te_{2.8}Se_{0.2}$; 3 – $Bi_{2.0}Te_{2.4}Se_{0.6}$; 4 – $Bi_{0.4}Sb_{1.6}Te_{3.0}$; 5 – PbTe; 6 – $(GeTe)_{0.96}$-$(Bi_2Te_3)_{0.04}$; 7 – $Si_{0.8}Ge_{0.2}$ (P 2.2 wt. %); 8 – $Si_{0.8}Ge_{0.2}$ (B 1.8 wt. %)), which in the in the working temperature range are not inferior to the best analogues [1].

One of the important problems of the designing of multisectional TE is optimization of the sizes of each section of the TE leg. In addition to the electrical parameters, height of the section (h) for the ΔT of TE determines the working temperature range of given section. Method and mathematical model for the calculation of sections sizes and software for its realization were developed, which allows optimizing TE structure [2].

Methods of chemical, electrochemical, vacuum deposition were developed and used for the fabrication of contact systems to each TM. Summarizing – structures of contact systems and methods and regimes of contact layers depositions were developed; electrophysical parameters and adhesion strength (adhesion strength was not less than 9.0 MPa) of contact layers were investigated [3, 4].

Application of developed TM and optimized TE structure allowed to create generators with multisectional TE, which demonstrated high efficiency of 10 %.

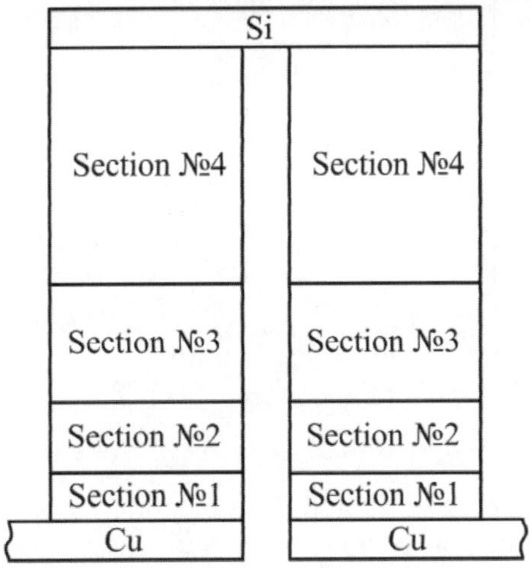

Figure 1. Scheme of the thermoelement.

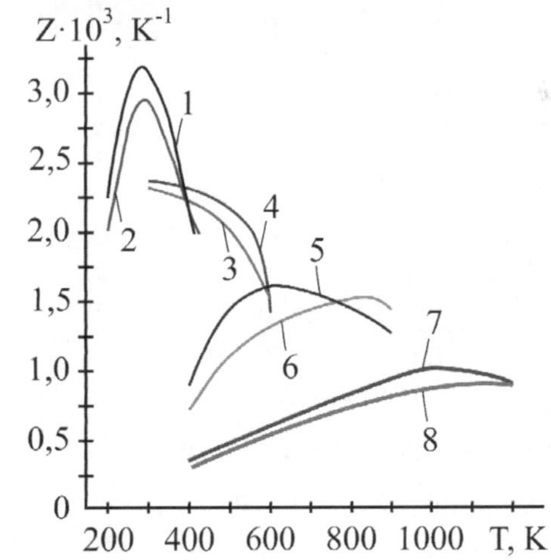

Figure 2. Temperature dependencies of Z for the thermoelectric materials.

One of the main problems of the high temperature TE designing is connected with the sufficient sublimation of TM at high temperatures. However, protection methods of TE legs is not developed [5, 6]. At practice for middle temperature materials (working temperature range up to 900 K) heat-resistant enamel KO-818 is used. For high temperature conditions protection options are not presented. In this work ways of protections of the TE legs working at the temperatures up to 1200 K are proposed. Technologies of protection coverings are developed, and samples with proposed coverings are fabricated. Samples with coverings were investigated.

Investigations of the sublimation of the developed TM for middle and high temperature leg sections were investigated before the formation of the protective coverings in the working temperature ranges characteristic for each section.

Investigations of the samples on the basis of PbTe, GeTe and SiGe by the differential scanning calorimetry (DSC) and thermogravimetric analysis (TGA) were carried out with using of TA Instruments SDT Q600 in the temperature range from 340 to 1200 K, argon flow of 100 ml/min and at the heating rate of 10 K/min.

To prevent TM evaporation of the leg sections following protective coverings were tested: enamel KO-818, Si_3N_4 and SiO_2 layers. Operations of cutting, grinding, polishing and cleaning were performed for all samples before the fabrication of protective covering.

Three layers of enamel KO-818 was manually put on every side of the sample. Draying at the temperature of 470 K during 2 h was carried out after the application of each layer. The total thickness of the layer was 0.1–0.2 mm.

Si_3N_4 and SiO_2 layers were deposited by the plasma chemical method (CORIAL D250) with the thicknesses of 1.0 and 0.25 μm, respectively. During the deposition gas pressure was 133 Pa, and temperature was 520 K. Weight losses of the samples with the protective coverings were controlled after the heat treatment during 1 h. Mettler Toledo XS205DU was used for the measuring of sample weight.

3. Results and discussions

According to the DSC and DTA measurements sufficient changes of the thermal properties and weight losses are observed for TM on the basis of PbTe at the temperatures higher than 950 K (figure 3). Meanwhile the weight losses exceed 15 %. For GeTe (figure 4) intensive endothermic effect was found on the DSC curve with the peak at the temperature of ~968 K, accompanied by the sharp weight losses, which reached ~80 %.

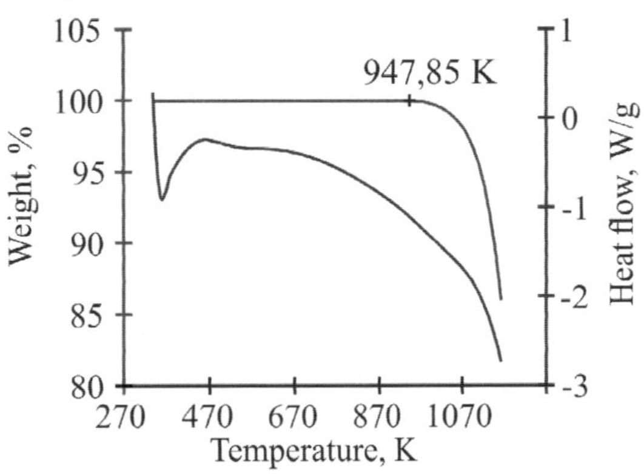

Figure 3. DSC and TGA curves for PbTe.

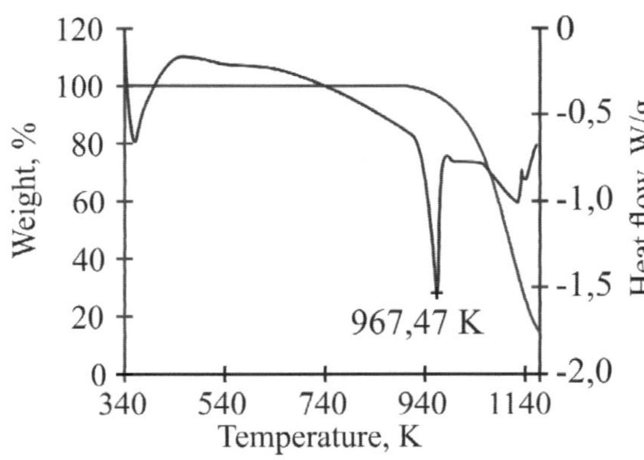

Figure 4. DSC and TGA curves for GeTe.

According to the DSC and TGA data there is no sufficient heat effects in the temperature range from room temperature to 1200 K for TM on the basis of SiGe. The weight losses are on the order of hundredths of a percent (figure 5).

Then estimation of the weight losses after the heating in the heat chamber of the samples with the protective coverings was carried out. For middle temperature TM on the basis of PbTe and GeTe testing temperature ranges were 670–900 K, while for the high temperature material SiGe-970 – 1200 K (figures 6–8).

4. Conclusions

It was established that enamel KO-818 is cracking already at the temperature of 770 K, and is not suitable for the protection of TM at higher temperatures. Using of Si_3N_4 as protective layer for TM on the basis of PbTe and GeTe lead to the weight losses of 0.18 and 0.14 %, respectively. For TM on the basis of SiGe weight losses were not found. In the case of using of SiO_2 as protective layer for TM on the basis of PbTe weight losses does not exceed 0.5 %, while for TM on the basis of GeTe and SiGe weight losses were not observed.

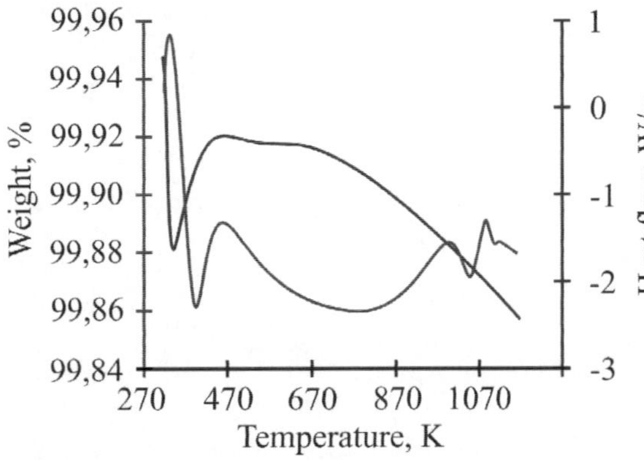

Figure 5. DSC and TGA curves for SiGe.

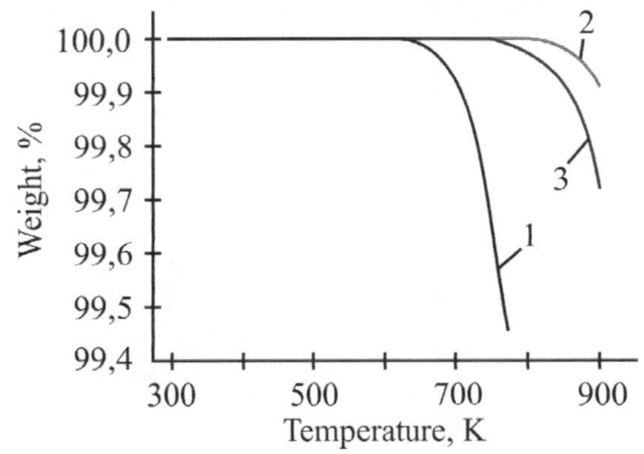

Figure 6. Temperature dependencies of the weight losses for PbTe sample with the protective coverings: 1 – KO-818; 2 – Si_3N_4; 3 – SiO_2.

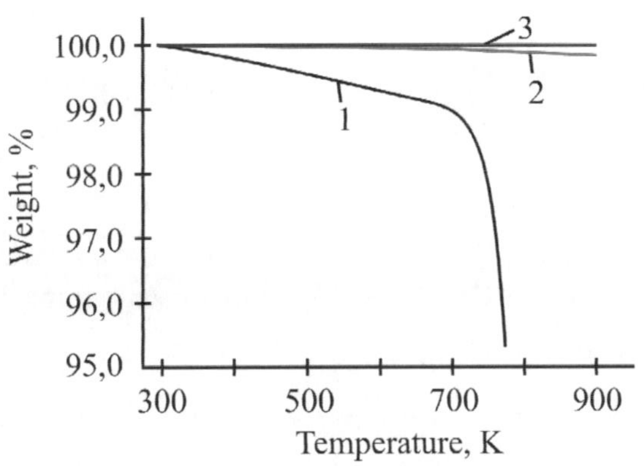

Figure 7. Temperature dependencies of the weight losses for GeTe sample with the protective coverings: 1 – KO-818; 2 – Si_3N_4; 3 – SiO_2.

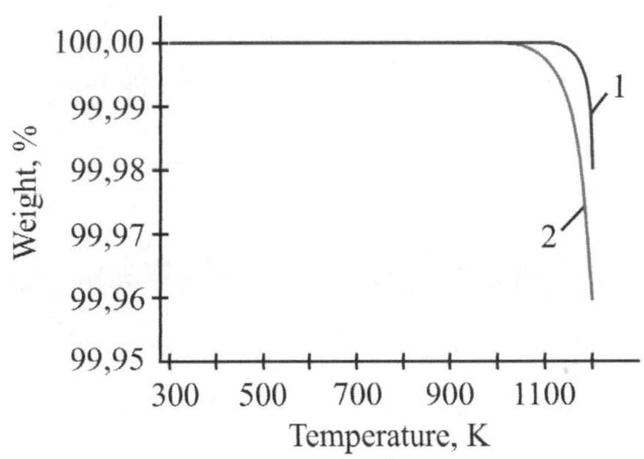

Figure 8. Temperature dependencies of the weight losses for SiGe sample with the protective coverings: 1 – KO-818; 2 – Si_3N_4; 3 – SiO_2.

The investigations showed that protective covering of enamel KO-818 can be used up to the temperature of 720 K. Si_3N_4 and SiO_2 protective layers fabricated by the plasma chemical method demonstrated effectiveness and stability in all temperature range.

Acknowledgments

This work was supported by Russian Science Foundation (project number 16-19-10625).

References

[1] Sherchenkov A A, Shtern Y I, Shtern M Y and Rogachev M S 2016 *Nanotechnologies in Russia* **11** 387–400
[2] Shtern I Yu, Gromov D G, Shtern M Yu, Sherchenkov A A and Rogachev M S 2017 *Proc. of the 2017 IEEE Russia Section Young Researchers in Electrical and Electronic Engineering Conf.* (Moscow: MIET) ed. S Lupin pp. 236–9
[3] Shtern Y I, Mironov R E, Shtern M Y, Sherchenkov A A and Rogachev M S 2016 *Acta physica polonica A* **129** 785–7

[4] Gromov D G, Shtern Yu I, Rogachev M S, Shulyat'ev A S, Kirilenko E P, Shtern M Yu, Fedorov V A and Mikhailova M S 2016 *Inorganic Materials* **52** 1132–6
[5] Park S H, Kim Y, Yoo C Y and Yoon G 2016 *J. Vac. Sci. Technol. A* **34** 061101-1–9
[6] Sadia Y, Ohaion-Raz T, Ben-Yehuda O, Korngold M and Gelbstein Y 2016 *J. Solid State Chem.* **241** 79–85

IOP Conf. Series: Journal of Physics: Conf. Series **857** (2017) 012042 doi:10.1088/1742-6596/857/1/012042

Electrodischarged Ti-(B, C, Si)-coating

A A Sivkov, I A Rakhmatullin, D Yu Gerasimov and D S Nikitin

Tomsk Polytechnic University, 634050, Tomsk, Russia

E-mail: riam@tpu.ru

Abstract: The possibility of using hyperhigh speed high current electrodischarged Ti-plasma, generated by coaxial magnetoplasma accelerator as a method for Ti-C, Ti-B and Ti-Si coatings is shown. It is estimated that the most ordered microstructure is formed for a sample obtained under W = 46.7 kJ and 2.0 g carbon load. The average level of nanohardness for this sample is 15.3 GPa.

1. Introduction

TiC-based coatings are highly wear resistance, superhard, strength, resistant to corrosion and oxidation [1], have high melting point temperature and thermally stable. They have low friction coefficient and high coefficient of electrical and thermal conductivity [2]. TiC-coatings could be synthesized by laser cladding [1], laser-assisted CVD [2] or plasma-assisted CVD [3], ion deposition [4] or PVD [5].

Titanium diboride is highly hard, has low electrical resistance, large cross section for thermal neutron capture, high wettability and high thermal shock tolerance. Due to this TiB-coatings could be used as electrode coatings in aluminum manufacturing, coating for turbine's blade, chemical reactors, crucibles, pump parts, thermocouples and cutting tools. Usually TiB-coatings obtained in solid-phase diffusion reaction of boron into titanium [6] or by laser cladding [7].

Titanium silicide is highly electro conductive and thermally stable material which could be applied as a contact layer between semiconductor and base structure [8].

Low adhesion, residual stress, long duration of coating deposition and preheating necessity of substrate until 1050 °C are still an aim of the search despite a wide spread of methods for Ti-based coatings obtaining [3].

Coaxial magnetoplasma accelerator (CMPA), generating hyperhigh high current electrodischarged plasma jet, could be a way for solving described problems. TiC/Ti-based coating with high adhesion level was deposited on a copper substrate using this CMPA at [9].

2. Experimental

High voltage high current CMPA with titanium electrodes was used for TiC-based coatings deposition [9]. Capacity energy storage was used as a power supply source. Interelectrode gap was filled with necessary precursor (B, C or Si). Electro erosion of titanium was occurred from the surface of electrodes after generation of high current plasma jet. Eroded titanium was reacted with loaded precursor under acceleration from electrode channel and impacting on substrate. The mass of loaded precursors was varied in a range 0.4–2.0 g. The distance between the end of electrode and substrate was 150 mm in all experiments. The level of charged voltage and capacity was 2.8 kV and 14.4 mF respectively.

Obtained samples were researched using X-Ray diffraction method (Shimadzu XRD6000S, CuKα). Microstructure of cross section grinds of samples was obtained using scanning electron microscope Hitachi TM3000 (SEM). Nanohardness of the samples was measured using NANO Hardness Tester NHT-S-AX under 300 mN loads.

3. Results and discussion

The level of TiC in coating increases with increasing of carbon mass load and level of energy allocated in process as shown by XRD. Titanium, iron and iron carbide are presented in coating beside titanium carbide. Carbon absence in coating is shown by XRD-pattern in figure 1. The presence of Ti caused by erosion of Ti-electrodes and loaded carbon deficit. Formation of iron in coating caused by contact of high temperature plasma jet with steel substrate and mixing with eroded material. Wherein the reaction of iron with accelerated carbon occurs with iron carbide formation.

Microstructure of cross section grinds of samples was researched using SEM. Thickness of coatings varies in a range 30–300 μm and increases with carbon load mass and allotted energy. The most ordered microstructure formed for a sample obtained under $W = 46.7$ kJ and 2.0 g carbon load as shown in figure 1.

Uniform distribution of TiC-grains in Ti-matrix could be seen according to the picture. This allows obtaining the average level of nanohardness for this sample up to 15.3 GPa.

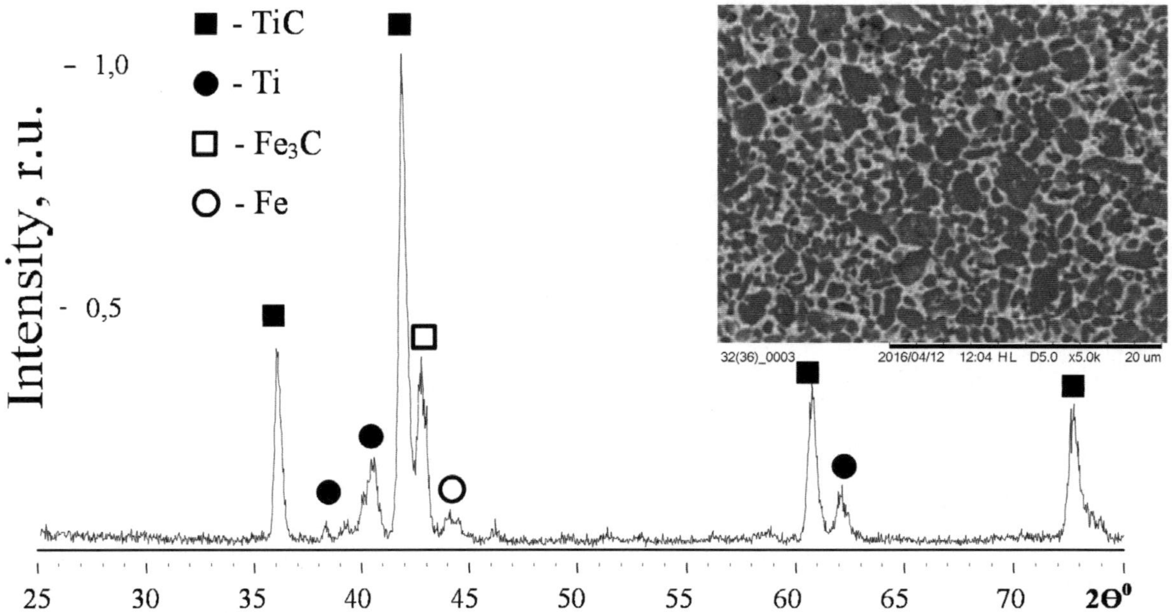

Figure 1. XRD-pattern for TiC-coating.

XRD-pattern of TiB-coating in figure 2 demonstrates formation of a few crystalline phases: TiB (ICDD #5-0700), TiB_2 (ICDD #7-0275), Ti (ICDD #1-1197), TiO_2 (ICDD #76-0326).

According to XRD-pattern in figure 3 Ti_5Si_3 (ICDD #78-1429) is a dominant phase in Ti-Si coating with small amount of $TiSi_2$ (ICDD #27-1402) and initial precursors if Ti and Si.

4. Conclusion

The results of the work establish the possibility of using coaxial magnetoplasma accelerator for TiC-coatings deposition on steel substrates in short time cycle up to 500 μs with coating thickness up to 300 μm and hardness level up to 15.3 GPa. The possibility of titanium borides and silicides coating deposition is shown.

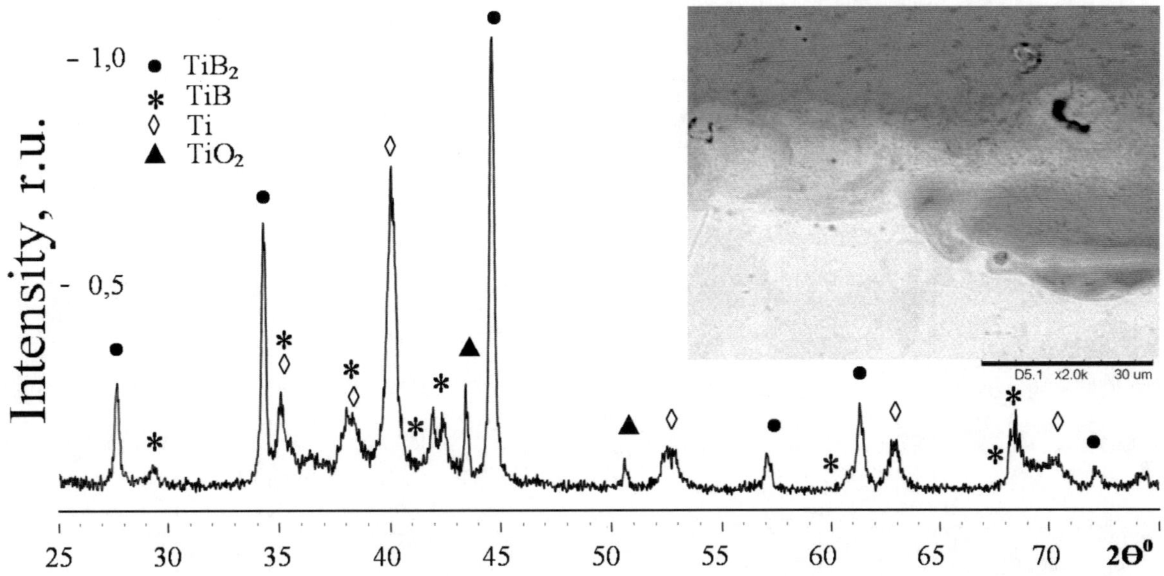

Figure 2. XRD-pattern for TiB-coating.

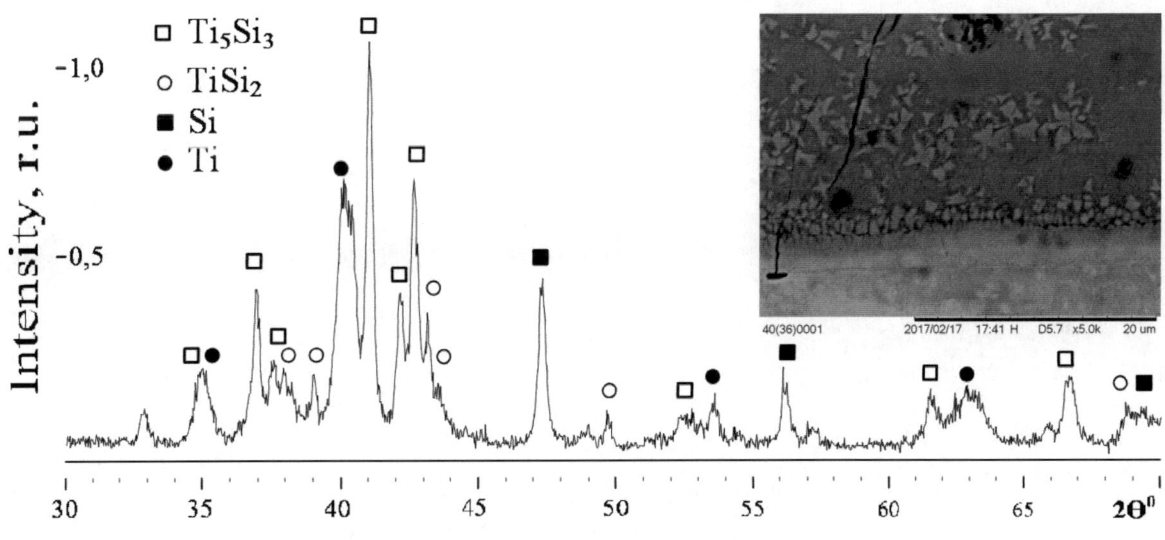

Figure 3. XRD-pattern for TiSi-coating.

Acknowledgment
Work was founded under Russian science fund project #15-19-00049.

References

[1] Savalani M, Ng C, Li Q and Man H 2012 *Applied Surface Science* **258** 3173–7
[2] Gonga Y, Tub R and Gotoc T 2013 *Materials Research Bulletin* **48** 2766–70
[3] Shanaghia A, Rouhaghdama A, Ahangaranic S, Chub P and Farahania T 2012 *Applied Surface Science* **258** 3051–7
[4] Zhang G, Li B, Jiang B, Yan F and Chen D 2009 *Applied Surface Science* **255** 8788–93
[5] Saoula N, Henda K, Kesri R, Shrivastava S, Erasmus R and Comins J 2012 *Acta Physica Polonica A* **121** 172–4
[6] Tikekar N, Chandran K and Sanders A 2007 *Scripta materialia* **57** 273–6

[7] Majumdar J, Chandra B, Nath A and Manna I 2006 *Surface and Coating Technology* **201** 1236–42
[8] Clevenger L, Harper J and Cabral Jr C 1994 *Journal of Applied Physics* **76** 7874–81
[9] Sivkov A, Gerasimov D, Shanenkov I and Shanenkova Yu 2016 *Surface and Coatings Technology* **291** 1–6

IOP Conf. Series: Journal of Physics: Conf. Series **857** (2017) 012043 doi:10.1088/1742-6596/857/1/012043

Structure and properties of particles deposited from vacuum arc discharge plasma on the way from the cathode spot up to the walls of a vacuum chamber

N A Smolanov

Mordovia State University, Saransk, Russia

E-mail: smolanovna@yandex.ru

Abstract. The author has analyzed the structure and properties of micro-particles, which are deposited from arc discharge plasma near the cathode on the substrate and walls of the vacuum chamber. Cathode was grade BT1.0 titanium spraying in nitrogen and acetylene. Particles and films formed in plasma flow with superposing of electric and magnetic fields. The dependence of physical properties of microparticles on the operation modes of the cathode pack has been analyzed. Possible mechanisms for forming fractal structures in vacuum arc discharge plasma have been considered.

1. Introduction

The cathode spot is the basis of arc discharges. Cathode spot products are electrons, ions and particles including liquid droplets [1–2]. Forming films from the cathode material or its compounds on substrate are resulted from a number of hard controllable processes. The analysis of these processes is complicated by the interaction of plasma particles moving in magnetic and electrical fields. Superposition of discharge and metal plasma with drops in it refers to dust plasma [3–4]. The structure of cathode plasma streams in vacuum arc in the presence of magnetic and electric fields is highly studied. However, the structure and properties of the particles deposited from plasma flow are still unstudied.

The purpose of the work is to study the structure, phase and elemental composition, and granulometry of fine powders.

2. Objects and methods of research

Powders with fine structure on the basis of titanium carbonitride were synthesized from arc discharge plasma on the unit NNV-6.6 I4. The material of the cathode is grade BT 1.0 titanium. Plasma forming gases, including coal-containing acetylene, were injected in the area of spraying titanium cathode. The powder deposited on the walls was extracted with a vacuum cleaner. To separate strong magnetic fraction we carried out magnetic separation of the powder in fields 0.2 T combined with simultaneous vibration. After separating with the use of the vibration shakers with the cell diameters 20, 60, 100, 140 and 180 μm, we determined and analyzed 5 groups of powders deposited on the walls of the vacuum chamber from arc discharge plasma. The dependence of mass fraction of dust particles on their size after separation on vibration shakers is shown in figure 1.

To study the structures and properties of micro-particles and films we used complex techniques and equipment: X-ray diffractometers DRONE-6 and Hecus S3-MICRO, electron microscope Quanta 200 i

IOP Conf. Series: Journal of Physics: Conf. Series **857** (2017) 012043 doi:10.1088/1742-6596/857/1/012043

3D FEI, particle size analyzer Shimadzu SALD -3101, analyzer TGA/SDTA851e, IR spectrophotometer IRAffinity-1 and ESR spectrometer PS 100. Magnetic properties were investigated at room and nitrogen temperatures.

We studied fine powders obtained by several operating modes of the cathode pack [5]. The main modes are the first mode (substrate potential is 250 V, arc current is 75 A, focusing coil current is 0.6 A) and the second mode (substrate potential is 0 V, arc current is 25 A, the walls of the vacuum chamber was an anode) without a focusing coil (0 A).

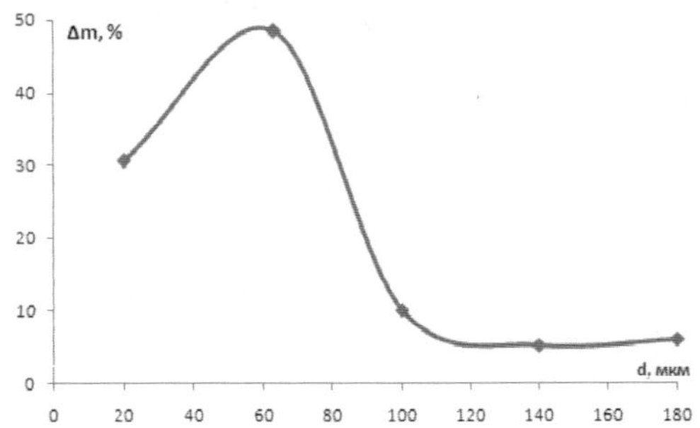

Figure 1. Dependence of dust particles from their size after separation with the use of the vibration shaker.

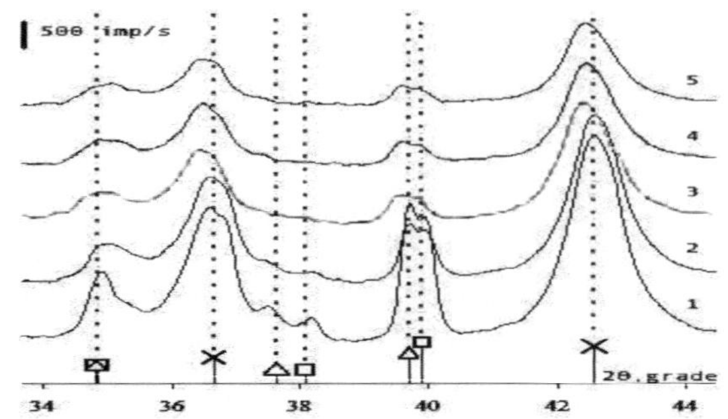

Figure 2. Plots of the X-ray diffraction pattern from powders of different sizes (1 – 20 μm; 2 – 63 μm; 3 – 100 μm; 4 – 140 μm; 5 – 180 μm). Bottom: bar X-ray data according to PDF-files (x – δ-TiN; □ – α-Ti; Δ – α-Ti 0.83 N 0.17) [9].

3. Results

The powders obtained from titanium-carbon-nitrogen low temperature plasma are multiphase mixtures of TiN (FCC-structure) and TiNC (HCP structure). The results of the radiographic analysis (figure 2) [9] and X-ray fluorescence analysis demonstrate complex phase and element compositions of obtained fine powders containing magnetic and nonmagnetic factions. Their elemental composition after magnetic separation varies and depends on the operation modes of the cathode pack. It has been determined that under the said deposition modes, the main part of obtained fine powder is titanium nitride, formed trough the carbide cycle reaction taking place during plasma particles movement [6].

The comparison of the dispersion of the particles obtained under different operation modes of the cathode pack (interelectrode potential, fractal focusing coil current) has been made. Uneven distribution of the participles has been found. The dependence of the distribution of particles on their diameters is determined by the deposition mode and is shown in figure 3.

The dependence of the titanium content in dust particles on their sizes is shown in figure 4. After magnetic separation, we detected iron in the powders, although there had not been iron in such numbers in the original material of the cathode (grade BT1.0 titanium).

4. Discussion

The results of analyzing carbonitride titanium based structures can be interpreted in terms of solid-state reactions. In analyzing these reactions, the interaction of ions in different states must be taken into account. The role of nitrogen in forming compounds in plasma chemical processes is not fully studied and raises a number of discussions. It is known [7] that the interstitial non metals atoms (C, N, O) in the lattice of transition metal atoms lead to the formation of strong chemical bonds between metal and non metal that significantly affects the chemical and physical properties. The paper [8] provides information of the nature of interstitial refractory phases as of solid solutions. The state of Me-Me-bonds, that form a lattice metal frame, plays the determining role. Some authors argue that three-dimensional lattices of NaCl-type, which are formed by the covalent bonds, are responsible for their high hardness.

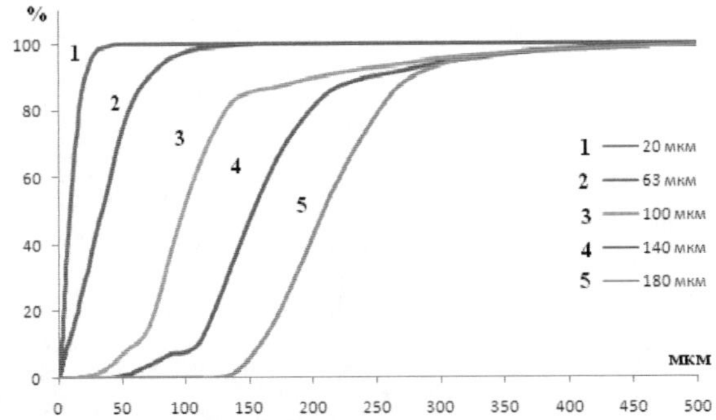

Figure 3. The dependence of the normalized number of particles on the diameter.

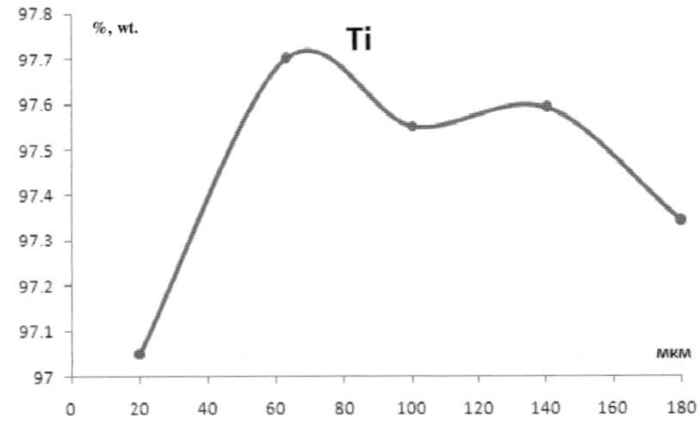

Figure 4. Dependence of the content of titanium on the dimensions of dust particles of particles.

Earlier, we used the method of small-angle X-ray scattering to determine the fractal structure of the arc discharge plasma particles [9]. The analysis of the conditions of forming these particles from the drop fraction of low-temperature plasma showed that the flow of ions of the moving cathode spot in arc discharge can be considered, at a specific point in time, as a current layer. The theory of current layers suggests that in plasma there are the magnetic field "special" lines, which are resulted from perturbations on distances exceeding the dimensions the neighborhood layer [10]. Growth front instability may be the reason for the formation of fractals. This condition is probably magnetic field dissipation in a small plasma area, in particular in the area of the cathode spot, where electrical and magnetic fields are parallel to each other [11].

References

[1] Anders A 2008 *Cathodic Arcs: From Fractal Spots to Energetic Condensation* (Berkeley: Springer Science) 540
[2] Lyubimov G A and Rakhovsky V I 1978 *UFN* **125** 665
[3] Smolanov N A 2016 *Journal of Physics: Conference Series* **669** 012055
[4] Fortov V E, Khrapak A G, Khrapak S A, Molotkov V I and Petrov O F 2004 *UFN* **174** 495
[5] Smolanov N A 2017 *Journal of Surface Investigation* **11** 353
[6] Chesnokov V V and Buyanov R A 2005 *Krit. tekhn. Membrany* **4** 75
[7] Samsonov G V 1969 *Nitridy* (Kiyev: Naukova dumka)
[8] Zhilyaev V A 2012 *Materialovedeniye* **4** 3
[9] Smolanov N A and Neverov V A 2015 *Pis'ma Mater.* **5** 179
[10] Frank A G 2010 *UFN* **180** 982
[11] Ledentsov L S and Somov B V 2015 *UFN* **185** 113

Shielding methods of quasilaminar jets outflowing from plasma torch with interelectrode insert

O P Solonenko and A V Smirnov

Khristianovich Institute of Theoretical and Applied Mechanics,
Siberian Branch of RAS, Novosibirsk, Russia

Abstract. Investigations of two methods of protection against mixing of surrounding air atmosphere with the free lengthy jets outflowing at low Reynolds numbers from plasma torches with interelectrode insertion (IEI) were carried out. The offered methods allow to significantly reduce the mixing of oxygen with plasma jets that gives a possibility of synthesis, treatment and spraying of powder materials, as well as melting of coatings critical to oxidation.

Up till now in practice of plasma spraying, synthesis and treatment of powder materials, as well as modifying of surface, including melting of coatings, the most popular were the plasma torches with self-stabilizing arc length generating turbulent plasma flows. Drawbacks of these plasma torches are discussed in paper [1]. In particular, they are characterized by high jet outflow velocity (up to 600 m/s and above) which results in small distances between the nozzle of plasma torch and the treated surface (~ 5–7 calibers), due to considerable dynamic pressure of the impinging jet the destruction of coating layer during its submelting can occur. At increase in the distance it is impossible to submelt the coating without significant thermal impact on the base, because due to intensive mixing of turbulent jet with the surrounding atmosphere, the extent of high-enthalpy core of the stream is small. Besides, this mixing leads to oxidation of the treated metal and cermet powders, as well as coatings at their submelting.

One of solutions to overcoming the specified drawbacks is the creation of stable quasi-laminar plasma flows of rather big extent outflowing from plasma torches at low Reynolds numbers. It allows to significantly reduce the mixing of surrounding gas both with the plasma jet and the area of its impingement on the surface. Plasma torches with IEI are perspective for implementation of such approach.

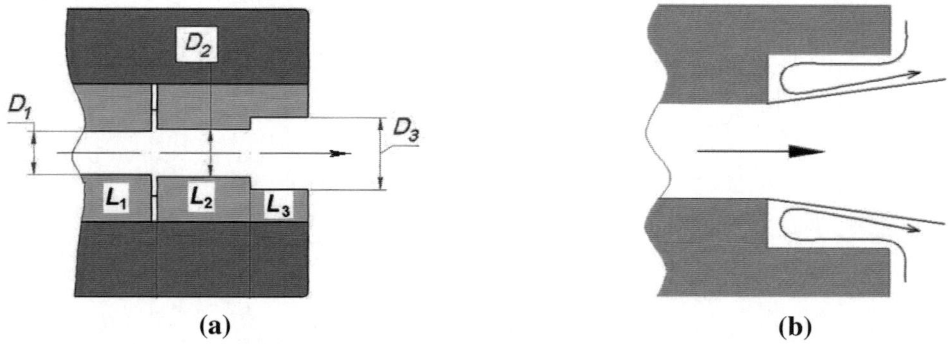

Figure 1. Scheme of the backward-facing step forming nozzle (**a**) and of the flow in it (**b**).

Results of the experimental studies of geometry optimization of the output backward-facing step forming nozzle (figure 1) providing the stable outflow of lengthy quasi-laminar jet from plasma torch with IEI [3] are presented in review papers [1, 2]. This type of jet flows is of interest for treatment and spraying of low-thermal-conduction oxide powders, and also post-treatment, including submelting, of coatings non-critical to oxidation.

In figure 2 the optical photo of argon-nitrogen plasma jet is presented, outflowing in the air atmosphere under the following conditions: arc current I = 200 A; voltage U = 200 V; main working gas – mixture of argon and nitrogen (G_{Ar} = 0.25 g/s, G_{N2} = 0.43 g/s); argon flow rate for anode protection – G_{Ar} = 0.07 g/s; D_1 = 9 mm; D_2 = 10 mm; D_3 = 18 mm.

Figure 2. Optical photo of argon-nitrogen quasi-laminar plasma jet.

However, the use of the forming nozzle in the form of the backward-facing step (figure 1) leads to ejection of air from the surrounding atmosphere in it.

Figure 3 illustrates the distributions of oxygen volume concentration in three control sections of the jet L = 100, 150 and 200 mm. From the presented results it can be seen that on jet axis the minimum volume content of oxygen is in the section L = 100 mm which usually meets the spraying distance, is not less than 5 % and grows with the distance from plasma torch nozzle exit.

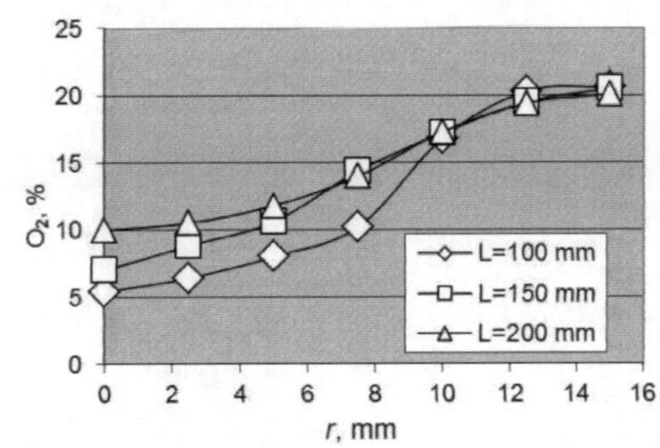

Figure 3. Distribution of the oxygen volume concentration in control sections of the jet.

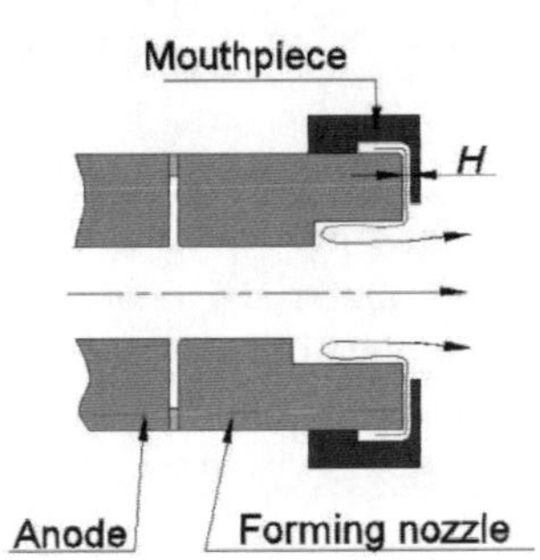

Figure 4. Schematic diagram of plasma jet shielding.

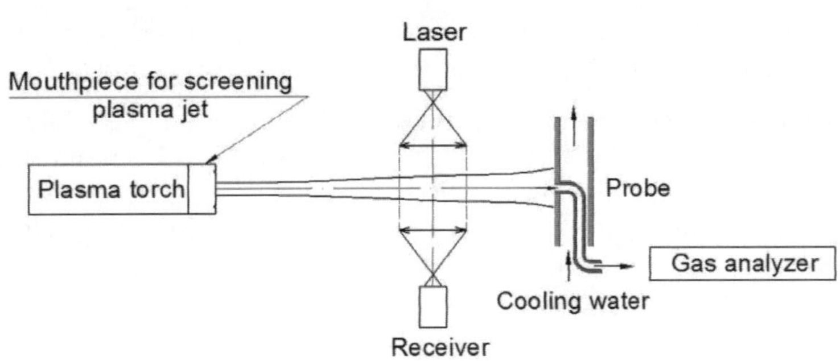

Figure 5. Scheme of the experimental setup for diagnostics of plasma jet.

In order to decrease the ejection of air in the backward-facing step from the surrounding atmosphere, and, therefore, the content of oxygen in plasma jet, a nozzle mouthpiece was developed, schematically presented in figure 4, and investigations on its optimization using the experimental setup were conducted which schematic diagram is presented in figure 5.

The nozzle mouthpiece for radial-annular slot shielding of plasma jet is schematically shown in figure 4 and includes a receiver with distributor in which, for the constant feed of the protective (shielding) gas, on the one side 12 holes with diameter of 1 mm were evenly and circumferentially made, and on the other side – 8 tangential grooves 3 mm wide and 0.15 mm deep for the homogeneous filling with shielding gas (argon) of circular slit of the given width H. As a probe for local sampling of gas the water-cooled tube with the external diameter of 3 mm was used in which the tube with the internal diameter of 0.5 mm was soldered. Sampling and analysis of the gas sample was provided by means of BONAIR TEST 1-6 gas analyzer at distances z = 100, 125, 150, 175 and 200 mm on jet axis at the operating conditions of plasma torch stated above.

When carrying out experiments the following parameters varied: width of the circular slit (H = 1 and 2 mm) and flow rate of shielding gas (argon). At the same time, flow rate of argon was set so that for slits H = 1 mm wide and H = 2 mm the weight-average flow rates of radial-converging jets for corresponding flow rates of argon were identical. In figure 6 the experimental data characterizing the distributions of oxygen volume concentration along the axis of plasma jets depending on width of slit and flow rate of shielding argon are submitted. As it can be seen from the submitted data, the use of shielding nozzle mouthpiece with radially converging near-wall protective argon jet allows, all other things being equal, to significantly reduce (by almost two times) the oxygen concentration on the axis of plasma jet for the control section L = 100 mm in comparison with its outflow without nozzle mouthpiece. It should be noted that the offered method of plasma jet protection, in comparison with commonly used axial schemes [4, 5], etc. does not demand increased flow rate of the shielding gas which leads to excessive turbulization of the jet with increase in distance from plasma torch nozzle exit.

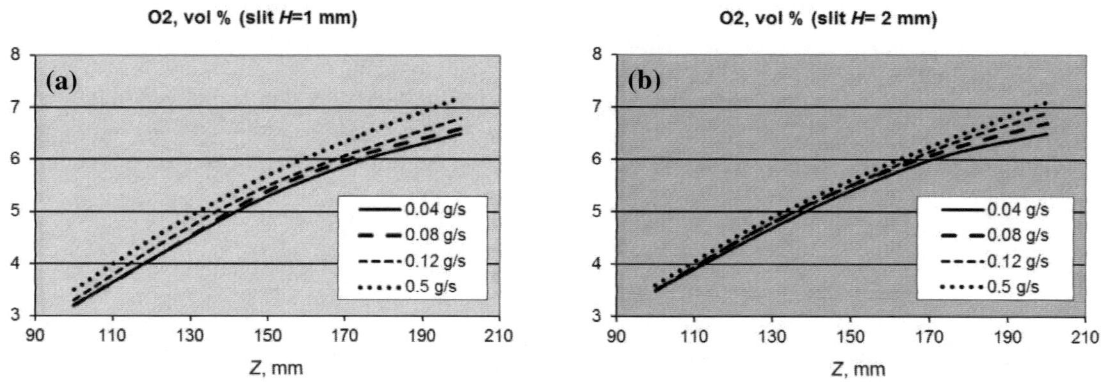

Figure 6. Distribution of oxygen concentration along the jet axis when using the shielding mouthpieces with the width of slit H =1 and 2 mm, **(a)** and **(b)**, respectively.

The considered above method of shielding of the plasma jet, outflowing at low Reynolds number from torch with IEI having forming nozzle in the form of backward-facing step, is of prime interest for surface treatment, including melting of coatings. For treatment of powder materials and plasma spraying, the use of the backward-facing step with ejection of gas from surrounding atmosphere is very problematic as at input of powder under the forming nozzle exit section the ejected gas will entrain the small particles of sprayed or treated powder in the backward-facing step that will lead to their unwanted deposition on the walls of forming nozzle.

In this regard, we have studied the second method of jet protection based on axisymmetric slit input of shielding gas (argon) into the channel of plasma torch just in front of the anode (figure 7(a)) behind which the forming nozzle with cylindrical or diffuser channel is installed.

However, unlike rather lengthy argon-nitrogen or nitrogen quasi-laminar plasma jets outflowing at low Reynolds numbers, it's quite problematic to achieve a lengthy laminar argon-helium jet without admixture of oxygen from the surrounding atmosphere.

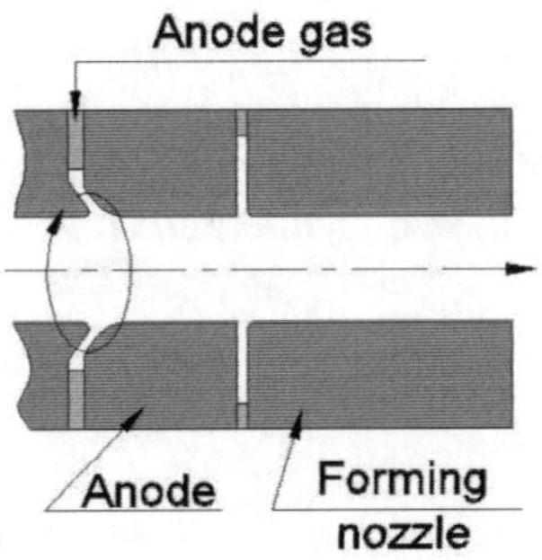

(a)

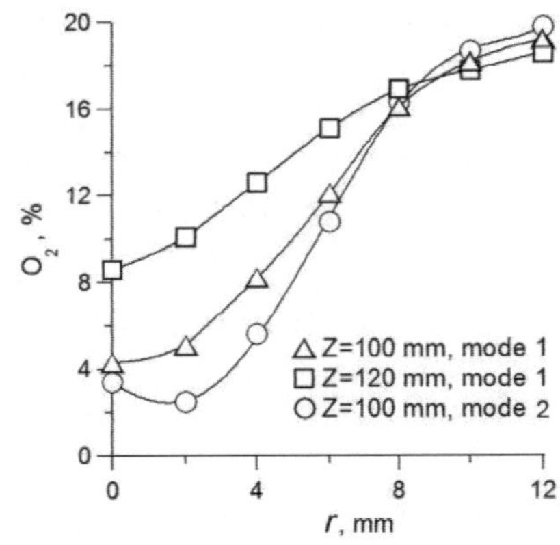

(b)

Figure 7. The scheme of axisymmetric slit input of shielding gas in front of anode **(a)** and distributions of oxygen concentration in the control sections of plasma jet **(b)**.

Investigations of behavior of argon-helium plasma jet were conducted at fixed composition and flow rate of the main working gas – mixture of argon and helium (G_{Ar} = 0.4 g/s, G_{He} = 0.2 g/s), but with variable flow rate of shielding argon supplied in front of the anode (figure 7(a)). The analysis of oxygen content in argon-helium jet shows (figure 7(b)) that at the flow rate of shielding argon G_{Ar} = 0.14 g/s (mode 1, Reynolds number Re ~ 150) the minimum oxygen content in the control section z = 100 mm is 4.1 %, and with increase in distance up to 120 mm it increases on the jet axis by almost two times. The corresponding measurements for argon-helium jet outflowing at the flow rate of shielding argon G_{Ar} = 0.42 g/s (mode 2) revealed feature connected with the minimum oxygen content observed not on the jet axis, but rather in zone with radius r ~ 2 mm where oxygen volume concentration equals 2.5 % (figure 7(b)).

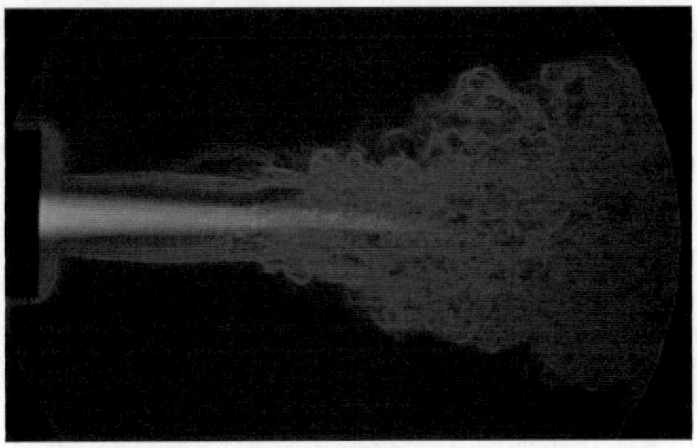

(a)

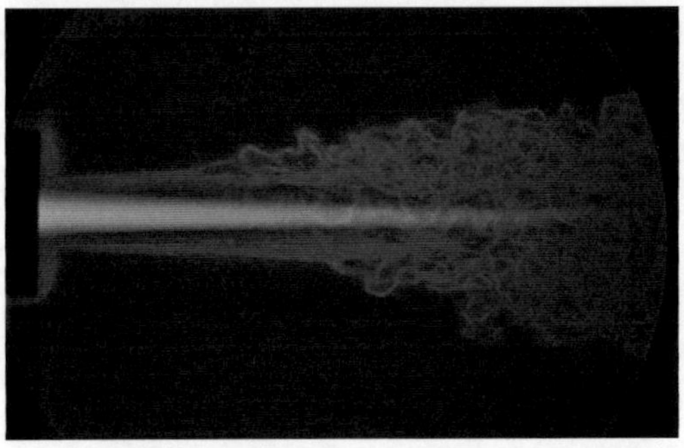

(b)

Figure 8. Shadowgraphs of flow in plasma jets depending on flow rate of shielding argon.

The analysis of the flow pattern in plasma jets outflowing from torch were carried out using the shadow device with adaptive visualizing transparency for recording phase heterogeneities [6]. The diameter of the observed area provided by the device was 120 mm and the frame exposure is 3 μs with image recording frequency of seven frames per second. Shadowgraphs of plasma jets at two flow rates of the argon supplied in front of the anode are provided in figure 8: G_{Ar} = 0.14 g/s (figure 8(a)) and G_{Ar} = 0.42 g/s (figure 8(b)).

References

[1] Solonenko O P, Nishiyama H, Smirnov A V, Takana H and Jang J 2015 *J. of Visualization* **18** 1
[2] Solonenko O P and Smirnov A V 2014 *J. of Phys.: Conf. Series* **550** 012017

[3] Solonenko O P and Smirnov A V 2007 *Multi-function plasma torch for generation of stable, lengthy and low-noise quasi-laminar jet* Design concept (know-how), Khristianovich Institute of Theoretical and Applied Mechanics SB RAS (ITAM SB RAS)
[4] Thomson I, Pershin V, Mostaghimi J and Chandra S 2001 *Plasma Chemistry and Plasma Processing* **21** 65–82
[5] Process and Apparatus for Shrouding a Turbulent Gas Jet 2010 *Air Products and Chemicals, Inc.*
[6] Shevchenko A, Kavun I, Pavlov A and Zapryagaev V 2006 *Proc. of the 12th Intern. Symp. on Flow Visualization* (Gottingen) 219

Modeling of bead formation process during laser cladding

S L Stankevich, R S Korsmik and E A Valdaytseva

Peter the Great Saint-Petersburg Polytechnic University, 195251, Saint-Petersburg, Russia

E-mail: s.stankevich@ilwt-stu.ru

Abstract. The article shows a numerical solution to the problem of heating of the powder particles and the substrate with the limited sizes by laser radiation. The powder transfer coefficient and the dimensions of the cladding bead are calculated based on estimates of the size of the molten pool.

1. Introduction

The cladding technology has found wide application in the field of repair, restoration, protective coatings and the creation of new products with special surface properties [1], due to its uniqueness and cost-effectiveness. Laser cladding is superior to other cladding technologies due to small areas of thermal influence, therefore microstructure and strength properties of the material remain unchanged.

Many parts of modern industrial equipment (parts of turbodrills, rotors of screw and centrifugal compressors, plungers and rods of artificial lifts, shafts, bearings, etc.) fail because friction and aggressive environment cause their surface to wear. This problem increases with specific load. This makes the works aimed at creation of functional coatings and devoted to reconditioning and reinforcing of the working surfaces of rapidly wearing parts very important [2]. Laser cladding consists of a few interrelated physical processes: the interaction of laser radiation with metal powder and substrate, heating and melting of the filler material and the substrate, etc. The bead is formed when the powder makes contact with the zone of the melt, is mixed with the base metal in a thin layer and undergo subsequent solidification. The powder jet is usually wider than the area of the melt so the powder of the additive does not fall into the molten pool and is blown away from its surface [3].

2. Heating of the powder particles

Consider the heating process of a gas-powder jet. The distance from the nozzle to the substrate is small while the velocity of powder particles in a stream of protective gas is high. To simplify the calculation, we assume that the trajectory of the particle is rectilinear, and the distribution of powder particles in the jet is uniform. The intensity distribution in the spot of the fiber optic laser is uneven and represents a Gaussian function. Knowing the distribution of the intensity and the trajectory of the powder particles, it is possible to estimate the heat flux affecting a particle of powder. To evaluate the heating of the powder particles we use the solution of the problem of temperature distribution in a homogeneous sphere of radius R with the initial temperature T_0 in the case of a constant heat flux with density q on the surface [4]:

$$T(r,t) = T_0 + \frac{qR}{\lambda}\left(\frac{3\chi t}{R^2} - \frac{3R^2 - 5r^2}{10R^2}\right) - \frac{2q}{\lambda R r}\sum_{i=1}^{\infty}\frac{\sin(\mu_i r)}{\mu_i^3 \cos(R\mu_i)}\exp(-\chi\mu_i^2 t), \quad (1)$$

where χ is thermal diffusivity; λ is thermal conductivity and μ_i is the positive root of the equation $tg(R\mu) = R\mu$.

The figure 1 shows the distribution of temperature field inside a powder particle with a diameter of 100 μm with physical properties of Inconel alloy 625. The figure shows that the surface melting of the powder particles is possible.

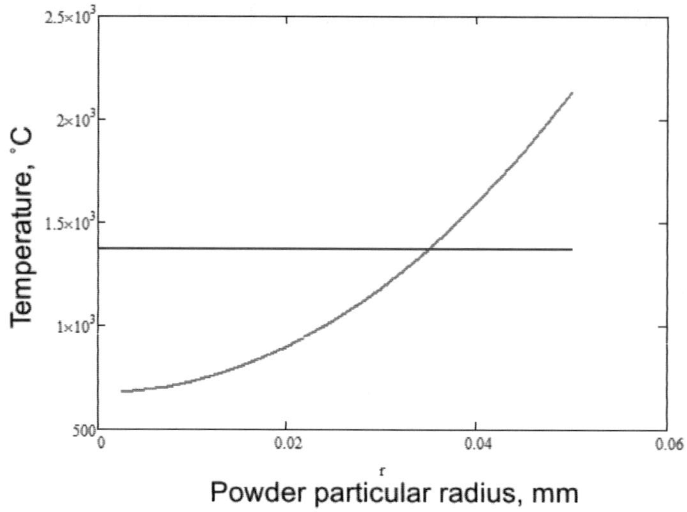

Figure 1. The distribution of temperature field within the powder particles after the interaction with laser radiation (red line shows the temperature change from the center to the periphery; the blue line is the melting temperature).

The temperature of the powder particles varies depending on the particle's path length in the zone of action of laser radiation as expected. The scheme of heating of particles in the gas-powder stream is shown in figure 2(a) and figure 2(b) shows surface temperature distribution of the powder particles at the time of contact with the substrate.

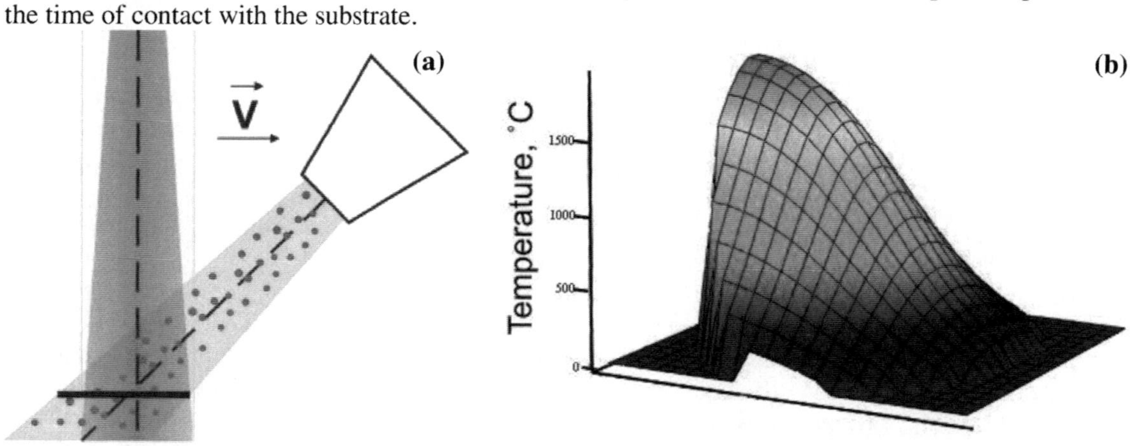

Figure 2. Scheme of the heating process of gas-powder jet by laser radiation (black line is the substrate level) **(a)**; surface temperature distribution of the powder particles at the time of contact with the substrate **(b)**.

3. Heating of the substrate by laser radiation

Since the defocused laser beam is a distributed heat source:

$$q(x, y) = k_{abs} I(x, y, z_0), \tag{2}$$

where k_{abs} is an absorption factor.

Heating of the substrate can be described as the ultimate state of the process of heat spreading from this source moving with constant velocity on the surface of a semi-infinite body relative to the moving *XYZ* coordinates [5]:

$$T(x,y,z)=\int_{-r}^{r}\int_{-r}^{r}\frac{q(x_i y_i)}{2\pi\lambda R(x-x_i,y-y_i,z)}\exp\left(-\frac{Vx}{2\chi}-\frac{VR(x-x_i,y-y_i,z)}{2\chi}\right)dx_i dy_i. \quad (3)$$

Taking into account the limited size of the substrate we will use the known method of introducing fictitious heat sources. Then the resulting thermal field in the product takes the form:

$$T_{\text{fin}}(x,y,z)=T(x,y,z)+T(x,y+2r,z)+T(x,y-2l,z)+T(x,y,z+2h). \quad (4)$$

4. Measurement of the cladding bead size

The cladding bead is formed by the molten powder mixed with the base metal into the molten pool. Assuming that powder transfer coefficient is the ratio of the surface area of the molten pool to the area of the fall of the gas-powder jet on the substrate it is possible to determine the amount of powder from which the bead is formed. We shall also take the width of the cladding bead as the width of molten pool. Then, considering that theoretical and practical [1] researches have shown that the cladding bead is ellipsoidal in cross-section, it is easy to calculate the height h of the bead:

$$h=\frac{2S}{w\pi}. \quad (5)$$

The resulting model was verified experimentally in the case of cladding of Inconel 625 powder on a substrate of heat-resistant alloy. Experimental and calculated values of the width of cladding bead and micro section are shown in figure 3. The experimental height of the bead is $h_r = 0.8$ mm and the calculated value is $h_t = 0.79$ mm.

Figure. 3. Metallographic sample of the cladding bead with parameters: power = 700 W; powder consumption = 4 g/min; speed – 5 mm/s; beam diameter on the substrate is 1.3 mm; (red line shows the calculated value of w_r and w_t experimental and calculated values of the width of the molten pool, respectively).

Acknowledgments

The work was carried out with financial support from the Ministry of Education and Science of the Russian Federation in frames of the complex project "Development of restoration of gas turbine engine working blades by laser cladding method". Contract No. 02.G25.31.0215 from 27.04.2016.

References

[1] Turichin G A, Somonov V V and Klimova O G 2014 *Applied Mechanics and Materials* **682** 160–5
[2] Turichin G A, Zemlyakov E V, Pozdeeva E Yu, Tuominen J and Vuoristo P 2012 *Metal Science and Heat Treatment* 54 139–44
[3] Erofeev V A and Kareninova I E *Computer simulation of laser powder surfacing* (Tula: Tula state University)
[4] Bitzadze A V and Kalinichenko D F 1985 *Collection of problems on equations of mathematical physics* 107
[5] Rykalin N N 2013 *Calculations of thermal processes in welding* (Ripol Klassik)

Study of Ni-Mo electrodeposition in direct and pulse-reverse current

Yu M Stryuchkova, N B Rybin, D V Suvorov, G P Gololobov, A B Tolstoguzov, D Yu Tarabrin, M A Serpova, V A Korotchenko and E V Slivkin

Ryazan State Radio Engineering University, 390005, Ryazan, Russia

E-mail: dmitriy_suvorov@mail.ru

Abstract. Process of electrochemical deposition of the coating based on a binary nickel-molybdenum alloy onto a nickel substrate under pulse mode with current reverse within the range of current density change from 2 to 9 A/dm^2 has been researched. Coating structure and its surface morphology have been studied. Method of X-ray energy dispersive spectroscopy has determined a percentage ratio of alloy components in the coating. Mode to obtain the densest and smoothest deposits has been identified under considered terms.

1. Introduction

Coatings on the basis of Ni-Mo alloy deposited by the electrochemical method have a high hardness, wear resistance, heat resistance, corrosion resistance in alkalis, chloride-containing environments and non-oxidizing acids (for example, HCl). Besides, Ni-Mo alloys are catalysts of the hydrogen electrochemical synthesis reaction that significantly decreases hydrogen evolution overpotential. At present the most available method to obtain coatings on the basis of nickel-molybdenum alloy is an electrochemical deposition from electrolytes containing nickel and molybdenum salts. This method does not require high energy and financial costs and also special complex equipment. Main issues restricting an area of galvanic Ni-Mo coating usage are coating defects in the form of cracks and also significant coating inhomogeneity and its high roughness. There are a lot of papers regarding optimization of conditions for electrochemical synthesis aimed at improvement of the surface structure and properties. Main electrolytes used for deposition of coatings based on Ni-Mo alloy are citrate and ammonia-citrate electrolytes [1–4]. Electrodeposition with direct current is traditionally used. Comparatively small number of papers dedicated to research of deposition of coatings based on Ni-Mo alloy under pulse mode is known [5–9]; very small part of them is dedicated to research of pulse mode with current reverse [10–13]. Paper [10] determined that under pulse-reverse mode of deposition surface roughness does not increase with rise of thickness in comparison with the mode of direct current, however coating surface images obtained under various modes have not been provided. Paper [11] published results of study of Ni-Mo alloy coatings under pulse-reverse mode of deposition. Obtained samples of coatings have an apparent globular structure with significant surface roughness. However, complex composition of the electrolyte complicates interpretation of obtained results. That is why, study of electrodeposition of Ni-Mo alloy under pulse-reverse current mode is important.

Purpose of the present paper is a research of influence of mode parameters of the pulse electrodeposition with reverse current on the coating structure and morphology on the basis of electrolytic Ni-Mo alloy under its deposition from the traditional ammonia-citrate electrolyte within current densities from 2 to 9 A/dm^2.

2. Experiment method

Nickel plates of the trademark NPA1 were a substrate for experiments. Final polishing of plates was performed using diamond paste NOM 0,5/0 up to appearance of surface high luster. After polishing, substrates were processed by ultrasound in 5 M solution HCl within 20 min. Then within 30 min washing was fulfilled in acetone and after – in alcohol within the same period of time. At the conclusive stage they were washed by distilled water. Before deposition of the coating the surface was cathodically activated in the solution containing nickel chloride and hydrochloric acid at the ratio 1:1 within 3 min under current density 1 A/dm^2. Electrodeposition of the nickel-molybdenum alloy was performed under pulse-reverse current with rectangular impulse shape where the reverse pulse followed after direct pulse. Ammonia-citrate electrolyte was used with the following composition [2]: $Na_2MoO_4{\cdot}2H_2O$ – 10 g/l; $NiSO_4{\cdot}7H_2O$ – 24 g/l; $NiCl_2\ {\cdot}6H_2O$ – 20 g/l; potassium citrate $K_3C_6H_5O_7$ – 70 g/l; NH_4OH – 60–80 g/l up to pH – 10.5. Two-electrode cell with the insoluble platinum anode and nickel cathode was used for deposition. Potentiodynamic current-potential curves were taken in order to determine areas of electrolysis current densities under deposition of the coating by nickel-molybdenum alloy on the nickel substrate. Three-electrode cell with the silver-chlorine reference electrode and platinum anode was used for polarization of the nickel electrode. The paper used a potentiostat IPC-Pro.3A. Researches of samples were performed on a scanning election microscope JEOL JSM-6610LV equipped for the elemental analysis by an Energy-dispersive X-ray spectroscopy microanalyzer INCA X-MAX made by Oxford Instruments.

3. Results and discussion

Before execution of experiments potentiodynamic current-potential curves have been obtained with sweep rate of sample potential ±10 mV/s under room temperature in solution of the ammonia-citrate electrolyte where the following electrodeposition of the alloy was performed. Significant hysteresis reflecting changes of the sample surface state as a result of its dissolution and following reduction is observed on polarization curves of direct and reverse currents. There is a pad of limiting current (approximately in the area of 1.4–1.6 V) on a polarization curve of direct route. So, deposition of compact coating of the nickel-molybdenum alloy from the ammonia-citrate electrolyte begins under current density being corresponding to this pad registered on the polarization curve and continues under higher current densities.

However, experiments have shown that under significant increase of the current density the deposited alloy becomes more porous and quality of the coating becomes worse. It is connected with presence of diffusion limitations by nickel and molybdenum under high current densities [2]. Figure 1(a) represents a typical SEM-image of the nickel-molybdenum alloy coating surface of one of the obtained samples deposited under direct current mode onto a nickel substrate under current density 40 A/dm^2. Time of electrodeposition was 10 min.

Experiments have determined that under considered terms increase of current density above 10 A/dm^2 leads to significant changes of structure and roughness of the deposited coating surface, i.e. porosity increases, cracks and delaminations appear. Change of current within density range from 2 to 9 A/dm^2 (more acceptable range from the point of view of current efficiency and possibility to carry out a reaction of electrochemical deposition under considered experiment conditions) has shown that a current value does not influence on structure and microrelief of the coating. Typical SEM-image of the nickel-molybdenum alloy coating surface of one of the obtained samples deposited under direct current mode onto a nickel substrate under current density 5 A/dm^2 and deposition time – 10 min is shown in figure 1(b).

At the following stage a number of alloy electrodeposition experiments under mode of pulse with current reverse under duration of direct pulse 50–500 ms and duration of the reverse pulse from 10 to 200 ms was executed.

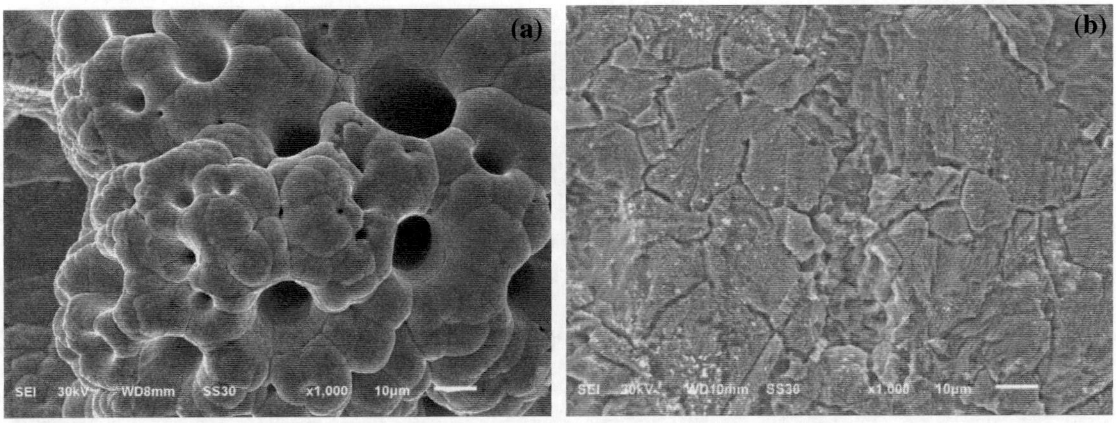

Figure 1. SEM-image of surface of the nickel-molybdenum alloy coating obtained under direct current mode onto a nickel substrate under various current densities: **(a)** – 40 A/dm^2; **(b)** – 5 A/dm^2.

Current density of the direct pulse was 5 A/dm^2, reverse one – 0.2 A/dm^2. Deposition time was 10 min in all cases. Analysis of the obtained sample surfaces has shown that within time interval of direct and reverses impulses being corresponding to tens of milliseconds (10–50 ms) a significant change of the deposited coating structure occurs: roughness and porosity decrease. "Refilling" of through cracks having a length of tens of micrometers is observed.

Figure 2 represents typical SEM-images of sample coating surfaces obtained under mode of pulse electrolysis with current reverse under various durations of reverse pulse being equal to 200 and 20 ms. Figure 3 represents SEM-images of Ni-Mo coating deposited under pulse-reverse electrolysis (5 A/dm2, 50 ms; –0.1 A/dm^2, 10 ms) at various surface magnifications. Time of electrodeposition was 10 min.

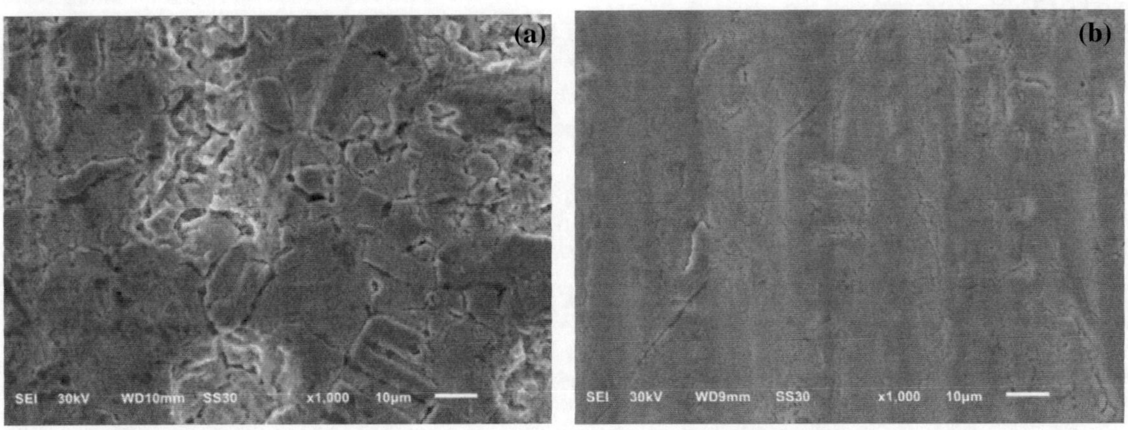

Figure 2. SEM-image of surface of the nickel-molybdenum alloy deposited under pulse electrolysis mode with current reverse onto a nickel substrate under various durations of reverse pulse: **(a)** – 5 A/dm^2, 50 ms; –0.2 A/dm^2, 200 ms; **(b)** – 5 A/dm^2, 50 ms; –0.2 A/dm^2, 20 ms.

It can be seen (figure 3) that the coating obtained within duration 10 ms of the reverse pulse has no defects (cracks), the surface of the coating is fairly smooth, characteristic dimensions of structural defects of the coating do not exceed 1 µm.

Method of X-ray energy-dispersive spectroscopy has determined a percentage ratio of nickel and molybdenum in deposits. Molybdenum content varied within 21–24 wt. %, nickel – 79–76 wt. %. Significant dependence of percentage content of alloy components on the electrolysis mode has not been detected.

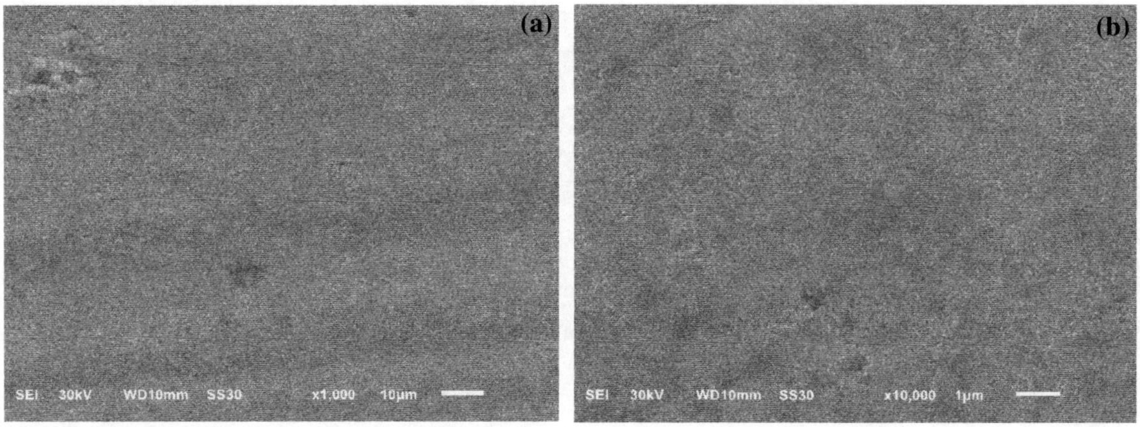

Figure 3. SEM-image of the surface of the nickel-molybdenum coating deposited under pulse-reverse current (5 A/dm^2, 50 ms; –0.1 A/dm^2, 10 ms) at various magnifications: **(a)** – ×1000; **(b)** – ×10 000.

4. Conclusions

The executed research has determined that under electrochemical deposition of the nickel-molybdenum alloy from the ammonia-citrate electrolyte, the good-quality coatings with minimum number of defects (cracks and pores) can be obtained using a non-stationary pulse mode of electrodeposition with reverse current, besides, step of reverse pulse should be within range 10–50 ms. Under such mode deposition of the alloy is possible within the whole researched range of current densities from 2 to 9 A/dm^2 without visible changes of coating structure and morphology. Besides, deposition mode does not influence on percentage content of Mo in coating.

References

[1] Beltowska-Lehman E, Bigos A, Indyka P and Kot M 2012 *Surface & Coatings Technology* **211** 67–71
[2] Vasko A T 1977 *Electrochemistry of molybdenum and tungstena* (Kiev: Naukova dumka) 120
[3] Huang P, Hou K, Wang G, Chen M and Wang J 2015 *International Journal of electrochemical science* **10** 4972–84
[4] Allahyarzadeh M H, Roozbehani B and Ashrafi A 2011 *Electrochimica Acta* **56** 10210–6
[5] Huang P, Hou K, Sheu H, Ger M and Wang G 2014 *Surface & Coatings Technology* **258** 639–45
[6] Nee C C, Kim W and Well R 1988 *Electrochemical Science and Technology* **5** 1100–3
[7] Beltowska-Lehman E and Indyka P 2012 *Thin Solid Films* **520** 2046–51
[8] Gonzalez-Buch C, Herraiz-Cardona I, Ortega E M, Garcia-Anton J and Perez-Herranz V 2013 *Chemical Engineering Transactions* **32** 865–70
[9] Donten M, Cesiulis H and Stojek Z 2005 *Electrochimica Acta* **50** 1405–12
[10] Chassaing E, Roumegas M P and Trichet M F 1995 *Journal of Applied Electrochemistry* **25** 667–70
[11] Yancheshmeh H S and Ghorbani M 2014 *Surface & Coatings Technology* **238** 158–64
[12] Ahmad Ya H, Mohamed A M A, Golden T D and D'Souza N 2014 *International Journal of electrochemical science* **9** 6438–50
[13] Sanches L S, Domingues S H, Carubelli A and Mascaro L H 2003 *J. Braz. Chem. Soc.* **14** 556–63

Electrodeposition of Ni-W gradient coatings

D V Suvorov, G P Gololobov, A B Tolstoguzov, S M Karabanov, D Yu Tarabrin, N B Rybin, Yu M Stryuchkova, M A Serpova and V A Korotchenko

Ryazan State Radio Engineering University, 390005, Ryazan, Russia

E-mail: dmitriy_suvorov@mail.ru

Abstract. The present paper has represented results of researches of the electrodeposition of Ni-W coatings having a gradient structure. The paper has determined conditions of synthesis providing a possibility to control W content in the coating within wide ranges (from 3.3 to 9.2 at. %). Influence of current density and pH value on coating characteristics has been studied. Morphological characteristics of the gradient coating have been researched.

1. Introduction

Binary alloy Ni-W has high wear resistance, hardness, corrosion resistance [1]. High resistance of Ni-W alloy to electrical erosion should be mentioned separately. One of "congenital" issues of coatings based on Ni-W alloy is an occurrence of the crack network significantly worsening coating corrosion resistance. One of more effective methods to decrease coating cracking significantly is a formation of the coating gradient structure where tungsten content increases from the substrate to the coating surface. Paper [2] researched a process of electrochemical deposition and characteristics of the Ni-W coating having a gradient structure. Pulse mode of electrodeposition was used in the citrate electrolyte. Tungsten content in the coating [2] increased from 1.5–1.8 to 4.5–5.0 % (by mass) in the direction from substrate to surface; coating has satisfactory characteristics of resistance to abrasive wear and corrosion resistance that explains a comparatively small content of tungsten in the alloy but it has a sparse crack network. Important task is a search of methods to form a gradient structure of the coating with high tungsten content and providing a small number of coating defects.

2. Experiment details

For electrochemical deposition the present paper used a pyrophosphate electrolyte developed and researched in [3,4] containing 0.2 M $NiSO_4$, 0.2 M Na_2WO_4, 0.6 M $K_4P_2O_7$, 0.15 M $(NH_4)_2SO_4$, electrolyte temperature 40 °C, pH level was varied from 8.5 to 9.5. For this electrolyte, stability area is a range pH 8.0–11, overrunning of this range leads to precipitation in solution (under pH < 8) or significant deterioration of the coating quality (under pH > 10). Plates of oxygen-free copper of the trademark OF-OK were used as substrates; they were sequentially processed by acetone in the ultrasonic bath and 1 M HCl solution. Galvanodynamic mode was used for deposition of coatings with variable tungsten content by coating thickness when current density was linear decreased from 10 to 1 A/dm^2 and from 5 to 1 A/dm^2. Time of electrodeposition of all samples was 40 min. Coating samples had metallic luster, good adhesion to the substrate. Obtained coating samples were researched using a scanning electron microscope JSM-6610LV (JEOL, Japan) with an EDS detector INCA X-

MAX (Oxford Instruments). Distribution of tungsten content along the coating depth was determined by multipoint scanning of the coating cross section by an EDS detector.

3. Experimental results and discussion

Figure 1 represents SEM-images of the coating surface obtained under linear measurement of the current density from 10 to 1 A/cm^2 at electrolyte pH value – 8.5 and distribution of tungsten content along the coating depth.

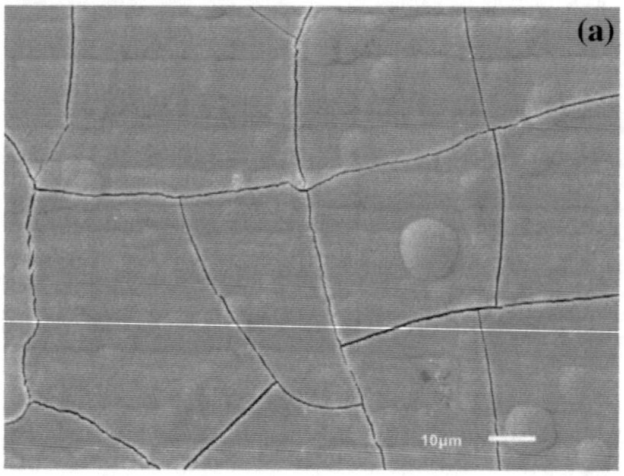

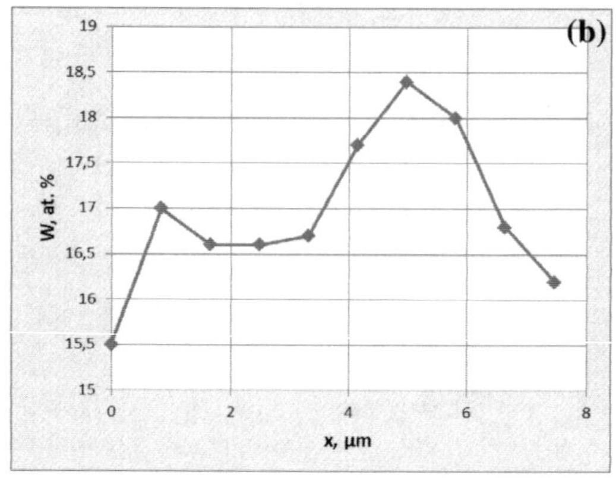

Figure 1. Characteristics of Ni-W coating obtained at linear decrease of the current density from 10 to 1 A/dm^2 (electrolyte pH value – 8.5): **(a)** – coating surface morphology; **(b)** – distribution of tungsten content along the Ni-W coating depth.

Coatings obtained at electrolyte pH being equal to 8.5 are enough smooth, have a small number of separate globules, comparatively small number of cracks (about 22 intersections for 1 mm), however distribution of tungsten concentration along the coating depth has a complicated profile and varies within narrow ranges from 15.5 to 18.4 at. %.

Figure 2 represents SEM-images of the coating surface obtained under linear measurement of the current density from 10 to 1 A/dm^2 at electrolyte pH value – 9.5 and distribution of tungsten content along the coating depth.

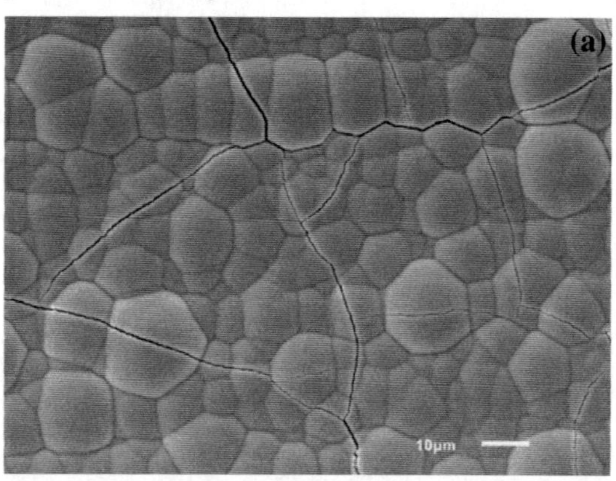

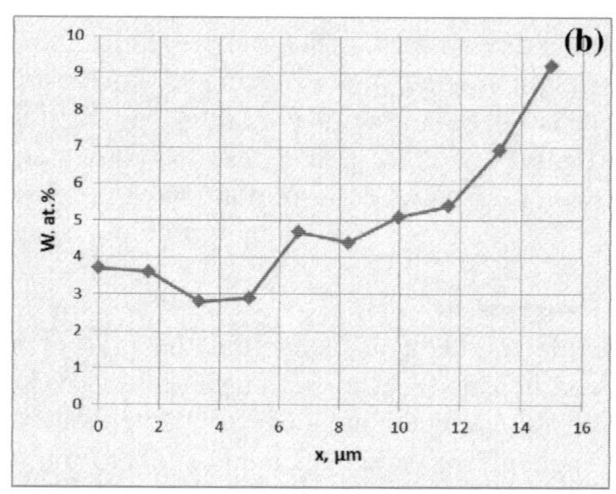

Figure 2. Characteristics of gradient Ni-W coating obtained at linear decrease of the current density from 10 to 1 A/dm^2 (electrolyte pH value – 9.5): **(a)** – coating surface morphology; **(b)** – distribution of tungsten content along the Ni-W coating depth.

Obtained coating has a prominent globular structure (globule size 5–30 μm) and comparatively sparse crack network (about 24 intersections for 1 mm). Crack network is formed by stochastic way and coincides with globule boundaries only in some fragments. Comparison of coating thicknesses obtained at various electrolyte pH has shown that at pH transfer from 8.5 to 9.5 integrated current efficiency increases practically in 2 times. Analysis of the obtained value of the gradient of elemental composition has shown that insignificant oscillations of tungsten concentration within 2.8–3.7 at. % are observed close to the substrate and then practically linear growth of the tungsten concentration up to 9.2 at. % is observed on the coating surface. Area of oscillations and low values of the tungsten concentration corresponds with high values of the current density 10–5 A/dm^2. Deposition was performed under the same conditions but with change of the current density from 5 to 1 A/dm^2 to achieve a linear gradient of the tungsten concentration in the coating (figure 3).

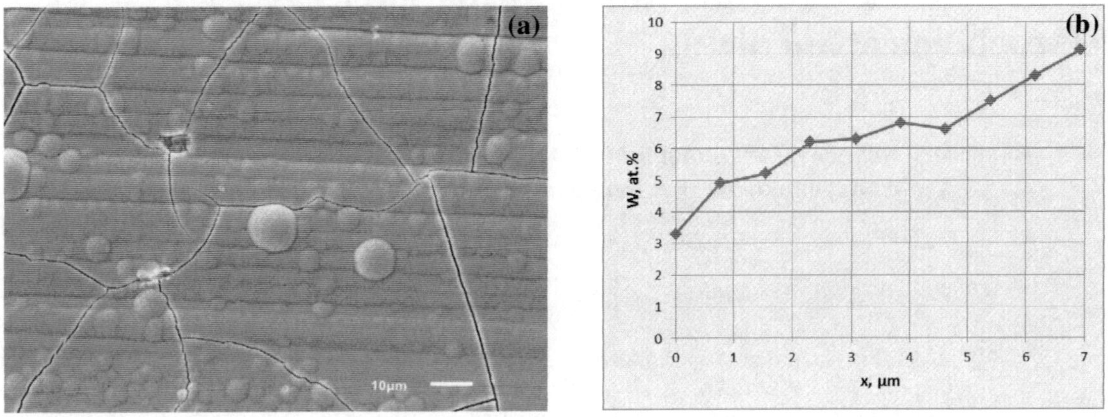

Figure 3. Characteristics of gradient Ni-W coating obtained at linear decrease of the current density from 5 to 1 A/dm^2 (electrolyte pH value – 9.5): **(a)** – coating surface morphology; **(b)** – distribution of tungsten content along the Ni-W coating depth.

Obtained coating is smooth enough, has some number of separately located large globules which size does not exceed 10 μm. Crack network is stochastic, enough analogous to the previous case. Enough linear gradient of the tungsten content in the coating increasing from 3.3 at. % on the substrate surface to 9.2 at.% on the coating surface has been obtained.

Analysis and comparison of obtained experimental data have shown that the pyrophosphoric electrolyte of mentioned content under pH value – 9.5 provides a possibility to control an elemental composition of the coating within 3.3–9.2 at. % by changing of cathode current density within the range from 1–5 A/dm^2. Besides, comparatively sparse crack network and small coating roughness are provided. High value of the current density leads to occurrence of the prominent globular structure. In electrolytes having pH being equal to 8.5 coatings with enough high tungsten content are formed but possibility to control a coating structure by changing of cathode current density is not provided.

4. Conclusion

Experimental researches have demonstrated a possibility of electrodeposition of gradient coatings based on Ni-W alloy having tungsten concentration practically linearly changing within 3.3–9.2 at. % in direction from a substrate to a surface of the coating. The paper has determined that for pyrophosphoric electrolyte at pH – 9.5 control of the coating composition is provided by changing of cathode current density. Besides, within the range from 1–5 A/dm^2 direct dependence of tungsten content in the coating on cathode current density is observed and enough smooth coatings are formed.

References

[1] Yamasaki T 2000 *Mater. Phys. Mesh.* **1** 127–32
[2] Allahyarzadeh M H, Aliofkhazraei M, Sabour Rouhaghdam A R and Torabinejad V 2016 *Canadian Metallurgical Quarterly* **55** 303–11
[3] Krasikov A V 2012 *Matters of material science* **1** 68–73
[4] Furmon M S, Krasikov A V and Drozdova N F 2012 *Matters of material science* **1** 74–9

IOP Conf. Series: Journal of Physics: Conf. Series **857** (2017) 012048 doi:10.1088/1742-6596/857/1/012048

Research of the TiO_2 nanotubes formation by AC-DC electrochemical anodizing

D V Suvorov, G P Gololobov, S M Karabanov, D Yu Tarabrin, E V Slivkin, Yu M Stryuchkova, M A Serpova and V A Korotchenko

Ryazan State Radio Engineering University, 390005, Ryazan, Russia

E-mail: dmitriy_suvorov@mail.ru

Abstract. The present paper has represented results of research of nanoporous TiO_2 formation process under electrochemical anodizing by direct current with an overlapped variable component. It has been shown that overlapping of the variable component with frequency 100 Hz–10 kHz leads to increase of the growth rate and qualitative change of the nanoporous TiO_2 coating structure – open inter-tube space appears, inter-tube "bridges" are generated and form of the nanotube section becomes "rounded".

1. Introduction

Interest to the titanium oxide is connected with its unique physical and chemical properties. Nanoporous titanium oxide is one of promising and required materials for photocatalysis and adsorption. Traditional method to synthesize nanoporous TiO_2 is electrochemical anodizing by direct current. TiO_2 structure depends on substrate morphology, electrolyte composition and temperature; however main parameter determining structure parameters is a value of anodizing voltage. Possibility to control TiO_2 structure is practically valuable because the structure determines physical and chemical properties of the coating (catalytic activity, etc.) [1–9]. So, search of new methods to control a structure of the nanoporous titanium oxide is important. Most researches of nanoporous TiO_2 formation have been executed under DC anodizing voltage. Papers [10–14] showed a possibility to control a structure by step change of voltage; besides typical time intervals of the voltage change is amounted to tens-hundreds seconds. Influence of AC voltage overlapping with frequency 1 Hz–100 kHz on DC anodizing voltage has not been researched in details. It explains importance and novelty of the performed work directed at study of nanoporous TiO_2 formation process under electrochemical anodizing by direct current with an overlapped AC component.

2. Experiments methods

Specialized test stand (figure 1) has been assembled for experimental researches of the nanoporous titanium oxide synthesis under overlapping of the AC component. The test stand includes two identical cells for performance of anodizing process. Only DC voltage was applied to one of cells and combined AC-DC voltage obtained from a source of DC voltage and generator – to other one. Control of anodizing current was performed by means of current shunts (resistors *R*1, *R*2) and digital oscilloscope. Control of voltage at the cell was performed from the resistor divider output of which was connected to the oscilloscope.

Electrolyte on the basis of ethylene glycol containing NH_4F 0.3 % (mass.), H_2O 0.5 % (mass.) was used. DC voltage component was varied within 30–80 V, AC component had a sinusoidal form, amplitude 2–10 V and frequency 1 Hz–100 kHz.

Titanium plates of the trademark VT1-0 were used to obtain experimental samples. Final polishing of plates was performed using diamond paste NOM 0,5/0 up. After polishing, plate surfaces were processed by ultrasound in 5 M solution HCl within 40 min. Then within 30 min washing was fulfilled in acetone and after – in ethanol within the same period of time. At the conclusive stage they were washed by distilled water. Obtained samples of the nanoporous TiO_2 coating were researched by a scanning election microscope JEOL JSM-6610LV.

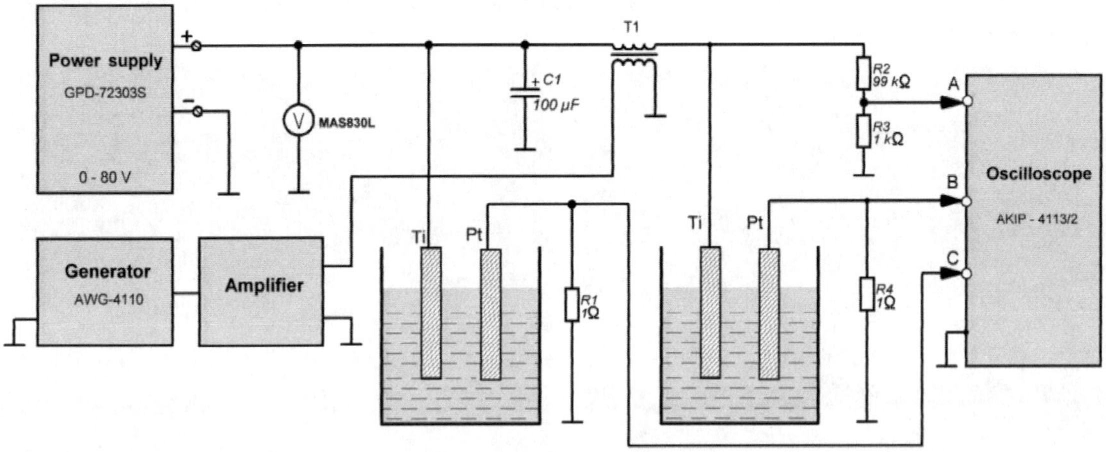

Figure 1. Electric circuit of the experimental test stand for obtaining of the porous TiO_2 under overlapping of the AC voltage component on to DC voltage one.

3. Results and discussion

Experimental researches have determined that maximum thickness of coatings under DC and AC-DC modes does not depends on a form of voltage anodizing and it is amounted to 10–12 μm for the present experiment (figure 2).

Figure 2. SEM-image of the nanoporous titanium oxide coating (anodizing DC voltage – 60 V).

Analysis of time dependencies of anodizing current for various modes has shown that the initial value of anodizing current under AC-DC mode is higher in 15–20 % in comparison with DC mode and "shelf" is observed with enough high current value (75 % of the maximum value) on time current diagram under AC-DC mode after which current sharply falls and becomes lower than an actual value at direct current. Change of the sample thickness has shown that after current fall growth of the coating terminates. Time to achieve TiO_2 level thickness up to a nominal value under AC-DC mode is less in 5–7 times that under direct voltage (under AC-component being equal to 10 % of DC-component). Figure 3 represents images of sample surfaces of nanoporous TiO_2 obtained under DC and AC-DC anodizing voltages. It is obvious that under overlapping of the AC voltage component typical change of the nanoporous titanium dioxide structure happens: form of nanotubes becomes "rounded", open inter-tube space appears, typical diameter in comparison with direct current slightly changes. Under anodizing voltage 60 V overlapping of the variable component in 10 V (1 kHz) leads to increase of a pore diameter in 25 %.

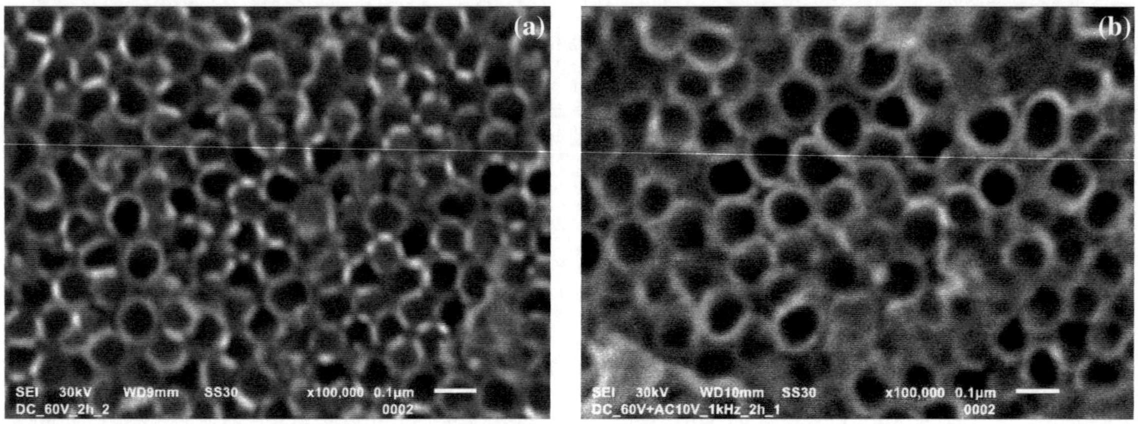

Figure 3. Structure of the nanoporous titanium oxide surface under various modes of anodizing: **(a)** – DC voltage, 60 V; **(b)** – AC-DC voltage (60 V DC + 10 V AC 1 kHz).

Image of a cross-section of the coating structure (figure 4) shows that under overlapping of the AC component inter-tube "bridges" are formed with thickness about 20 nm at distance 50–100 nm from each other.

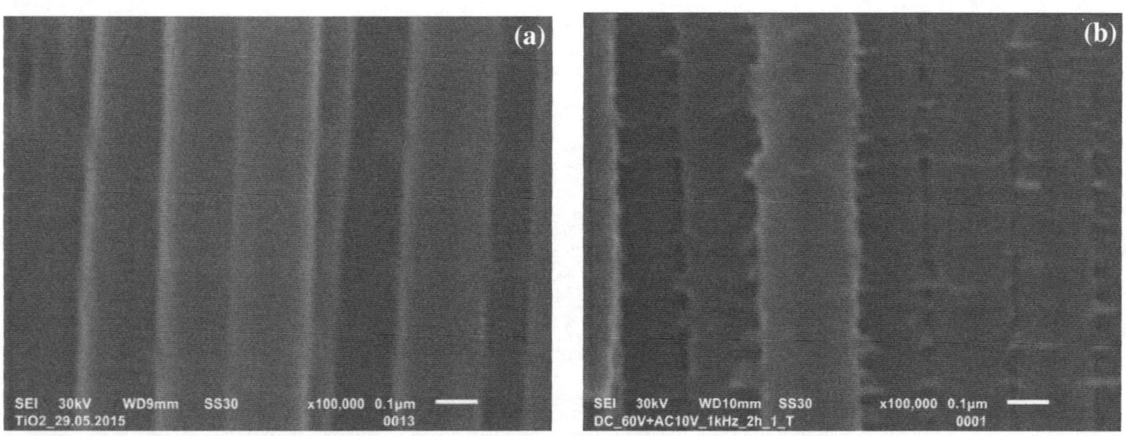

Figure 4. Image of a cross-section of the coating structure of the nanoporous titanium dioxide structure: **(a)** – DC voltage, 60 V; **(b)** – AC-DC voltage (60 V DC + 10 V AC 1 kHz).

Visual influence of frequency on characteristics of the coating structure appears within the frequency range 100 Hz–10 kHz. Under lower and higher values of frequency influence of the variable component is insignificant and geometrical characteristics of the coating structure are not practically different from characteristics under direct current.

Results can be interpreted as increase of the pore growth rate to the deep of titanium under overlapping of the current variable component. Additionally inter-tube space begins to be etched and opened that leads to rounding of the pore form and qualitative change of geometrical parameters of the growing nanoporous titanium oxide surface – appearance of horizontal «bridges» in space between tubes. Explanation of the observed occurrence can be a supposition that capacitive current passing through the lower barrier level of the pore causes a change (increase) of its transport properties in relation to ions penetrating through it that determines an increase of pore growth process rate.

4. Conclusions

Data on influence of overlapping of the anodizing voltage AC component within frequency range 100 Hz–100 kHz on growth rate, surface morphology, structure parameters of the nanoporous titanium oxide have been obtained. Conditions for titanium anodizing have been determined when formation of inter-tube «bridges» with thickness 20 nm at distance 50–100 nm from each other connecting separate tubes in the direction being perpendicular to their axis happens. It has been shown that under overlapping of AC-component (amplitude 10 % of DC-component value) rate of TiO_2 level formation increases in 5–7 times. Visible influence of AC-component appears within frequency range 100 Hz–10 kHz. The paper has determined that within mentioned range of frequencies the form of TiO_2 nanotubes becomes “rounded”, open inter-tube space appears. So, the paper has shown that overlapping of the AC component provides a possibility to control characteristics of the coating structure. Obtained results have high practical importance for development of technologies to synthesize new types of catalytically active coatings based on the nanoporous titanium oxide.

References

[1] Yu X, Li Y, Wlodarski W, Kandasamy S and Kalantar-zadeh K 2008 *Sensors and Actuators B* **130** 25–31
[2] Berger S, Hahn R, Roy P and Schmuki P 2010 *Phys Status Solidi B* **247** 2424–35
[4] Lu X, Wang G, Zhai T, Yu M, Gan J, Tong Y and Li Y 2012 *Nano Lett.* **12** 1690–6
[5] Yuan S, Yu L, Shi L, Wu J, Fang J and Zhao Y 2009 *Catalysis Communications* **10** 1188–91
[6] Tian G, Fu H, Jing L and Tian C 2009 *Journal of Hazardous Materials* **161** 1122–30
[7] Fujishima A, Zhang X and Trykc D 2008 *Surface Science Reports* **63** 515–82
[8] Paulose M, Shankar K, Varghese O, Mor G and Grimes G 2006 *J. Phys. D: Appl. Phys* **39** 2498–2503
[9] Firouzdor V, Brechtl J, Hauch B, Sridharan K and Allen T R 2013 *Applied Surface Science* **282** 798–808
[10] Luan X, Guan D and Wang Y 2012 *J. Phys. Chem. C* **116** 14257–63
[11] Guan D and Wang Y 2012 *Nanoscale* **4** 2968–77
[12] Luan X and Wang Y 2014 *Materials Science in Semiconductor Processing* **25** 43–51
[13] Lin C, Chen S and Cao L 2013 *Materials Science in Semiconductor Processing* **16** 154–9
[14] Jia Y, Zhang M, Cui J, Lin K, Zheng H, Zhu J and Samia A 2012 *Nano Energy* **1** 796–804

IOP Conf. Series: Journal of Physics: Conf. Series **857** (2017) 012049 doi:10.1088/1742-6596/857/1/012049

Investigation of structural and optical properties of solid solution films in PbF_2-YF_3 system

A N Tropin[1,2]

[1] JSC NII GIRICOND, 194223, Saint Petersburg, Russia
[2] Department of Physics, Saint Petersburg State University of Aerospace Instrumentation, 190000, Saint Petersburg, Russia

E-mail: 216@giricond.ru

Abstract. This work deals with the physical chemistry of metal fluorides and their thin films applicable in thin-film optics and is related to the investigation of the structural and optical properties of solid solution films in the system PbF_2-YF_3 for the creation of interference coatings for the near- and mid-infrared spectra.

1. Introduction

Nowadays, the thin-film optical coating technology requires films based on low-refractive film-forming materials transparent in the spectral range with wavelengths exceeding 10–12 μm [1].

Traditionally, films with a low refractive index are obtained from fluorides of the following metals: calcium (CaF_2), barium (BaF_2), lead (PbF_2), yttrium (YF_3) and some others [2, 3]. However, there are a number of constraints to the successful application of the films prepared from these materials [1].

The problem of practical application of films with a low refractive index arises from the fact that the production of coatings operating in the mid-infrared region requires sufficiently thick layers with the physical thickness up to 1 μm or more. As a result, severe mechanical stresses build up both in the layers themselves and in the multilayer structures based on these layers, which often causes the coating to peel.

For example, the coatings containing YF_3 layers with the overall thickness of several micrometers become extremely unstable with changes in the temperature and are prone to peeling from the surface of optical parts due to mechanical stresses arising in YF_3 films.

In this case, using mixtures and solid solutions of metal fluorides can probably provide an effective solution. In particular, the prospects of using solid solution films in the PbF_2-YF_3 system seem good. However, data on studying these films in terms of their application in the optics of thin-film coatings have not been reported in the scientific literature.

In this regard, the purpose of the work was to study the properties of the film-forming materials and their thin films in the PbF_2- YF_3 system for creation on their basis multilayer interference coatings applicable in the near- and mid-infrared spectral ranges.

The object of the study: film-forming materials of the fluoride system PbF_2-YF_3 and their thin films prepared by electron-beam vacuum evaporation onto substrates of single-crystal silicon, germanium and silicate glass.

Subject of the research: physicochemical properties of the film-forming materials in the PbF_2-YF_3 system, optical and structural properties of the thin films formed from the investigated materials.

IOP Conf. Series: Journal of Physics: Conf. Series **857** (2017) 012049 doi:10.1088/1742-6596/857/1/012049

Research methods and instruments:
– X-ray diffraction analysis (a diffractometer "Dron-2,0», Cu-Kα);
– Scanning Electron Microscopy (a complex SUPRA 55VP-25);
– Fourier Transform Infrared Spectroscopy (an IR Fourier spectrometer FSM 1201).

2. Experimental setup

PbF_2 and YF_3 as well as specially synthesized solid solutions in the system PbF_2-YF_3 having different compositions were used as the starting film-forming materials. Data on the composition of the samples and the results of X-ray diffraction analysis of initial film-forming substances are listed in table 1.

Table 1. Results of X-ray diffraction analysis of film-forming materials.

Sample number	Composition: mole fraction	Structure
1	≈100 % PbF_2	Corresponds to the structure β-PbF_2, cubic lattice
2	75 % PbF_2 + 25 % YF_3	β- PbF_2 based solid solution
3	50 % PbF_2 + 50 % YF_3	Corresponds to the structure $Pb_{0.52}Y_{0.48}F_{2.48}$, tetragonal lattice. The main diffraction lines of this structure coincide with the lines of PbF_2 but with a reduced parameter
4	25 % PbF_2 + 75 % YF_3	Corresponds to the structure $Pb_{0.52}Y_{0.48}F_{2.48}$ + YF_3
5	≈100 % YF_3	Corresponds to the structure YF_3, orthorhombic lattice

The films of the materials under the investigation had thickness of ~1 μm and were produced by E-beam evaporation in the vacuum unit VU-2M. The pressure in the vacuum chamber was not lower than $5 \cdot 10^{-5}$ Torr and the temperature of substrates during deposition was 100 °C. Single crystal silicon (111), germanium and silicate glass served as substrate materials. The rate of film deposition was about 2 nm/s. The thickness of the layer in the process of deposition was monitored by a broad-band optical monitoring system IRIS 1017 operating in the reflection mode.

3. Results and discussion

Dispersion of film optical constants was studied on samples obtained on the germanium substrates. X-ray diffraction analysis was performed on the film samples formed on single-crystal silicon and silica glass substrates. Analysis of structural features showed that the properties of films formed on single crystal silicon substrates are identical to those of the films formed on silicate glass substrates. Table 2 lists the results of X-ray diffraction analysis of the films. The number of each film sample corresponds to the number of the film-forming material sample from table 1.

Table 2. Results of X-ray diffraction analysis.

Film number	Structure
1	Corresponds to the structure β-PbF_2, texture along the plane (111). The orthorhombic phase of PbF_2 can be present in small quantities
2–4	The diffraction patterns from the films having such compositions practically coincide with each other and correspond to β-PbF_2 with the texture along the plane (111), but with a modified parameter. This change in the lattice parameter can be due to both the stressed state of the film and the formation of the structure $Pb_{0.52}Y_{0.48}F_{2.48}$
5	The film is X-ray amorphous

The analysis of the film microstructures indicates that film № 1 is polycrystalline and has the grain size of 40 to 100 nm, 60 nm on the average. Film № 2 is more homogenous in the grain size with the

IOP Conf. Series: Journal of Physics: Conf. Series **857** (2017) 012049 doi:10.1088/1742-6596/857/1/012049

average value there of 50 nm. The films on samples № 3–5 are fine-grained and an average grain size is of 20 nm.

The elemental composition of films № 1–3 corresponds to PbF_2 and the composition of film № 5 corresponds to YF_3. Film № 5 exhibits drop-shaped defects with sizes of about 5 μm (figure 1). The elemental composition of the defects corresponds to the elemental composition of the film.

Figure 1. Microphotograph of a YF_3 film defect.

The elemental composition of film № 4 (atomic %): F – 74.36 %, Y – 13.94 %, Pb – 11.69 % is similar to that of $Pb_{0.52}Y_{0.48}F_{2.48}$.

Dispersions of the film refractive index $n(\lambda)$ and the absorption coefficient $k(\lambda)$ were calculated from the transmittance and reflection spectra of films formed on the germanium substrate with known optical constants. Calculation results are presented in figure 2.

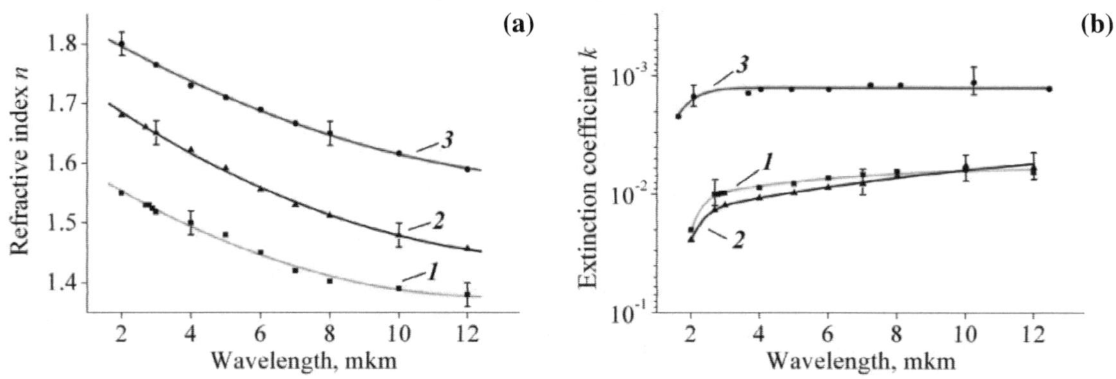

Figure 2. Refractive index (**a**) and extinction coefficient (**b**) curves of investigated films: *1* – YF_3; *2* – $Pb_{0.52}Y_{0.48}F_{2.48}$; *3* – PbF_2.

The values of refractive indices for the examined PbF_2 and YF_3 films are close to the known values [2, 3]. The refractive index for $Pb_{0.52}Y_{0.48}F_{2.48}$ film has some intermediate values.

4. Conclusions

We have studied the structural and optical properties of films obtained from new film-forming materials, namely from solid solutions with different compositions in the PbF_2-YF_3 system. Our results

IOP Conf. Series: Journal of Physics: Conf. Series **857** (2017) 012049 doi:10.1088/1742-6596/857/1/012049

have shown that the structure and composition of the films deposited by E-beam evaporation in the vacuum from the starting materials containing 75 % PbF_2 + 25 % YF_3 and 50 % PbF_2 + 50 % YF_3 correspond to those of films obtained from PbF_2. The structure of the films prepared from the material containing 25 % PbF_2 + 75 % YF_3 corresponds to $Pb_{0.52}Y_{0.48}F_{2.48}$.

Dispersions of optical constants in the spectral range from 2 to 12 μm have been calculated for PbF_2, YF_3 and $Pb_{0.52}Y_{0.48}F_{2.48}$ films.

Acknowledgments
The author thanks S V Shtelmakh and S P Teslenko for their assistance in carrying out X-ray diffraction analysis and microstructural studies.

References
[1] Tropin A N 2016 *Uspekhi Prikladnoi Fiziki* **4** 206–11
[2] Kruschwitz J D T and Pawlewicz W T 1997 *Applied Optics* **36** 2157–9
[3] Pellicori S F and Colton E 1992 *Thin Solid Films* **209** 109–15

Influence of intermetallic coatings of system Ti-Al on durability of slotting tool from high speed steel

E L Vardanyan, V V Budilov, K N Ramazanov, R N Khusnimardanov and R Sh Nagimov

Department of industrial and manufacturing engineering, Ufa State Aviation Technical University, Russian Federation, Ufa, st. K. Marxa 12

E-mail: vardanyaned@gmail.com

Abstract. The operation conditions and mechanism of wear of slotting tools from high-speed steel was researched. The analysis of methods increasing durability was carried out. The effect of intermetallic coatings deposited from vacuum-arc discharge plasma on the physical-mechanical high-speed steel EP657MP was discovered. The pilot batch of the slotting tool and production tests were carried out.

1. Introduction

For today aircraft engine construction sets itself the tasks of modernizing and development modern turboshafts, which have high efficiency coefficient and engine thrust due to the new heat-strength, heat-resistant and high-strength materials. For the treatment of internal slots in different products of turboshafts uses a slotting operation. In the working process, the slotting tool experiences alternating dynamical loads, wherein the greatest wear exposed the backside of tool [1].

At present time physical deposition processes [2] for different components, including cutting tool, are increasingly used due to high reliability, uniqueness, the possibility of obtaining coating of various composition, structure with high productivity of tool. For increase durability of metal-cutting tools widely applied coatings of the Ti-Al-N system [2]. For most of bulk materials with high hardness is characteristic a large value of elastic modulus E, therefor such materials are brittle. To evaluate the resistance of materials to elastic deformation uses the ratio of hardness to the elastic modulus H/E, which called the plasticity index of material [3]. It follows that to increase the resistance to elastic deformation, the material must have a high hardness with a low modulus of elasticity. Earlier, the authors [2, 4] developed a new method for obtaining coatings based on the intermetallic compound of the Ti-Al system, which based on the synchronous deposition of Ti and Al from two electric arc evaporator, by heating substrate to a temperature of 700 K. As a results of conducted researches, has been established, that with the increase the substance intermetallic phases in the coating, mechanical properties are increase.

In this way, the aim of this work is to research the influence of intermetallic coatings of the Ti-Al system on the durability of slotting tools from high-speed steel.

2. Objects and methods of research

As the object of research was chosen slotting tool made of EP657MP steel, which is a gear-cutting tool with the shape of a gear wheel. Depending on the working rime of tool, its wear can be divided on

IOP Conf. Series: Journal of Physics: Conf. Series **857** (2017) 012050 doi:10.1088/1742-6596/857/1/012050

three steps (figure 1), which consistent to the three sites of wear. Area I – initial wear – burn-in period. Area II – period of normal wear. Area III – period of intense wear. In the working process slotting tool is incured a thermal wear, formed as a result of friction of the backside of the teeth of cutting tool to the surface of the workpiece.

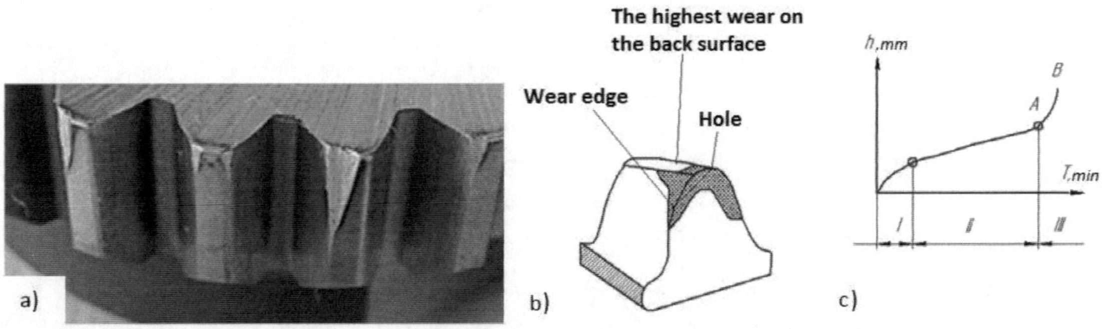

Figure 1. Wear a gear-cutting: **(a)** – catastrophic wear of the teeth of a slotting tool; **(b)** – the nature of the tooth wear; **(c)** – dependence of wear on the tool life.

The samples of EP657 material were coated intermetallic coatings of the Ti-Al system by developed technology [4]. Samples in the vacuum chamber are located as well as in parallel and perpendicular, to determine the difference of the thickness on vertical and horizontal surfaces. Coatings were covered by simultaneously spraying of two one-component cathodes of Al and Ti, and also rotating the table around itself axis. During the deposition coatings, when the samples were rotated in the chamber, intermetallic phases Ti-Al system are formed on the surface. Technological modes of the coating are varied the next limits: the pressure in the chamber $P = 10^{-1}$–10^{-2} Pa; arc current I = 60–120 A; processing time 90 min. The thickness of the samples was measured by device called CALOTEST. The microhardness, modulus of elastics were measured by using a nanodidomer called NANOVEA. The microstructure and chemical composition of the surface were determined by scanning electron microscope.

3. Experiment results

The thickness of the coating on the samples which were located parallel to the flow was 3 μm, perpendicular – 5 μm. The microstructure of the coated samples is shown in figure 2.

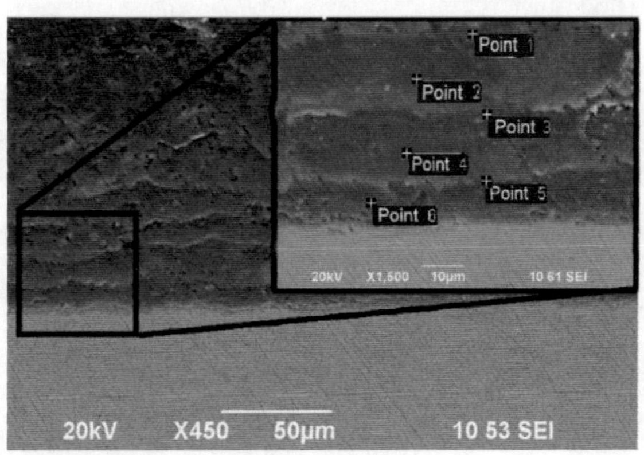

Figure 2. Microstructure of samples processed by the developed technology.

Also, using the scanning electron microscope, was researched the chemical composition of the coating layers. The microstructure of the explored area is shown in figure 2. The results on the chemical composition of the coating layers is showed in table 1. The results of measurements of the chemical composition of the layers are shows that the nitrogen content is maximal in the dark areas and decreases on the light layers, which is correspond to the intermetallic layers. In the transition from the

IOP Conf. Series: Journal of Physics: Conf. Series **857** (2017) 012050 doi:10.1088/1742-6596/857/1/012050

upper layers to lower, increases the content of iron and chromium, because the depth of the layer since measuring the chemical composition is up to 3 μm.

Table 1. Chemical composition of the layers.

Point	Atomic ratio, %				
	N	Al	Ti	Cr	Fe
1	35.12	22.8	40.7	0.6	0.4
2	27.27	26.7	45.4	0.64	0.4
3	39.77	22.2	36.6	0.55	0.5
4	27.4	22.1	48.5	0.75	0.8
5	34.34	21.5	42.9	0.63	0.7
6	33.15	21.5	43.1	0.69	0.9

Also were studied the mechanical properties of coated samples. The loading and unloading curves are shown in figure 3, the hardness of the coating H was 34.75 GPa, the modulus of elasticity E = 204.17 GPa. Herewith, the elastic recovery value W_e is 62 % and the plasticity index of the material is 0.17, which made possible to provide increased mechanical properties and wear-resistant characteristics of the cutting edges of the tool.

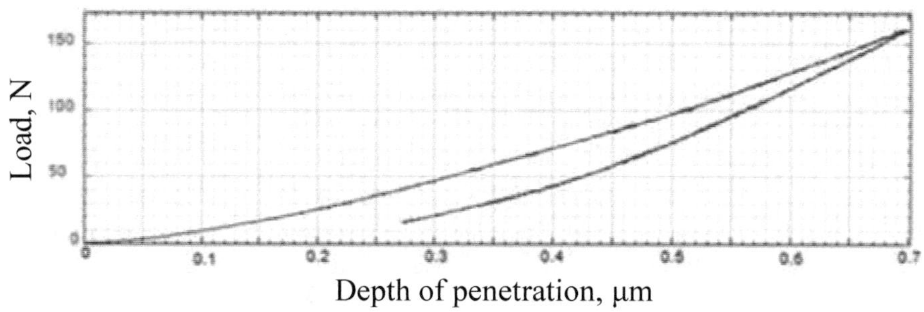

Figure 3. Microstructure of samples processed by the developed technology.

Based on the results of researches the experimental batch of the instrument was processed (figure 4) and production test were carried out.

Figure 4. Slotting tool with intermetallic coating.

The results of the tests established, that the deposition of composite coatings based on intermetallic Ti-Al system on metalcutting slotting tool for involute splines allowed to process detail in size with the drawing requirements, with using a single instrument and thereby improving durability compared with regulatory 6 times (figure 4), the tool wear was not more than 1 mm, which will use the tool in future (after regrinding and repeated coating), as opposed to catastrophic wear of the teeth (figure 1(a)).

4. Conclusion

Operation conditions and the wear mechanism of slotting tools form high-speed steel are investigated. A research of mechanical properties of samples with intermetallic coatings showed, that the microhardness is 34.75 GPa and the modulus of elasticity is 204.17 GPa. Also it was found that the value of elastic recovery W_e is 62 % and the plasticity index of the material was 0.17, which allowed the increased mechanical properties. Based on the results of the researches tests it was found that the application of intermetallic coatings of the Ti-Al system made it possible to increase the tool life in comparison with the standard one by 6 times.

References

[1] Kalashnikov S N and Kalashnikov A S 1983 *Machine building* 264
[2] Budilov V V and Vardanyan E L 2016 *Journal of Surface Investigation: X-ray, Synchrotron and Neutron Techniques* **10**(4) 728–31
[3] Shtanskiy D V et al. 2006 *Physics of the solid state* **48** 1231–8
[4] Vardanyan E L et al. 2016 *Proceed. XXVII Intern. Symp. Discharges and Electrical Insulation in Vacuum* pp. 739–41

Active elements of film sources of sound-thermophones

B P Vasiliev, F F Legusha, K V Rezrezova and G V Chizhov

St. Petersburg State Maritime Technical University, St. Petersburg, Russia

E-mail: bp_vas@mail.ru, legusha@smtu.ru, kv_neveselova@mail.ru

Abstract. The report examines the device and factors affecting the acoustic efficiency of thermoacoustic sound sources – thermophones. Film thermophones are studied, the structure of which is made in the form of a system of layers. Thermophones can be considered as an example of the practical application of films and coatings to ensure the operation of physical instruments. A great importance for the work of thermophones is the quality of the production of thin films, which are their active elements.

1. Introduction

For the first time the description of a film thermophone as a generator of sound waves appeared in 1999 [1]. Since then, interest in such sources of sound has not died out. This can be judged, for example, according to the list of references given in [2]. The amplitude-frequency characteristic (AFC) of the film thermophone is uniform over a wide frequency range up to 150 kHz. Such thermophones are the only acoustics generators of sound waves, corresponding to the definition of a piston sound source.

2. Research object

The layered structure of the film thermophone is clearly visible in the cross-sectional image shown in figure 1. This image was obtained by electron microscopy in [3]. Each of the layers has its own functional load. A thin metal film, having a thickness of the order of 10 μm, serves as its active element (AE), the flow of an alternating electric current through which generates sound waves.

In turn, the AE is deposited on the surface of a heat-insulating layer – a substrate having a thickness of 47 μm. The overall mechanical strength and heat dissipation of the stationary heat flux from the AE is provided by a carrier plate having a thickness of 0.6 mm. To produce AE, pure metals, various metal alloys, layers of carbon nanotubes, etc. are used. The thickness h_2 of the AE is usually in the range from 20 nm to 20 μm. To select the thickness of the AE, one can use condition

$$h_2 \le 0.1\delta_2' = 0.1\sqrt{2a_1/\omega}\,, \tag{1}$$

where δ_2' is the thickness of the thermal boundary layer in the material of the AE; $\omega = 2\pi f$ is the frequency; $a_2 = \chi_2/\rho_2 C_2$ is the thermal diffusivity; ρ_2 is the density and C_2 is the heat capacity of the AE substance. When condition (1) is satisfied, the amplitude of the variable temperature at any point in the volume of the AE synchronously follows the changes in the amplitude of the alternating current flowing through the AE. At high frequencies, when condition (1) is not satisfied, the acoustic efficiency of the thermophone is reduced.

IOP Conf. Series: Journal of Physics: Conf. Series **857** (2017) 012051 doi:10.1088/1742-6596/857/1/012051

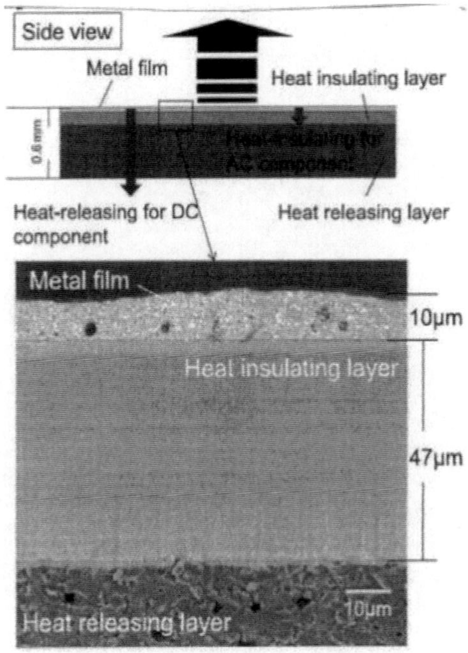

Figure 1. Cross-section of the film thermophone.

As a substrate, it is proposed to use a layer of porous silicon. The thicknesses of the substrate h_3 are in the range from 40 to 100 μm. To ensure reliable thermal insulation of the AE, the thickness of the substrate is chosen from the condition

$$h_3 \ge \lambda_{T3} = 2\pi\delta_3' = 2\pi\sqrt{2a_3/\omega}, \tag{2}$$

where λ_{T3} is the length of the thermal wave and $a_3 = \chi_3/\rho_3 C_3$ is the thermal diffusivity of the substrate substance. Condition (2) is usually not performed at low frequencies, which leads to a decrease in the levels of sound emitted by the thermophone.

3. Thermophone sound radiation

As shown in [4–7], sound emission in a thermophone is provided by two surfaces parallel to each other. The first radiating surface, located at a distance $x = \lambda_{T1}$, where λ_{T1} is the length of the thermal wave in the air, from the AE surface and generates a traveling acoustic wave p_1 due to the thermoacoustic effect. In addition, the substance of a solid body, on whose surface the AE is applied, is affected by a variable heat flux. This leads to the appearance of mechanical vibrations of the free surface of the body due to the thermal expansion of the substance and, as a consequence, to the radiation of the traveling sound wave p_3, from the surface of the AE ($x = 0$). The central axes of the combined sound sources coincide, the total amplitude of the sound pressure at the point x on the central axis of the thermophone is given by

$$p(x) = p_1(x) + p_3(x) = 2U_{01}\rho_1 c_1\left|\sin\left(\frac{k\alpha_1}{2}\right)\right| + 2U_{03}\rho_1 c_1\left|\sin\left(\frac{k\alpha_3}{2}\right)\right|, \tag{3}$$

where $\alpha_1 = \sqrt{(x-\lambda_{T1})^2 + r_0^2} - x$; $\alpha_3 = \sqrt{x^2 + r_0^2} - x$; r_0 is the equivalent radius of the radiating surface; ρ_1 is the density; c_1 is the speed of sound in the air and $k = \omega/c_1$ is the wave number.

The formulas for calculating the amplitudes of the vibrational velocities entering into expression (3) have the form, here, a_1 is the thermal diffusivity of the gas; $U_{01} = \sqrt{a_1\,\omega}\beta_{V1}T_m'$; $U_{03} = \sqrt{a_3\,\omega}\beta_{V3}T_m'$, where a_1 is the coefficient of thermal volumetric expansion of the gas; T_m' is the amplitude of the variable temperature of the AE; $\beta_{V3} = 3\alpha_{L3}$; α_{L3} is the coefficient of thermal linear expansion of the substrate substance. To calculate the amplitude of the variable temperature of the AE surface, one can use the following expression [6]

$$T'_m = \frac{q_e}{\sqrt{\omega\left(K_{H1} + K_{H3}\right)^2 + \left(\omega\varepsilon\right)^2}}, \tag{4}$$

where $q_e = p_e/S$ is the specific peak power of heat release AE; $p_e = R_e I_m^2$ – power consumed by the thermophone; R_e – total electrical resistance of the AE; I_m is the amplitude of the alternating current; S is the surface area of the AE; $K_{H1} = \sqrt{\chi_1 \rho_1 C_{p1}}$; $K_{H3} = \sqrt{\chi_3 \rho_3 C_{p3}}$; ε is the heat capacity per unit area of the AE of thickness h.

The main difficulty that arises in the calculation of thermophones is the lack of data on the physical parameters of the AE substance and the substrate. Table values of these parameters, available in the literature, are not suitable for this purpose, since they change in thin layers due to dimensional effects. We show that the physical system studied by us makes it possible to obtain experimentally the values of the parameters necessary for carrying out the calculations. To do this, we turn to expression (4). Film thermophones operate in a wide frequency band, and this allows us to distinguish between low-frequency and high-frequency excitation modes of thermophones. At low frequencies, the inequality $\omega(K_{H1} + K_{H3})^2 >> (\omega\varepsilon)^2$ is satisfied. Formula (4) is simplified and takes the form

$$T'_m = q_e \big/ \sqrt{\omega}\left(K_{H1} + K_{H3}\right). \tag{5}$$

At high frequencies, the inequality $\omega(K_{H1} + K_{H3})^2 << (\omega\varepsilon)^2$ is fulfilled and, consequently, we have

$$T'_m = q_e / \omega\varepsilon. \tag{6}$$

4. Conclusion

If formulas (5) and (6) are used to calculate the sound pressure (3), then we obtain the linear dependences $p(p_e)$. The magnitude of the sound pressure amplitude can be measured quite accurately with modern microphones. This allows using the simplest calculations to obtain the parameter $(K_{H1} + K_{H3})$ at low frequencies, and ε at high frequencies. This method was used by us in [5, 7] to determine the value of the parameter K_{H3} of laminated plastics. Given the unique technical characteristics of film thermophones, it can be argued that they can be used as sound sources, both for conducting scientific research, and for creating technical devices for various purposes. Reported features of the work of thermophones can be used to create acoustic methods for measuring certain thermophysical parameters of a substance.

References

[1] Shinoda H, Nakajima T, Ueno K and Koshida N 1999 *Nature* **400** 853–5
[2] Asadzadeh S S, Moosavi A, Huynh C and Saleki O 2015 *Journal of Applied Physics* **117** 095101
[3] Nishioka T, Teshima Y, Mano T, Sakai K, Asada T, Matsukawa M, Ota T and Hiryu S 2013 *Proceedings of Symposium on Ultrasonic Electronics* **34** 45–6
[4] Legusha F F and Neveselova K V 2014 *Marine Intelligent Technologies (MIT)* **3** 71–7
[5] Legusha F F and Neveselova K V 2015 *Marine Intelligent Technologies (MIT)* **4** 60–5
[6] Neveselova K V 2014 *Marine Intelligent Technologies (MIT)* **4** 110–5
[7] Vasiliev B P, Legusha F F, Razrezova K V and Chizhov G V 2016 *Marine Intelligent Technologies (MIT)* **4** 118–23

Anti-friction properties of thin layered nanosized protective films "Epilam": synergy effect

A S Vokhidov

Avtostankoprom Ltd., 190020, Saint Petersburg, Russia

E-mail: rvokhidov@list.ru

Abstract. In the general concept of temperature-resistant antifriction coatings, greate attention is paid to the growing use of multifunctional films. Epilams are polymer compositions, studying organic chemistry, namely to compositions (liquids), which, when applied to a solid surface, form on it nanosize multifunctional molecular film of a certain way of oriented molecules of surface-active agents (surfactants, SAA), which significantly changes the surface energy of the base (substrate).

1. Introduction

In recent times new materials: alloys, composite metals, ceramics, textolite and other that require a thin protective coating. The use of modern film-forming lubricating composition for producing a thin layer (nanosize thickness up to 100 nm) of a protective coating that preserves the geometrical, and electrical parameters of the surface, while reducing energy surface, friction [1], adhesion (sticking of the particles) in the absence of corrosion activity, aging, wear can be characterized by the synergy effect of resource that is imperative for the preservation of technical characteristics, durability and reliability.

In addition to handling mechanisms and bevel pairs, where there is friction, for example the bed slideways of machine tools with CNC, conical pairs of gears, the compositions can also be used as an additive in internal combustion engines, transmission, transmission (including robotic). Methods of thermodispersion or added in the form of additives, which are caused by intermolecular forces (van der Waals) between the fluids, oils, greases, provide durability of the applied coating, but in chemical connections (chemisorption) the molecule triboactive and irreversible adsorbate form chemical connections.

After activation of the polymer composition (production process) the fluorine-containing macromolecule acquires properties entirely different from properties of those composite units from which they are built. Under the influence of certain rays of the fluorine molecules are activated and begin to "ordered" vertically-directed, i.e. the hydrophilic ends. Charged molecules can reduce surface energy of the material to be adsorbed, hemosorbents hydrophilic tail (polar layer), while the hydrophobic tail remains on the surface (non-polar layer). There are different activation methods, for example ultrasonic. In the works of V E Panin and group of authors the use of ultrasound accelerates diffusion and adhesion processes in the coating with different materials [2]. In this source indicates that the analysis of the influence of ultrasound on the structure is determined by the penetration of ultrasound (and charges) in a layer of liquid phase coating, thus changing the structure properties of the alloy (material). According to the authors [2, 3] the mechanism of activation structures of the composition under the influence of ultrasonic vibrations is a process that affects the structure of the composition. Vibrations of the ULTRASONIC instrument are excited by the transformation of mechanical fluctuations of current or voltage. In practice, the role of the system of airless mixing with the use of Doctorow (ejectors), based on the Venturi effect, carry out internal combustion engines, gearboxes and other automatic stirring mechanisms.

2. Research of a nanosized monomolecular surface-active agents films

The process of obtaining a film characterized by monomolecular layer. Are determining parameters such as substrate temperature, deposition rate, indicators of charge of the molecules, the degree of activation of the surfactant, pressure and humidity of the working atmosphere, the porosity and surface finish.

The study profilograms (figure 1) can be carried out with the help of modern devices nanoscopy, study on the level of neopane (adhesion) – an instrument adhesimeter where to achieve maximum grip of the focusing device with the surface of the test specimens, they are left for 7 days (the standard methods). After 7 days it stops around the perimeter of its circumference is cut, and then the pull test. For example, the results of tests on the bond strength of the focusing parts adhesimet with plates covered in the nanotechnology center of the Institute Gosniti (Moscow) composition "Epilam SFC-05" received following data for adhesion: 0.55 MPa, which confirmed the very low adhesion of the fluoropolymer, which is 6 times less than on polyethylene, 11 times – than glass, in 13 than steel and 15 times than ceramics.

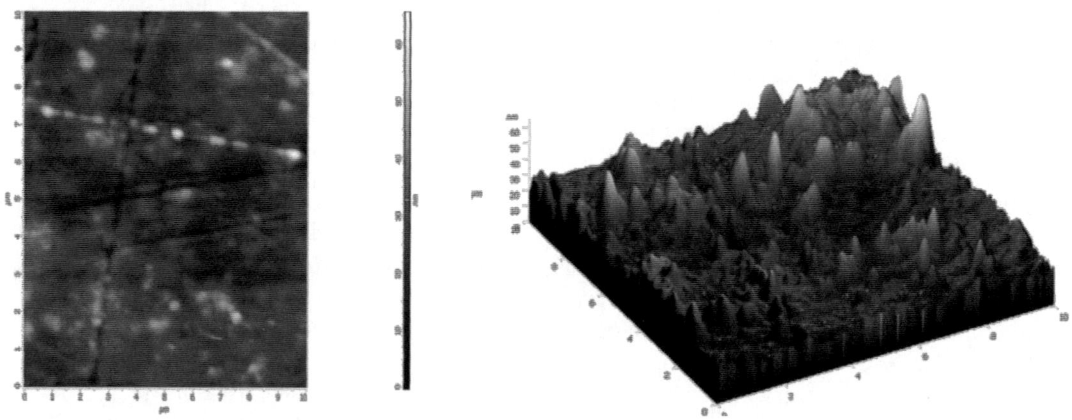

Figure 1. The profilogram the surface of a fiberglass dielectric material after covering by Epilam SFC-05 nanofilms.

The impact of the strong links between metal ions and surfactants the formed complexes increase the polarization of the hydrophilic and hydrophobic ends. The nature of the coverage on average is consistent with theoretical notions about surfaces Langmuir–Blogett. The height of the "pile" (palisade Langmuir) of 1–3 nm, the coating on the plot is uniform, without visible spikes. The surface roughness of the coating film: the RMS is 2 nm, the arithmetic mean is 1.6 nm (figure 2).

Study of parameters of the microrelief of the substrate surface can be a significant source of information on the modifying effects of a film-forming composition prior to application and after the application of a uniform nanosized layer coatings [4]. A plurality of functional impact (antifrictional, antiadhesives, bacteriostatic) in total creates a synergistic result.

Theoretical and practically synergy effect is achieved through colloidal processes, where the properties of colloidal systems depend not only on particle size but by the presence of interfaces with significant free surface energy. In the source, it was concluded that in addition to the kinetic stability depending on the size of the particles has a resistance of particles to mutual adhesion (aggregate stability). Stabilizing the coefficient of surface energy is achieved not only smoothing of the surface roughness in the range of 10.6 nm (exposure contributes to the conductivity of nanofilms), but also to improve the water resistance of semiconductors, etc., providing a clean surface. Moreover, the surface layer can withstand temperatures ranging from –160 to +420 °C, providing thermal protection of the surface layer depending on the reduction or increase of temperature.

Ostwald maturation neutralized due to the acceleration of coalescence and reduction of surface energy, which directly affects the angle of wettability of the deposited and developing droplets of moisture and corrosive liquids on the surface. High adsorption ability at the interfaces liquid–solid and determined by the rate of adsorption of nano-films and can be described by the equation [5]:

$$\alpha = 0.856 \cdot 10^{-11} + 8.900 \cdot 10^{-7}\ °C, \quad (1)$$

which corresponds to linear dependence in the region of Henry isotherms, i.e. the initial stage, in which is formed the first adsorption layer.

The diffusion coefficient in the aggressive air environment has higher value and can reach 10^{-6} m^2/s, which means that the movement of fine particles of different sizes when applied by aerosol (selective spraying) will be more intense according to the criterion of relativity.

Figure 2. The profilogram the surface after covering by Epilam SFC-20 nanofilms and chemical thermoactivation.

3. Conclusions

The process of obtaining a uniform layer due to the fact that macromolecules of this type of surfactant, polar, ordered (the activation process) hydrophilic/ hydrophobic ends and have a magnetic dipole moment. The polar portion of the macromolecule, getting into the field of the part surface, oriented in the directions of the force fields details [6], its polar part of the adsorption bonds is fixed on details, and non-polar part remains outside over the item. Thanks to the strong links between metal ions and surfactants the formed complexes increase the polarization of the hydrophilic and hydrophobic ends, which leads to a duration of thin coatings that reduce friction coefficient and low adhesion surface.

Coverage (including epilam) insulate the surface of the material of the part, prevent the introduction of superficial layers of the working surfaces of agents of the aggressive environment and thus protect the material from wear, corrosion and fracture.

References

[1] Ahmatov A S 1963 *Molecular physics of boundary friction* 314–7
[2] Panin V E, Klimenov V A et al. 1993 *The new materials and technologies* 114–8
[3] Dunaev A V 2013 *Non-traditional tribology. Modification of friction surfaces* 82–7
[4] Vokhidov A S 2014 *Publications ITMO Univ.* 165–72
[5] Ryabinin N A 2002 *Publications of 4 International Conference* 47–50
[6] Kanarev F M 2007 *The beginning of physical chemistry of microworld* **1** 259–76

IOP Conf. Series: Journal of Physics: Conf. Series **857** (2017) 012053 doi:10.1088/1742-6596/857/1/012053

The regularities of phase formation in $Fe_3Si(111)/Si(111)$ structure at vacuum annealing

M N Volochaev[1,2], **I A Tarasov**[1,2], **Yu Yu Loginov**[1], **A G Cherkov**[3] **and I V Kovalev**[1]

[1] Siberian State Aerospace University n. a. academician M. F. Reshetnev, 660014, Krasnoyarsk, Russia
[2] Institute of Physics n. a. L. V. Kirensky FRC KSC Siberian Branch, Russian Academy of Sciences, 660036, Krasnoyarsk, Russia
[3] Novosibirsk State University, 630090, Novosibirsk, Russia

E-mail: volochaev91@mail.ru

Abstract. The regularities of phase formation and thermal stability in $Fe_3Si(111)/Si(111)$ structure at stepped vacuum annealing (350, 450 and 550 °C) were investigated. The layer of 32 nm Fe_3Si was deposited onto Si(111) substrate by molecular beam epitaxy at 260 °C. Transmission electron microscopy (TEM) measurements demonstrated that the film thickness increases by ~19 % at 350 °C annealing without changing the phase composition. The polycrystalline ε-FeSi sublayer was formed on the interface at 450 °C annealing. Further annealing at 550 °C led to the ~ 80 nm polycrystalline film formation containing the crystallites of ε-FeSi, Fe_5Si_3, and β-$FeSi_2$ phases.

1. Introduction

Recently, one of the perspective directions in the field of electronics is the study of ferromagnetic/semiconductor heterostructures and the creation of integrated circuits based on them [1, 2]. The basic characteristics of the ferromagnetic layer in ferromagnetic/semiconductor heterostructures are high Currie temperature and high spin polarization level. The iron silicide Fe_3Si with DO_3 structure (Currie temperature is 840 K [3], spin polarization is about 40 % [4]) is an attractive ferromagnetic material for spintronic devices [5, 6]. There is a series of papers [7–10] discussing the growth conditions optimization, studying the structure and physical properties of epitaxial Fe_3Si films and Fe_3Si-based multilayer structures grown on various substrates (Si, Ge, GaAs).

However, the problem of phase formation and thermal stability in the $Fe_3Si(111)/Si(111)$ structure shown in the present study was not previously discussed.

2. Experiments

The Fe_3Si thin films were deposited by molecular beam epitaxy technique in ultrahigh vacuum ($1.3\cdot10^{-8}$ Pa) onto atomically pure *n*-type Si(111) substrate. The substrate preparation, growth and growth control technology is identical for the films studied in [6, 10], except that the growth temperature was raised to 260 °C. The thickness of as-grown films was 32 nm according to the ellipsometry data. When deposition was finished $Fe_3Si(111)/Si(111)$ structures were annealed at different temperatures: 350, 450 and 550 °C, respectively for 1 h at 10^{-7} Pa.

IOP Conf. Series: Journal of Physics: Conf. Series **857** (2017) 012053 doi:10.1088/1742-6596/857/1/012053

As-grown and annealed structures were characterized by transmission electron microscopy (Hitachi HT-7700 microscope operating at 100 kV) at "plan-view" and "cross-section" mode. The "plan-view" specimens were prepared by mechanical lapping and polishing, followed by Ar+ ion milling according to standard TEM techniques. The "cross-section" samples were prepared by Hitachi FB-2100 single-beam focused ion beam (FIB) system using FIB technique described in [10].

3. Results and discussion

Figure 1 shows the TEM study results of as-grown $Fe_3Si(111)/Si(111)$ specimen.

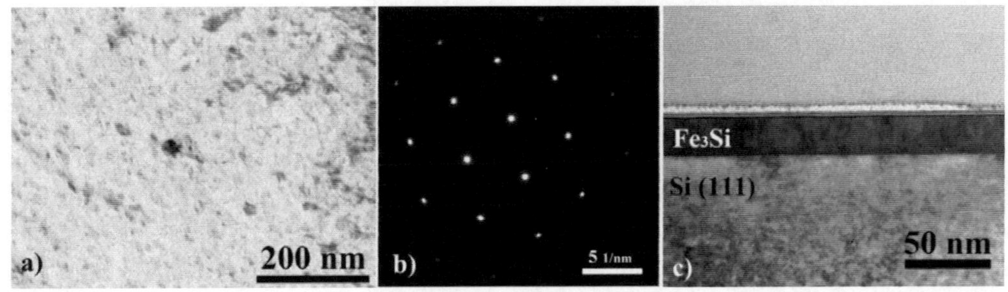

Figure 1. TEM images of as-grown $Fe_3Si(111)/Si(111)$ specimen: **(a)** – "plan-view" image; **(b)** – the SAED pattern corresponding to image (a); **(c)** – "cross-section" image.

The silicide film thickness is 32 nm. The "cross-section" image (figure 1(c)) demonstrates linear and sharp interface between Fe_3Si and Si(111) without a reaction sublayer, which indicates a high quality of the grown silicide film. The selected area electron diffraction (SAED) pattern corresponds to the Fe_3Si phase with the DO_3 structure.

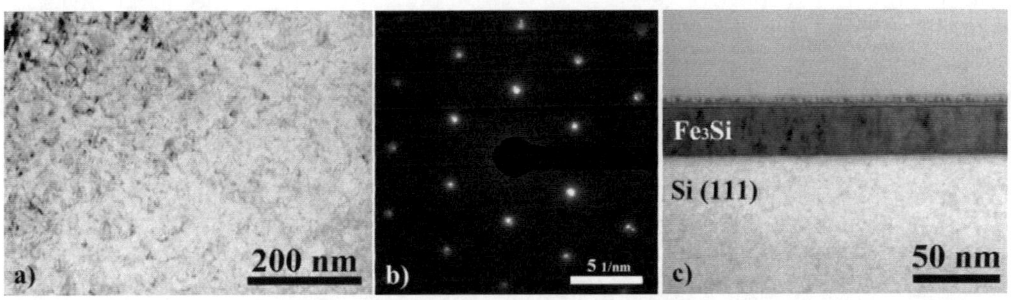

Figure 2. TEM images of 350 °C annealed $Fe_3Si(111)/Si(111)$ specimen: **(a)** – "plan-view" image; **(b)** – the SAED pattern corresponding to image (a); **(c)** – "cross-section" image.

The structure and phase composition of $Fe_3Si(111)/Si(111)$ specimens are not changed at 350 °C annealing (figure 2).

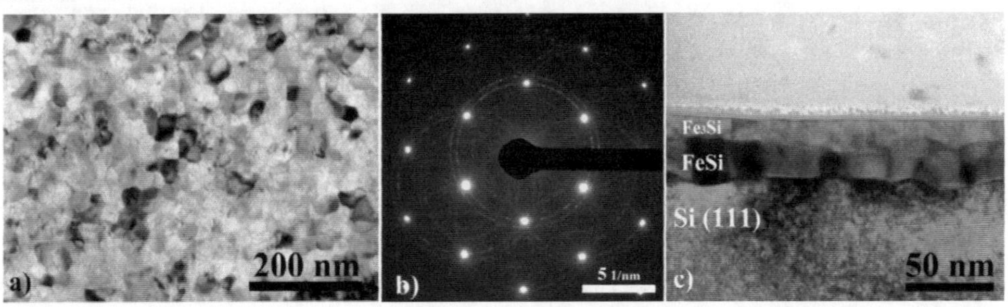

Figure 3. TEM images of 450 °C annealed $Fe_3Si(111)/Si(111)$ specimen: **(a)** – "plan-view" image; **(b)** – the SAED pattern corresponding to image (a); **(c)** – "cross-section" image.

The interface between the film and the substrate is still linear and sharp (figure 2(c)). However, the film thickness increased by 6 nm. This fact indicates the diffusion processes occurring on the interface. Figure 3 demonstrates the formation of polycrystalline sublayer at the $Fe_3Si(111)/Si(111)$ interface after 450 °C annealing. The SAED pattern (figure 3(b)) corresponds to single-crystal Fe_3Si and polycrystalline ε-FeSi phases. The ε-FeSi layer thickness varies from 10 to 45 nm, the average thickness is about 30 nm with linear crystallite sizes of 20 to 50 nm (figure 3(a), (c)).

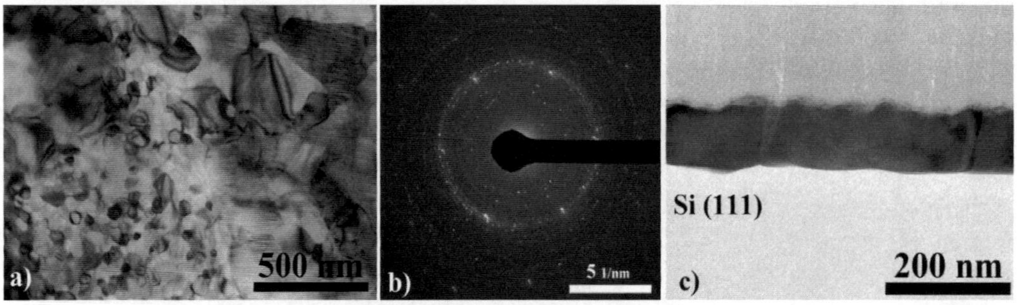

Figure 4. TEM images of 550 °C annealed $Fe_3Si(111)/Si(111)$ specimen: **(a)** – "plan-view" image; **(b)** – the SAED pattern corresponding to image (a); **(c)** – "cross-section" image.

At 550 °C annealing a polycrystalline film of iron silicides was formed (figure 4). The TEM studies results showed the presence of ε-FeSi phase "islands" with crystallite sizes from 30 to 100 nm surrounded by a matrix composed of the Fe_5Si_3 and β-$FeSi_2$ crystallites with crystallite sizes from 100 to 500 nm (figure 4(a), (b)). The average thickness of the film increased to 80 nm.

4. Conclusions

The regularities of phase formation in $Fe_3Si(111)/Si(111)$ structure at stepped vacuum annealing (350, 450 and 550 °C) were investigated. During the annealing at 350 °C the film thickness increased by ~19 % due to diffusion. The structure and phase composition not changed. The polycrystalline ε-FeSi phase sublayer with the thickness of 10 to 45 nm, and crystallite sizes from 20 to 50 nm was formed on $Fe_3Si(111)/Si(111)$ interface at 450 °C annealing. After annealing at 550 °C the polycrystalline film composed of β-$FeSi_2$ and Fe_5Si_3 phases crystallites with crystallite sizes from 100 to 500 nm and ε-FeSi phase "islands" with crystallite sizes from 30 to 100 nm was formed. The film thickness increased to ~80 nm.

References

[1] Wolf S A, Awschalom D D, Buhrman R A et al. 2001 *Science* **294**(5546) 1488–95
[2] Sugahara S and Tanaka M 2004 *Appl. Phys. Lett.* **84**(13) 2307–9
[3] Hong M, Chen H S, Kwo J et al. 1991 *J. Cryst. Growth* **111** 984–8
[4] Odkhuu D, Yun W S and Hong S C 2011 *Thin Solid Films* **519** 8218–22
[5] Mantovan R, Georgieva M, Fanciulli M et al. 2008 *Phys. St. Solidi (A)* **205**(8) 1753–7
[6] Yakovlev I A, Varnakov S N, Belyaev B A et al. 2014 *JETP Lett.* **99**(9) 527–30
[7] Hamaya K, Ueda K, Kishi Y et al. 2008 *Appl. Phys. Lett.* **93** 132117
[8] Ueda K, Sadoh T, Ando Yu et al. 2008 *Thin Solid Films* **517** 422–4
[9] Hsu Y L, Lee Y J, Chang Y H et al. 2007 *J. Cryst. Growth.* **301-302** 588–91
[10] Volochaev M N and Loginov Yu Yu 2016 *Vestnik SibSAU* **17**(3) 792–6

IOP Conf. Series: Journal of Physics: Conf. Series **857** (2017) 012054 doi:10.1088/1742-6596/857/1/012054

Control of optical properties of metal-dielectric planar plasmonic nanostructures by adjusting their architecture in the case of TiAlN/Ag system

D L Wainstein[1], V O Vakhrushev[1,2] and A I Kovalev[1]

[1] Surface Phenomena Researches Group, I. P. Bardin Central Research Institute for Ferrous Metallurgy, 23/9, bdg. 2 Radio str., Moscow, 105005, Russia
[2] Moscow Technological University (MIREA), 78 Vernadsky Avenue, Moscow, 119454, Russia

E-mail: gareq1211@gmail.com

Abstract. The multilayer Ag/($Ti_{34}Al_{66}$)N metal-insulator-metal (MIM) heterostructures with different thicknesses of individual layers varied from several to several hundred nanometers were fabricated by DC-magnetron sputtering on the surfaces of Si single crystal wafers. The coatings structure was determined by STEM. The phase composition and crystallography of individual layers were studied by X-ray diffraction. The reflection indexes were measured in the photons energies range from 1 to 5 eV, or from 1240 to 248 nm. The spectroscopy of plasmon losses and plasmon microscopy allowed us to measure the plasmons losses characteristic energies and their surface distribution. The energies of plasmons peaks and their locations are strongly depending on Ag layers thickness in the MIM nanocomposite. The surface plasmon with energy about 4 eV was observed in the middle of 20 nm Ag layer. The plasmons were localized at the metal/dielectric interface for Ag layers 5 nm and less. The reflectance spectral profiles edges positions at long and short waves are correlated with plasmons energies and features of their spatial distribution. The MIMs based on the TiAlN/Ag can find applications as optical filters, photovoltaic energy conversion devices, etc.

1. Introduction

Modern antireflection optical coatings are designed as combinations of dielectrics and metal thin layers. Metal layers are acting through the bulk reflectance and absorption properties of the material. Dielectric thin films are using optical interference. The reflectance and transmittance in these coatings are achieved due to pumping of plasmon oscillations in the multilayer coating by impinging light.

Excitation of plasmons at optical frequencies and coupling them back into freely propagating light are processes of great interest for the manipulation with transmitted and reflected light. Plasmonics involving manipulation of bulk and surface plasmon resonance in a metal allows changing the wavelength and intensity of electromagnetic radiation at its absorption and reemission. These processes are occurring in the metal-dielectric metamaterials with dimensions of structural elements comparable with electromagnetic wavelength. Multilayer designs can deliver higher performance and enable operation over a wider range of wavelengths filtering and focusing electromagnetic waves. Physical properties and electronic structure of a metal and a dielectric in the nanocomposite are controlling energy and intensity of free electrons resonance as well as electromagnetic field localization on interfaces.

IOP Conf. Series: Journal of Physics: Conf. Series **857** (2017) 012054 doi:10.1088/1742-6596/857/1/012054

Plasmonic structures can manipulate light at a subwavelength scale because large spatial frequency components associated with metallic structures with subwavelength-sharp features [1–3]. The attention to plasmon metamaterials design paid in recent years is stipulated by ideas to apply their exciting optical functions such as field localization, refractive index engineering, phase and amplitude control, and many more [4].

2. Materials and methods

The multilayer $(Ti_{34}Al_{66})N/Ag$ heterostructures with different thicknesses of individual layers (see table 1) and two reference 580 and 1000 nm TiAlN coatings were fabricated by DC-magnetron sputtering. The thicknesses of individual layers were varied from several to several hundred nanometers. All samples were covered with TiAlN layer to protect silver layer from oxidation.

Table 1. Multilayer coatings composition parameters.

Sample	TiAlN layer thickness, nm	Ag layer thickness, nm	TiAlN/Ag bilayers	Total thickness, nm
1	40	40	6	480
2	40	20	8	480
3	20	20	12	480
4	20	10	16	480
5	10	5	33	495

Light reflection measurements (figure 1(a)) were acquired in the total reflection mode (incidence angle 5°) with a Shimadzu UV-VIS-NIR Spectrophotometer SolidSpec-3700. The reference signal was taken from an aluminum coated mirror. One can see that each multilayer coating has its own band of maximal reflection suppression.

To restore the optical constants of MIMs we used the Drude-Lorentz (DL) model and Transfer Matrix Method (TMM) to fit the light reflection experimental data. Such fitting was based on Levenberg-Marquardt algorithm.

According to DL model the complex dielectric function $\varepsilon(\omega) = \varepsilon_1 + i\varepsilon_2$ could be written as:

$$\varepsilon = \varepsilon_\infty - \omega_p^2 \sum_j \frac{f_j}{\omega_j^2 - \omega^2 + i\omega G_j}; \ \omega_0 = 0, \tag{1}$$

where ω_p is a bulk plasmon frequency, eV; ω – primary light frequency, eV; ω_j – resonant frequency, eV; G_j – dumping coefficient (attenuation time), eV.

The complex refractive index $N = n + ik$ of planar MIM is an important attribute of absorption of light propagating through the film.

Refraction (n) and extinction (k) coefficients were calculated as:

$$n = \sqrt{\frac{\sqrt{\varepsilon_1^2 + \varepsilon_2^2} + \varepsilon_1}{2}}; \quad k = \sqrt{\frac{\sqrt{\varepsilon_1^2 + \varepsilon_2^2} - \varepsilon_1}{2}}, \tag{2}$$

The surface reflectivity coefficient was calculated as:

$$R(T_e, T_p, \omega) = ((n-1)^2 + k^2)/((n+1)^2 + k^2). \tag{3}$$

3. Results and discussion

The reflective index $n(\omega)$ can be presented as a function of angular plasmon frequency ω_p, total scattering rates v_m, angular light frequency ω and extinction coefficient k as:

$$n^2 = \frac{1}{2}\left[\left(1 - \frac{\omega_p^2}{\omega^2 + v_m^2}\right)^2 + \left(\frac{v_m}{\omega}\frac{\omega_p^2}{\omega^2 + v_m^2}\right)^2\right]^{\frac{1}{2}} + \frac{1}{2}\left(1 - \frac{\omega_p^2}{\omega^2 + v_m^2}\right), \tag{4}$$

where bulk plasmon frequency ω_p is defined by the material

$$\omega_p = \sqrt{Ne^2/\varepsilon m_{eff}}\,, \tag{5}$$

N, ε, m_{eff} are the density, dielectric permittivity and the effective mass of the conduction electrons, respectively. Influence of the ionic lattice on the conduction electrons can be accounted using the effective electronic mass m_{eff} instead of m_e. According to DL model, the refractive index depends on plasmon frequency.

Features of plasmons energy and spatial distribution in thin Ag layers as well as plasmon-polariton interactions on metal-dielectric interfaces in TiAlN/Ag MIM coatings under study were reported earlier in [5]. There was found that plasmon frequencies are rising while their intensities are attenuating with Ag layer tightening. It is paradoxical because decreasing of plasmons intensities due to density of conduction electrons reduction shall be accompanied by plasmon frequency (i.e. energy) decreasing according to equation (4). The contradiction can be resolved accounting features of electronic structure of silver as a *d*-element.

The calculated *d*-screened surface plasmon energy in Ag layer depends on the bulk plasmon energy ω_p in presence of *d*-electrons characterized by the frequency dependent dielectric function $\varepsilon_d(\omega)$ as [6]:

$$\omega_s = \frac{\omega_p}{\sqrt{1+\varepsilon_d(\omega_s)}} = 3.62\ \text{eV}. \tag{6}$$

Measured plasmon energies were 3.55, 3.45, 3.39 and 3.20 eV for 5, 10, 20 and 40 nm Ag layers, correspondingly [7]. The plasmon energies are bigger in thinner nanolayers because plasmons in thin Ag layer with *s*- and *d*-bands crossing the Fermi level are strongly damped by interband transitions. Increased binding energy of the plasmon with ion core (localized 5*s* electrons) provides its bigger excitation energy as we see with Ag layer thinning. The similar growth of the surface plasmon energy at nanostructuring was demonstrated in [8, 9] for silver clusters.

Figure 1(a) demonstrates complex influence of Ag/TiAlN coatings architecture on their reflectance. So, the coating with 40 nm layers of Ag and TiAlN (ML-40/40×06) possesses reflectance more than 30 % in whole investigated band of photon energies. Its antireflectance properties are enough poor.

Figure 1(a) demonstrates also that each multilayer coating has its own band of maximal reflection suppression in region of frequency $\omega_0 = \lambda_0/\hbar c$ corresponding to resonant adsorption by surface plasmon resonance. In the long-wave region at $\omega < \omega_0$ reflectance is growing with ω decreasing, and threshold of reflectance is shifted according to ω_0 variations. The resonance frequency of adsorption goes to longer waves at increasing number of bilayers of same thicknesses.

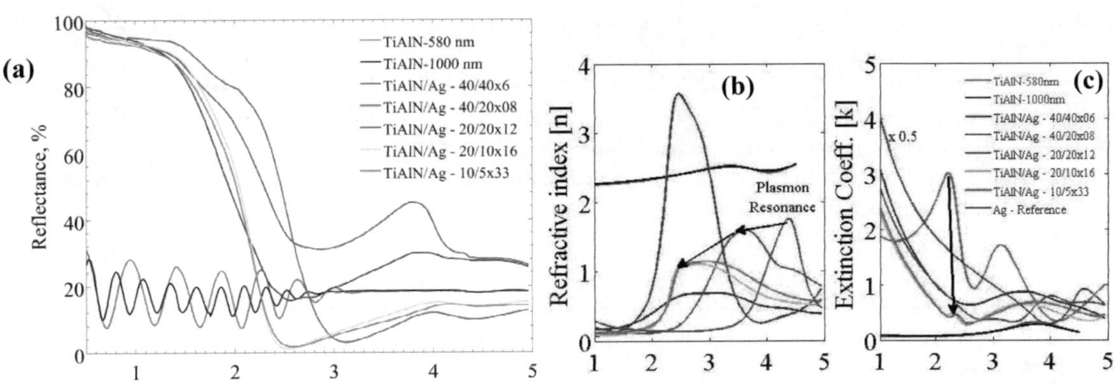

Figure 1. Reflectance spectra of single layer TiAlN (580 and 1000 nm) and TiAlN/Ag coatings **(a)**, refractive index *n* **(b)** and extinction coefficient *k* **(c)** of the coatings and bulk Ag calculated by TMM.

IOP Conf. Series: Journal of Physics: Conf. Series **857** (2017) 012054 doi:10.1088/1742-6596/857/1/012054

It is clearly seen from figure 2(a) that there are at least four characteristic excitations playing an important role in MIMs optical properties formation. The first group is responsible for absorption of electromagnetic waves by plasmons of metallic layers. These plasmons were detected by REELS and attributed as surface plasmons [5]. The energy of this group of excitations increases with individual layers thinning. It means that nanostructuring in MIM should cause the shift of light absorption edge toward IR region. But as we can see in figures 1(b), (c) there are some more features at the *n* and *k* dependencies. There is a peak around 2.5 eV for finest architectures (10/5 and 20/10) of the nanocomposite that correspond not only to plasmons excitations but also its further propagation of the plasmon into volume of TiAlN thin layer.

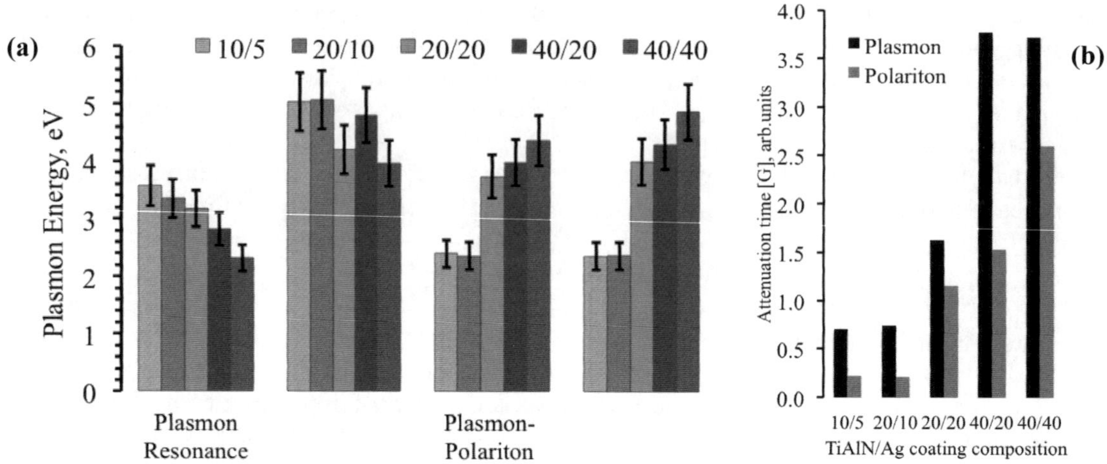

Figure 2. Plasmon resonance in TiAlN/Ag multilayer coatings: resonant frequency **(a)** and damping coefficient (attenuation time) **(b)** for plasmon-polariton excitations.

So, the second group of excitations (figure 2(b)) reflects the plasmon-polariton penetration (PPP) from metallic to insulator layer. Frequencies of plasmons that resonate with Ti 3*d* or Ag 5*s* electrons in TiAlN depend on the thicknesses of either insulator or metallic individual layers. Such kind of resonance gives the two absorption edges (e.g. Lorentz terms) with close resonant frequencies ω_i but with significantly different attenuation times G_i. According to our calculations both plasmon and polariton attenuation time (figure 2(b)) is 4–5 times lower for 10/5 and 20/10 samples than for 40/40 one. As attenuation time of excited polariton decreases, the depth of plasmon penetration into nitride layer is decreasing also. The reflection of light in this region (see figure 2(b)) of wavelengths sharply goes down.

So we can conclude that while characteristic size of TiAlN/Ag MIM nanolayers are lower than 20/20 nm, the plasmon-polariton resonance happens to play the prevalent role in optical properties formation instead of plasmon excitation for relatively thick MIM structures.

The results on plasmon losses measured by electron spectroscopy techniques and restored from optical data are in the perfect agreement.

4. Conclusions

Plasmon nanocomposites of Ag/TiAlN system could be applied as broadband optical coatings with high light transmission and low reflection at selected wavelength range of electromagnetic spectrum with practically total reflection of IR radiation and relatively high reflection of UV. We found interrelations between plasmon resonance and light absorption band for MIM nanocomposites with metal layers thickness from 20 to 5 nm. The IR and UV cut-off positions are shifted through the electromagnetic spectrum depending on the plasmonic system parameters determined by thickness and number of layers in the nanolaminate. Basing on these regularities, the reflection and transmission bands of coatings can be tuned by architecture of the plasmonic system varying number and thickness of dielectric and metal layers as well as their chemical composition. The design of these optical band filters should take also into account the features of polaritonic excitations rising in dielectric layers.

Acknowledgements
The research was carried out with financial support of RSF project No. 14-12-00170.

References

[1] Barnes W L, Dereux A and Ebbesen T W 2003 *Nature* **424** 824–30
[2] Ozbay E 2006 *Science* **311** 189–93
[3] Atwater H A 2007 *Sci. Am.* **296**(4) 56–63
[4] Liu Z, Liu G, Liu X, Huang S, Wang Y, Pan P and Liu M 2015 *Nanotechnology* **26**(23) 235702
[5] Kovalev A I, Rashkovskiy A Yu, Wainstein D L, Gago R, Soldera F and Endrino J L 2016 *Curr. Appl. Phys.* **16** 459–68
[6] Pitarke J M, Silkin V M, Chulkov E V and Echenique P M 2007 *Rep. Prog. Phys.* **70** 1–87
[7] Kovalev A I, Wainstein D L, Rashkovskiy A Yu, Gago R, Soldera F, Endrino J L and Fox-Rabinovich G S 2015 *Journ. of Nanomaterials* **2015** 876247
[8] Ealet B, Elyakhloufi M H, Gillet E and Ricci M 1994 *Thin solid films* **250** 92–100
[9] Tiggesbiumker J, Koeller L, Meiweis-Broer K H and Liebsch A 1993 *Phys. Rev. A* 48 R1749

Regularities of electronic structure transformations in nanomaterials with decreasing their characteristic size

D L Wainstein and A I Kovalev

Surface Phenomena Researches Group, I. P. Bardin Central Research Institute for Ferrous Metallurgy, Radio str., 23/9, bdg. 2, 105005, Moscow, Russia

E-mail: d_wainstein@sprg.ru

Abstract. The set of nanomaterials (PbS equiaxial nanocrystals, ZnO films and whiskers, Ag films in TiAlN/Ag nanolaminate multilayer coatings) was studied in a wide range of characteristic sizes. Parameters of their electronic structure as functions of the characteristic size of the nanosystems were analyzed using X-ray photoelectron spectroscopy (XPS) and High resolution electrons energy losses spectroscopy (HREELS) methods. There were determined critical sizes manifesting transition to nanostate. Basing on dependences of donor and acceptor XPS lines size shifts, the two-stage mechanism of electronic structure transformations is proposed: the electronic structure starts to transform below D_0 due to surface/volume atoms ratio growth while D_q is associated with preferred role of quantum effects in solids.

1. Introduction

During last decades nanomaterials are attracting attention of science and industries due to their properties sharply distinguished from traditional "bulk" materials. Specific properties of such small objects are obviously connected to growing role of free surface or interfaces with decreasing their characteristic dimensions firstly due to growth of surface energy in the system and secondly through radical transformations of electronic structure and consequently properties due to quantum phenomena. The limit cases of such systems are 2D materials possessing only surface: graphene and carbon nanotubes as well as dichalcogenides of transitional metals, MoS_2 or $TiSe_2$ for example.

According to ISO [1] and IUPAC [2] definitions, the "nanomaterials" term means materials with structure elements with size 100 nm and below at least in one dimension. Such round number rightfully raises doubts in its physical signification. So this research was aimed to study systematically regularities of electronic structure transformations in various materials (metals, semiconductors) with decreasing their size.

2. Materials and methods

There were studied equiaxial PbS nanocrystals, ZnO films and whiskers, Ag layers in TiAlN/Ag coatings. Characteristic dimensions of these systems determined by SEM or STEM were in range from several hundred to units of nanometers.

The electronic structure was investigated using ESCALAB MK2 (VG) spectrometer with hemispherical energy analyzer. The samples surface was cleaned by one of two standard ways: Ar^+ ions etching at vacuum 10^{-5} Pa, accelerating voltage 8 kV, ion current 20 μA either heating in vacuum 10^{-8} Pa to 100–120 °C. The cleaning quality was estimated by Auger electron spectroscopy (AES).

Internal levels and valence band were studied by X-ray photoelectron spectroscopy (XPS) using a monochromatic Al K_α source with a spot size 1×8 mm. The energy resolution (Au 4*f* line) was 0.6 eV, energy scanning step was 0.05 eV. The samples charge was suppressed by low energy electrons beam. Binding energies of electronic levels E_b were calculated using basic XPS equation:

$$E_b = h\nu - E_{kin} - e\varphi, \quad (1)$$

where $h\nu$ = 1486.6 eV is an excitation quant energy; E_{kin} is a measured energy of electrons and $e\varphi$ is a work function assumed as a constant equal 3.81 eV for the spectrometer.

The intraband and interband transitions were studied by high resolution electrons energy losses spectroscopy (HREELS) using EMU-50 monochromatic electrons source (FWHM = 17 meV, E_0 = 9.5–13.5 eV). The analyzer was operating in high resolution mode with energy scanning step 0.025 eV. Fine structure of spectra was analyzed using UNIFIT 2007 [3] software.

3. Results and discussion

We have investigated in this work the regularities in changes of internal electronic levels positions, or electrons binding energies, densities of states (DOS) in valence bands, band gaps widths, work functions of materials as functions of characteristic size: diameters of crystals and rods, thickness of films. Earlier we reported in [4] and [5] about studies of PbS nanocrystals of various size. The size dependent transformations in electronic structure of ZnO thin films and whiskers were presented in [6]. The paper [8] was dedicated to tracing fine electronic structure changes in thin metal Ag layers in multilayer nanolaminate TiAlN/Ag coatings.

XPS lines Zn $2p_{3/2,\,1/2}$ and O 1*s* recorded from ZnO nanofilms with different thicknesses are presented in figures 1(a), (b). One can see that donor and acceptor lines are moving to higher binding energies due to size shift. These movements have the same values in accuracy limits of the experiment that is about ± 0.05 eV.

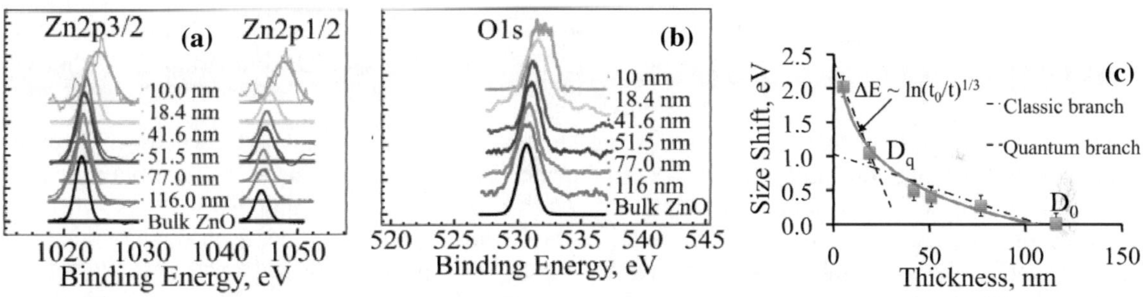

Figure 1. ZnO nanofilms. XPS lines Zn $2p_{3/2}$–$2p_{1/2}$ **(a)**, O 1*s* **(b)**, size shift vs. film thickness **(c)**.

The dependence of size shift vs. a nanostructure size is described well by empiric proportion:

$$\Delta E(D) \sim \ln((D_0/D)^{n/3}, \quad (2)$$

where $\Delta E(D)$ is size shift of the electronic level binding energy, eV; D is a characteristic size of the nanosystem, nm; n is number of nanodimensions: 3 for equiaxial particles, 2 for nanorods, 1 for thin films; approximation parameter D_0 corresponds to minimal size of nanocomponents when binding energies of XPS lines are located at their standard values. In relatively big nanosystems the size shifts are driven by classic mechanism due to growth of surface atoms in the system. The linear "classic" branch on figure 1(c) corresponds to this phenomenon. While characteristic size is small, we observe accelerated growth of such shifts due to quantum size effects. The second tangent line (a quantum branch) corresponds to this part of the plot. The crossing point of these right lines corresponds to the characteristic size D_q manifesting transition to nanoregion. D_0 and D_q have the following values for substances under investigation: PbS nanocrystals on neutral Si – 430 and 155 nm; PbS nanocrystals on *n*-doped Si – 270 and 155 nm; ZnO whiskers – 106 and 22 nm; ZnO nanofilms – 106 and 21 nm; Ag nanolayers in TiAlN/Ag – 200 and 24 nm. It is interesting that quantum threshold in PbS is significantly bigger than IUPAC limit.

Regularities described by proportion (2) are fulfilled also for outer fully occupied levels but sifts of these levels are significantly bigger than for inner ones. Figure 2 demonstrates size shifts dependencies for the set of donor and acceptor levels in PbS nanocrystals. The analogous results on electrons levels shifts in Pd nanoparticles with various size were published in [7].

Generalizing obtained experimental data on measured energies of various levels in the nanomaterial with characteristic size $D < D_q$, we can plot the differences between energies of various levels in this object comparing to the bulk material reference. Such sketch is presented in figure 3.

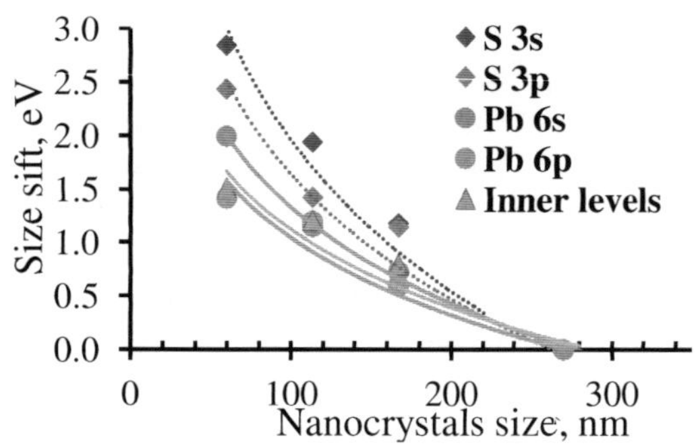

Figure 2. Size shifts of various levels in PbS nanocrystals.

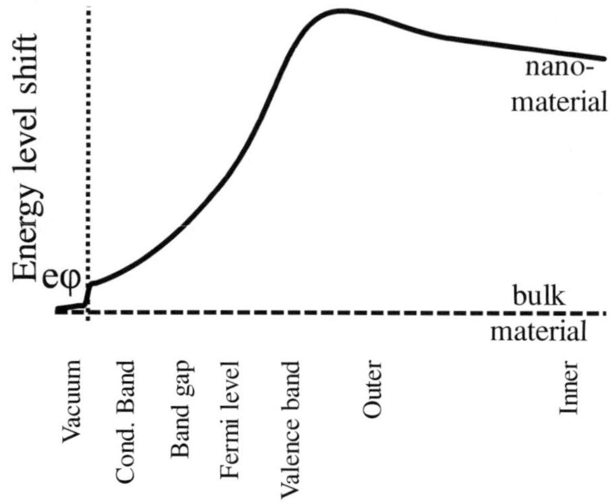

Figure 3. Increasing of binding energies of various levels in a nanomaterial comparing to a "bulk" reference.

Increasing of energies on this scheme means the shift of corresponding levels down from vacuum level. One can see that outer fully occupied levels located below valence band have maximal increase of binding energies. Shifts of deeper levels are lower due to screening effects. We observe also widening of band gap with its alignment in various directions of reciprocal space in semiconductors. Metal (Ag) demonstrates increasing intensity of Ag 5*s* XPS peak in consequence of localization electrons from conduction band at metal layers thinning [8]. Some input into observed growth of binding energies is provided by work function increasing. This part is enough small as can reach approximately 0.3 eV comparing to 3.0 eV shift of S 3*s* level.

4. Conclusions

1. There were determined common regularities of electronic structure transformations in metallic and semiconductor nanomaterials with different composition and morphology in wide ranges of characteristic dimensions from several hundred to units of nanometers.

2. As binding energies of different electronic levels are changing non-uniformly with a nanosystem size decreasing, we can conclude that such transformations are stipulated by rearrangement of the substance's electronic system rather than by "final state" phenomena in atoms after photoelectrons emission.

3. The two-stage mechanism of electronic structure transformations with decreasing of characteristic size of nanosystem was demonstrated. The threshold values D_0 corresponding to electronic structure transformations start and D_q manifesting beginning of strong changes driven by quantum size effects were determined for materials under investigation.

IOP Conf. Series: Journal of Physics: Conf. Series **857** (2017) 012055 doi:10.1088/1742-6596/857/1/012055

Acknowledgements
The research was carried out with partial financial support of RSF project No. 14-12-00170 (in part of fabrication and characterization of TiAlN/Ag coatings) and RFBR project No. 16-08-00527.

References
[1] ISO/TS 800004-1 2011 *Nanotechnologies. Vocabulary. P. 1: Core terms* (Geneva)
[2] McNaught A D and Wilkinson A 1997 *IUPAC. Compendium of Chemical Terminology* (Oxford: Blackwell Scientific Publications)
[3] UNIFIT for Windows software (http://www.uni-leipzig.de/~unifit)
[4] Kovalev A I, Wainstein D L, Rashkovskiy A Yu, Osherov A and Golan Y 2010 *Surf. Interf. Anal.* **42** 850–4
[5] Kovalev A, Wainstein D, Rashkovskiy A, Golan Y, Osherov A and Ashkenasy N 2012 *Vacuum* **86** 638–42
[6] Mosquera A, Horwat D, Rashkovskiy A, Kovalev A, Miska P, Wainstein D, Albella J M and Endrino J L 2015 *Sci. Rep.* **3** 1714
[7] Aruna I, Mehta B R, Malhotra L K and Shivaprasad S M 2008 *J. of Appl. Phys.* **104** 4308
[8] Kovalev A I, Wainstein D L, Rashkovskiy A Yu, Gago R, Soldera F, Endrino J L and Fox-Rabinovich G S 2015 *J. of Nanomat.* **2015** 876247

IOP Conf. Series: Journal of Physics: Conf. Series **857** (2017) 012056 doi:10.1088/1742-6596/857/1/012056

Thin-layer films and coatings

S N Yakupov and N M Yakupov

Institute of Mechanics and Engineering, Kazan Science Center Russian Academy of Sciences, 420lll, Kazan, Russia

E-mail: tamas_86@mail.ru

Abstract. The coatings are widely used in all industries. There is a need to establish an effective tool for evaluating the mechanical properties of coatings and adhesive. The process for determining the mechanical characteristics of coatings and adhesive. Modern developments in the field of coatings trends.

1. About coverings

There is no sphere of human activity in which technical and economic problems aren't tried to solve on the basis of films and coatings. In the future, this trend will increase even more. The rapid development of nanotechnology and nanomaterials promotes effective problem solving. Variety of coatings and adhesive components are created. Necessary qualities of coatings are provided by development of thin-layer complex composite structures and adhesives which are formed directly on the surface of the structure.

To create effective tools for diagnosing the state of coatings and adhesive, and to evaluate the mechanical properties of produced or acquired coatings and adhesive there is need of deep knowledge of the known methods and models of determine the mechanical properties of coatings and adhesive properties of coating to substrate [1].

2. Determination of mechanical properties of coatings

For studies of coatings with complex structure, including samples with defects or holes physical methods are not always applicable. Particularly, the method of "indentation" [2] – is not effective in the study of surfaces which have complex surface structure.

The molecular approach of the study of the mechanical properties of thin structures is in the embryonic state. The study of the mechanical characteristics of coatings by standard uniaxial testing gives a large variation of test results. It is impossible to study mechanical properties of coatings having a complex structure, defects and injuries by the standard uniaxial method.

By this theme there are domestic inventions (AS № 1742671, 1458766, 601599A, 1441243A1, 765697A) and foreign inventions – United States Patent 5764068A and Japanese Patent JP 8313422A. It should also be noted the Russian Federation patents: № 1756786, 2184361, 2296976.

An effective approach of determining the mechanical properties of shell coating is experimental-theoretical method [3, 4]. This method allows determining the mechanical properties of coating and can be used for the study of nano-coating [5].

Experimental-theoretical method for testing the mechanical properties of plane [3, 6, 7] and spherical surfaces [8] is based on a synthesis of experimental data and theoretical relations derived from the nonlinear theory of thin shells, the theory of elasticity and plasticity. Scheme of the

IOP Conf. Series: Journal of Physics: Conf. Series **857** (2017) 012056 doi:10.1088/1742-6596/857/1/012056

installation is shown in figure 1. Using the experimental data it is possible to build curve "deflection H – pressure p". Using the obtained results of measurements produce a theoretical data processing, this make possible to define, in particular, the elastic modulus or conditional elastic modulus, build curves of deformation and make a conclusion of the degree of deterioration of the material.

The method is also developed for the study of mechanical properties of biological membranes. This test methods and devices which use in the experimental-theoretical methods have the RF patents for inventions: № 2184361, 2296976, 2310184, 2387973, 2403556.

In particular, the influence of solar radiation on mechanical properties of polymer films of thickness 0.1, 0.12 and 0.18 mm with black and white substrates was studied. There are the curves "deflection – pressure p" (figure 2) and "stress intensity σ_i – deformation intensity ε_i" (figure 3) for films of thickness 0.12 mm. It also was found that with film thickness increasing the effect of influencing of solar radiation increase too.

By the separate studies of the substrate and the coated substrate (titanium oxide) the mechanical characteristics of the nano-coatings are defined in the system "coating-substrate" – figure 4 [1, 4, 9].

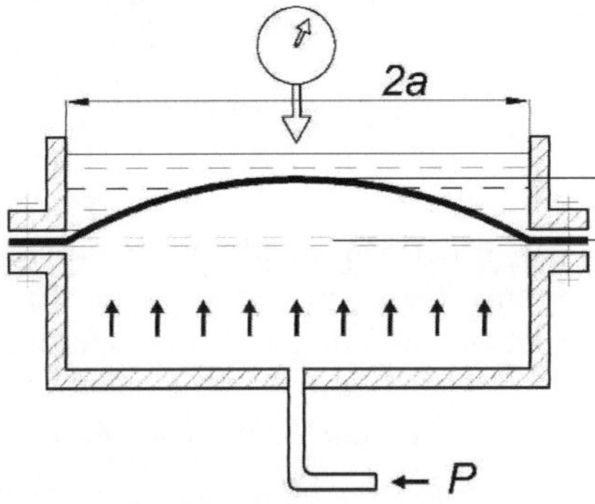

Figure 1. Plant layout.

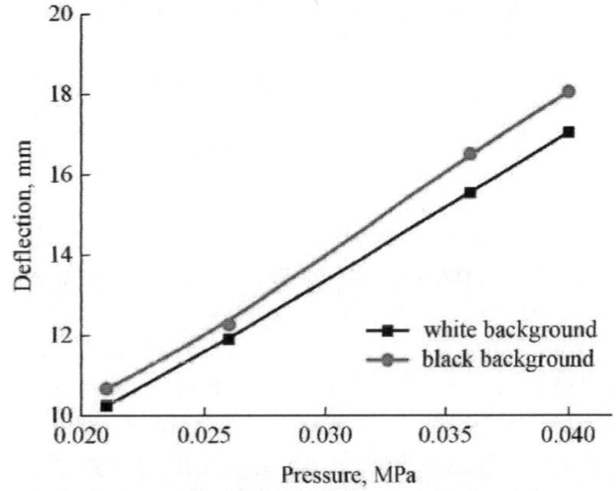

Figure 2. Curve $H - p$.

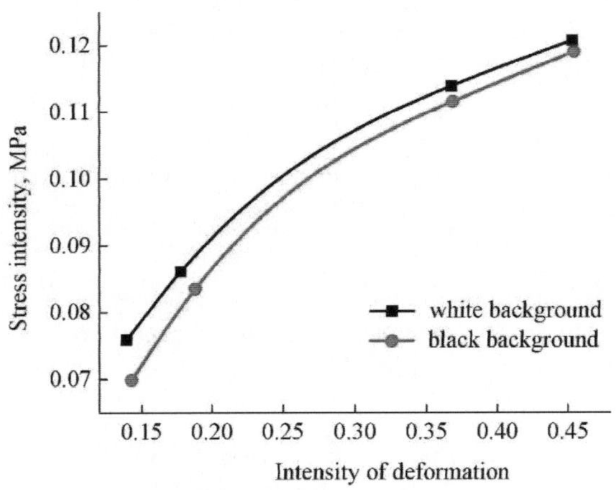

Figure 3. Curve $\sigma_i - \varepsilon_i$.

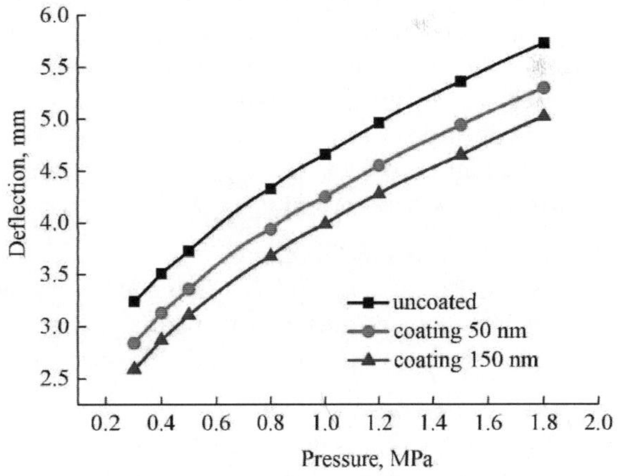

Figure 4. Curve $H - p$.

3. Determination of coating adhesion

The method for determining the adhesion of film to substrate using the parameters of the "bubble" [10] is known, it also has a number of disadvantages. Scheme of installation for determining the adhesion by the patent [11, 12] is shown in figure 5. A tensile load is applied to the covering through the hole, it is possible to watch the pressure change and formed shape of the dome, measure the change in diameter of base of the dome in the process of peeling of coating. Then the experimental data is used

IOP Conf. Series: Journal of Physics: Conf. Series **857** (2017) 012056 doi:10.1088/1742-6596/857/1/012056

in the ratio of the non-linear theory of shells. The method allows studying the adhesive properties of hard coatings and getting consistent results.

Method of determining the adhesion by the patent [12] improves accuracy of determining of the parameters of adhesion taking into account the mechanical properties and material thickness of coating, as well as height and elliptical form of base of the dome (figure 6). Polymeric coating adhesion to the metal substrate is defined.

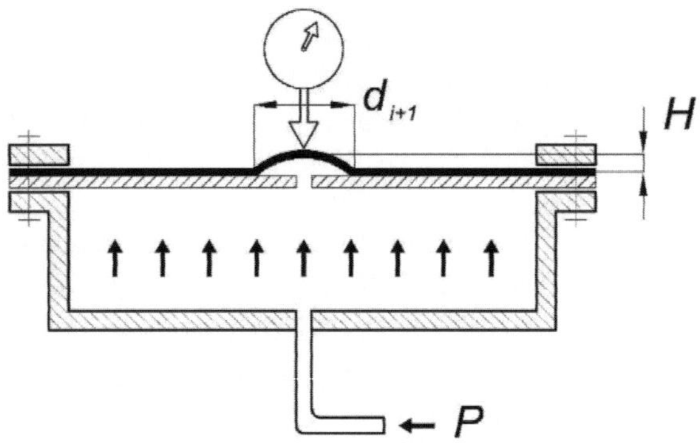

Figure 5. Plant layout.

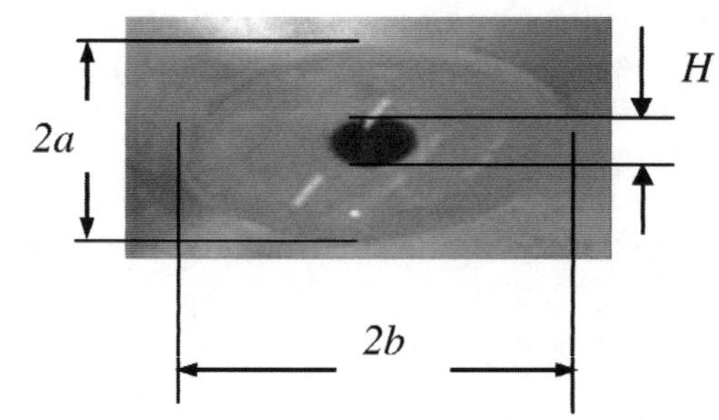

Figure 6. Parameters bubble.

4. About current trends

The most interesting data about surfaces and the trends in field of functional coatings for protection against corrosion of metal surfaces are noted. With the development of new technologies for the treatment of surfaces of construction elements it becomes possible to change the properties of materials and create new functional materials which have various applications [13–15], and others. Encapsulation of functional substances reacting on polymeric or inorganic supports, allows the creation of a new generation of intelligent coatings.

4.1. Self-treat covers

Protection from corrosion can be realized by "treatment" of defects in polymer coating by the addition polymerizable agents, or delay of corrosion due to the presence of corrosion inhibitors. Including the capsules with functional substances in coating is considered by Dry and Sottos [16]. Capsules are destroyed during the development of cracks in coating material and outputting the substances for polymerization of destroyed volume of polymer coating. Important moment for corrosion protection is an antifouling effect of coating. New biocides group are developed, in particular encapsulated antifouling substances are produced [18].

4.2. Super-hydrophobic coatings

Hydrophobicity is achieved by the encapsulation of functional substances [19] or by changing composition of outer surface of coating layers. Super-hydrophobic composite coating of a styrene copolymer, methyl methacrylate and silica nanoparticles was reviewed by Huang et al. [20].

4.3. Modification of coatings

Modification of coatings with carbon nanotubes and nanofibers make possible to produce nanostructured surfaces with increasing the mechanical properties [21]. Epoxy paints containing siloxane, allow to obtain coatings with the desired properties [22]. Siloxane allows creating self-cleaning and not covered by ice coatings. Conductive polymers – polyaniline, polypyrrole and thiophene – can be used to change the conductivity of anti-corrosion barrier coatings [13, 23].

5. Conclusions

The effective methods of evaluating the mechanical properties of coatings and adhesive which allows studying the perspective newly established modern functional coatings are developed.

References

[1] Yakupov N M, Yakupov S N and Rynkovskay M I 2016 *2nd International Congress on Technology – Engineering & Science* 143–5
[2] Oliver W and Pharr G 1997 *J. Mater. Res. Soc. Symp. Proc.* **473** 57
[3] Yakupov N M, Galimov N K and Leontiev A A 2000 *Mechanics of Composite Materials and Structures* **6** 238–43
[4] Yakupov S N 2010 *Mechanics of Composite Materials and Structures* **16** 436–44
[5] Yakupov N M, Kupriyanov V N, Nurullin R G and Yakupov S N 2010 *The patent for the invention of the Russian Federation № 2387973* Bull. **12**
[6] Yakupov N M, Nurgaliev A R and Yakupov S N 2008 *Factory Laboratory. Diagnosis materials* **74** 54–6
[7] Yakupov N M, Nurullin R G and Yakupov S N 2009 *Journal of Mechanical Engineering* **6** 44–7
[8] Galimov N K, Yakupov N M and Yakupov S N 2011 *Solid Mechanics* **3** 58–66
[9] Yakupov S N 2015 *XI All-Russian Congress on fundamental issues of Theoretical and Applied Mechanics* 4323–5
[10] Taplin J 1979 *Fracture Mechanics. The destruction of the materials* (Moscow: Publishing House "Mir") 222–4
[11] Goldstein R V, Yakupov N M, Nurullin R G, Yakupov S N and Yakupova R N 2011 *The patent for the invention of the Russian Federation № 2421707* Bull. **17**
[12] Yakupov S N 2016 *The patent for the invention of the Russian Federation № 2572673* Bull. **2**
[13] Montemor M F 2014 *Surface & Coatings Technology* **258** 17–37
[14] Garcia J, Fischer H R and van der Zwaag S 2011 *Prog. Org. Coat* **72** 211–21
[15] Taylor S R, Shiflet G J et al. 2009 *Nanotechnol. Appl. Coat* 126–55
[16] Dry C M, Sottos N 1995 *Smart Materials, Proc. SPIE* **1993** 438–44
[17] Taryba M, Snihirova D, Lamaka S V, Ferreira M G S, Montemor M F, Wijting W K, Toews S and Grundmeier G 2011 *Electrochim. Acta* **56** 4475–88
[18] Lejars M, Margaillan A and Bressy C 2012 *Chem. Rev.* **112** 4347–90
[19] Wang Q, Li J, Zhang C, Qu X, Liu J and Yang Z 2010 *J. Mater. Chem.* **20** 3211–5
[20] Huang Y F, Huang C, Zhong Y L and Yi S P 2013 *Surf. Eng.* **29** 633–6
[21] L'darkhanova F I, Mironova G A et al. 2012 *Prot. Met. Phys. Chem. Surf.* **48** 796–802
[22] Ahmad S, Gupta A P, Sharmin E et al. 2005 *Prog. Org. Coat.* **54** 248–55
[23] Zhang X, He Q, Gu H, Colorado H A, Wei S and Guo Z 2013 *ACS Appl. Mater. Interfaces* **5** 898–910

Formation of modified layers on the outer surface of cylindrical samples from E110 alloy under the influence of radial argon ions beam in the mode of multilayered films ion mixing

A S Yashin, D A Safonov, B A Kalin, P V Fedotov and N V Volkov

National Research Nuclear University MEPhI (Moscow Engineering Physics Institute), 115409, Moscow, Russia

E-mail: yashin_itf@mail.ru

Abstract. Results of experiments of modified layers formation on the long-length cladding tubes from E110 alloy (Zr – 1 % Nb) are presented in this paper. Samples were modified by surface polishing by radial Ar^+ beam with wide-range energy spectrum, series of films from alloying elements deposition and their subsequent processing in the ion mixing mode. The concentration of implanted atoms is 1–5 at. %, with a maximum penetration depth of 1–2 μm.

1. Introduction

Modification of materials by charged particles fluxes has a positive effect on the operational properties of the material at various scales – from nanomaterials and elements of microelectronics to the constituent parts of engineering machinery. One of the promising areas for the use of ion plasma technologies may be modification of materials for the core of nuclear reactors to improve the corrosion and wear resistance of fuel claddings [1, 2]. In this case, it is important to carry out local and multicomponent doping of the material.

2. Materials and methods

This paper presents the results of experiments performed on the ILUR-03 installation, which is a system for complex ion plasma treatment of long-range cylindrical samples. Feeding and moving of the samples takes place by means of an automatic manipulator through the gateway chamber (position 1 on figure 1) with a differential pumping system, behind which there is a preliminary chamber with an outlet to a high-vacuum pump, a chamber with an ion source and a camera with a magnetron system for depositing films (positions 2-4 on figure 1, respectively).

The ion source provides the formation of a radial beam with a wide energy spectrum of 0.5–5 keV and is a cylindrical set anode placed in a magnetic field created by a system of inductors and top magnetic cores. Argon was used as a working gas 10^{-2}–10^{0} Pa pressure in the chamber and 10^{-4} Pa residual gas pressure.

Cladding tubes from E110 (Zr – 1 % Nb) alloy with outer diameter of 9.15 mm and a length of up to 500 mm were used as the samples. Modification of the near-surface layer of the sample material by an ion beam was carried out in the modes of surface cleaning and polishing before the films deposition and in formed "substrate-multilayer film" system ion mixing mode for local doping of Fe, Mo, Y, Cr, Ni, Al, Mg [3]. The total irradiation dose was $(1–10)\cdot 10^{18}$ ion·cm^{-2}.

Figure 1. Installation ILUR-03: 1 – gateway vacuum chamber; 2 – preliminary evacuation chamber; 3 – zone radial ion beam action; 4 – zone of films deposition. The arrows show the pipelines of the vacuum system.

3. Results and discussion

As can be seen from figure 2, the ion-modified surface reveals a higher topography homogeneity, especially in the micrometer and submicrometer scale. Large technological scratches are smoothed, the boundaries of crystallographically differentially oriented crystallites are not observed.

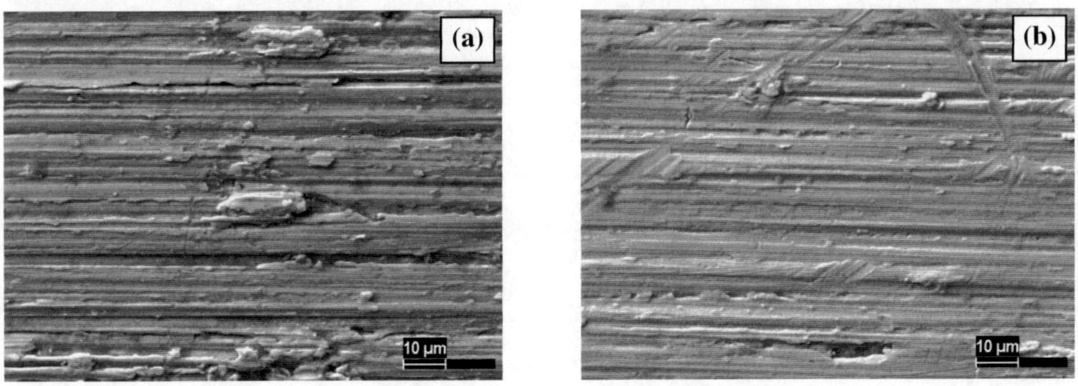

Figure 2. Images of the outer surface of the samples in initial state **(a)** and after ion polishing **(b)** obtained by the SEM method with ×1000 magnification.

Deposited film from various alloying elements atoms repeats the topography of the initial surface and has a finely dispersed structure with characteristic size of ~100 nm (figure 3). After "substrate-multilayer film" system mixing by ion beam, surface has a structure similar to that before the alloying elements films deposition, which can be seen in figure 4.

Depending on the film thicknesses and their relative arrangement ratio, the distribution of the elements near the modified surface may have domed or sloping shape with a characteristic drop to the equilibrium concentrations value (figure 5). The maximum concentration of the implanted elements is ~1–5 at. %, which exceed their equilibrium solubility limit in the alloy.

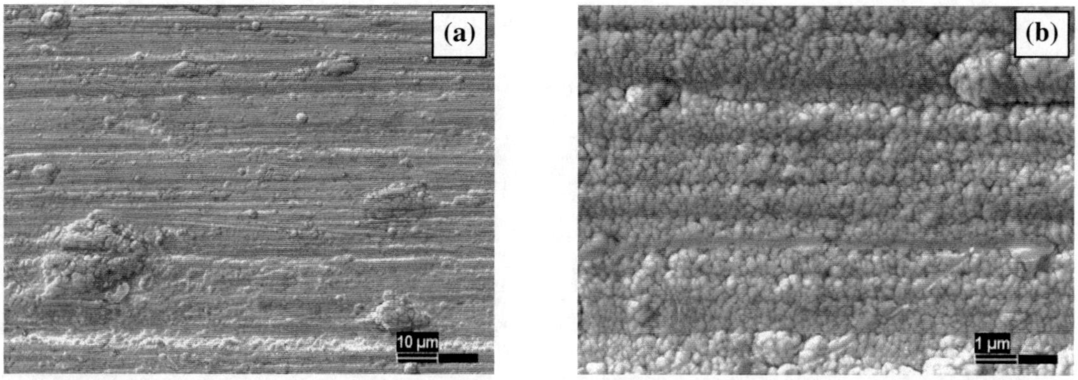

Figure 3. Images of the sample with a deposited film from FeCrNi atoms outer surface with ×1000 **(a)** and ×10 000 **(b)** magnification, obtained by SEM.

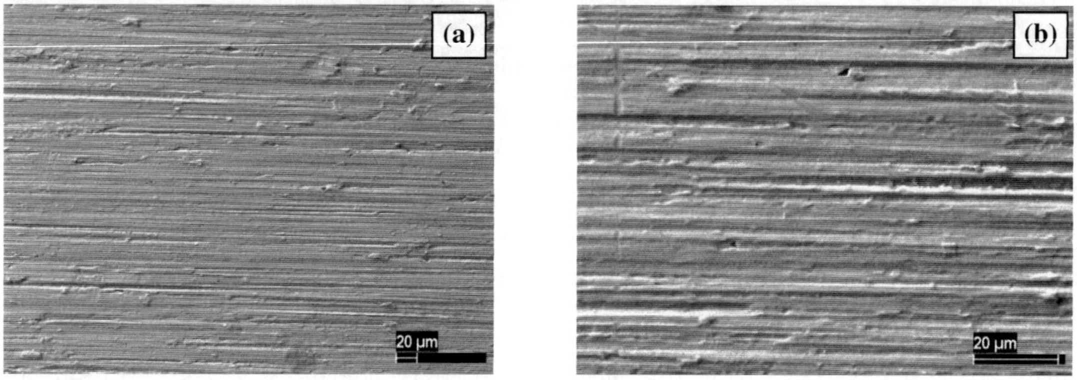

Figure 4. Images of the alloying by FeCrNi atoms sample outer surface with ×250 **(a)** and ×1000 **(b)** magnification, obtained by SEM.

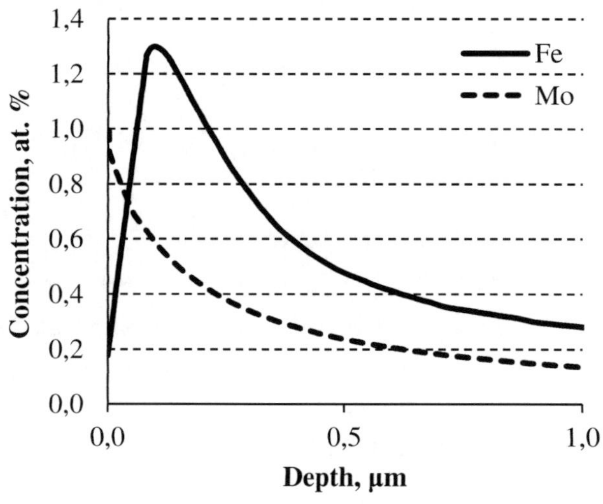

Figure 5. Distribution of elements in the near-surface layer of material doped as a result of Fe and Mo films ion mixing by Ar^+.

5. Conclusions

Modified layers with a content of alloying elements of Fe, Mo, Y, Cr, Ni, Al, Mg up to 1–5 at. % and maximum penetration depth up to 1–2 μm were obtained near the surface of cylindrical samples from sponge based E110 alloy under the action of radial beam of Ar^+ ions as a result of polishing and ion mixing of multilayer films.

IOP Conf. Series: Journal of Physics: Conf. Series **857** (2017) 012057 doi:10.1088/1742-6596/857/1/012057

References

[1] Kalin B A, Volkov N V, Valikov R A and Yashin A S 2016 *Physics and Chemistry of Materials Processing* **4** 5–11
[2] Kalin B A, Volkov N V, Valikov R A, Yashin A S and Yakutkina T V 2016 *IOP Conf. Series: Materials Science and Engineering* **130** 012006
[3] Kalin B A, Volkov N V, Valikov R A and Yashin A S 2016 *Physics and Chemistry of Materials Processing* **3** 5–8

IOP Conf. Series: Journal of Physics: Conf. Series **857** (2017) 012058 doi:10.1088/1742-6596/857/1/012058

Morphology of powders of tungsten carbide used in wear-resistant coatings and deposition on the PDC drill bits

E S Zakharova, I Yu Markova, A L Maslov, N I Polushin and A I Laptev

National University of Science and Technology «MISiS», Moscow, Russia

E-mail: elenazakharova93@gmail.com

Abstract. Modern drill bits have high abrasive wear in the area of contact with the rock and removed sludge. Currently, these bits have a protective layer on the bit body, which consists of a metal matrix with inclusions of carbide particles. The research matrix of this coating and the wear-resistant particles is a prerequisite in the design and production of drill bits. In this work, complex investigation was made for various carbide powders of the grades Relit (tungsten carbide produced by Ltd "ROSNAMIS") which are used as wear-resistant particles in the coating of the drill bit body. The morphology and phase composition of the chosen powders as well as the influence of a particle shape on prospects of their application in wear-resistance coating presented in this work.

1. Introduction

Constant improvement of the equipment for oil and gas production requires creation of modern materials. One of the most important points of the drill bit is significant abrasive wear in the area of contact with the rock and generated sludge. Increased wear of the bit body leads to premature crumbling of polycrystalline diamond and consequently reducing the drill bit life. Therefore, wear-resistant coatings applied on their surface are widely used to protect the drill bit body from abrasive wear [1, 2]. Today the method of manual flame spraying is widely used according to its high coverage rate, local heating of the coated material, and the simplicity of working with equipment. Today the surface materials are mostly consisting of a Ni-based alloy doped by Cr, Ti, etc. [3, 4]. Materials based on tungsten carbides have been widely used as wear-resistant particles. Along with a complex research as already applied wear-resistant coatings and the matrix, an important point is the complex research of initial powder.

The wear resistance of the surface material for the drill bit bodies strongly depends on the structural state of the matrix and the strength of confinement of a particle in the matrix during bit operation [5]. Location of carbides and their shape influence on the properties of the coating [6]. A close-grained structure and a highly-developed surface of the particles significantly contributes to their confinement of a particle in the matrix. Therefore, the aim of this work is the complex study of the following powders Relit LKV "Z"-2, Relit LKV "Z"-6, Relit LKV "c", VSNGN-88 (nickel-based alloys and cast tungsten carbides). These powders have high hardness, wear resistance and can be used as additives in surfacing wear-resistant coating.

2. Materials and equipment

As the test materials was chosen four powder cast tungsten carbide produced by Ltd "ROSNAMIS": Relit LKV "Z"-2, Relit LKV "Z"-6, Relit LKV "c", VSNGN-88 (nickel-based alloys and cast tungsten carbides). It is known that powders of a cast tungsten carbide consist of several phases. Therefore it is important the presence of the phase having the highest hardness. In table 1, for clarity, presents data on the microhardness of various phases of cast tungsten carbide [7]. Passport data of powders by chemical composition are presented in table 2. The dispersity and microhardness of the initial carbide particles are shown in table 3. Data in tables 2 and 3 is provided by Ltd "ROSNAMIS".

Table 1. Elemental analysis of the matrix coatings.

Carbide	Type of crystal lattice	Microhardness, MPa	Density, g/cm^3	Melting point, °C
WC	hexagonal	17000	15,7	2600
W_2C	hexagonal	30000	17,15	2850

Table 2. Chemical composition of the investigated tungsten carbide powders.

Grade	Chemical composition, %								
	Free carbon	Total carbon	W	Fe	Magnetic impurities	Si	Ni	Cr	B
Relit LKV "Z"-2	0.03	3.97	95.98	0.02	0.02	–	–	–	–
Relit LKV "Z"-6	0.02	3.98	95.97	0.02	0.02	–	–	–	–
Relit LKV "c"	0.05	4	98.89	0.11	–	–	–	–	–
VSNGN-88	–	5.18	83.44	0.4	–	0.45	9.1	1.1	0.33

Table 3. Properties of the investigated tungsten carbide powders.

Grade	Dispersity, μm	Microhardness, HV
Relit LKV "Z"-2	40–180	2340
Relit LKV "Z"-6	500–800	2340
Relit LKV "c"	40–180	2550
VSNGN-88	20–125	–

Investigation of the microstructure of wear-resistant coatings was carried out on a scanning microscope JSM-6480LV, located at the Center for collective use in NUST "MISiS". X-ray-graphical measurements were performed in NUST "MISiS" in the lab of superhard materials on a DRON-3M device. X-ray photography was conducted on cobalt radiation.

3. Research results of the microstructure of the coatings

Figure 1 presents the electronic image of the investigated powders. Powders of grades Relit LKV "Z"-2 and Relit LKV "Z"-6 represent particles of a fragmental form. According to the results of scanning electron microscopy, the particle size of the sample Relit LKV "Z"-2 in the range of 40–180 μm, the sample Relite LKV "Z"-6 – 500–800 μm. Analysis VSNGN-88 powder morphology showed that it consists of small and large particles (20–130 μm) with the internal block structure. Powder of the grade Relit LKV "C" is a spherical particle size of 40–180 μm.

The elemental composition and particle size distribution corresponds to the stated characteristics of the samples powders. Results of X-ray analysis and quantitative ratio of phases of the powders presented in figure 2 and table 4.

W_2C (hardness of 30 GPA) has a higher hardness and wear resistance compared to WC (hardness of 17 GPA). Therefore, it is considered that the more W_2C in the powder composition, the higher is wear resistance. The leading position on the content of polyamide tungsten is a powder Relit LKV "c". Carbide grains are spherical-shaped. Therefore, it is possible to synthesized coatings with a maximum

IOP Conf. Series: Journal of Physics: Conf. Series **857** (2017) 012058 doi:10.1088/1742-6596/857/1/012058

content of the carbide phase, which should favorably affect the durability the wear resistance of the bit body. An additional positive factor for the use of Relit LKV "c" in wear-resistant coatings is the reduction of the coefficient of friction in the treated rock. Spherical grains are more easily crumbled from the matrix under intensive loads, which should reduce the wear resistance of the coatings. Consequently, the resource of these coatings depends on the operating conditions.

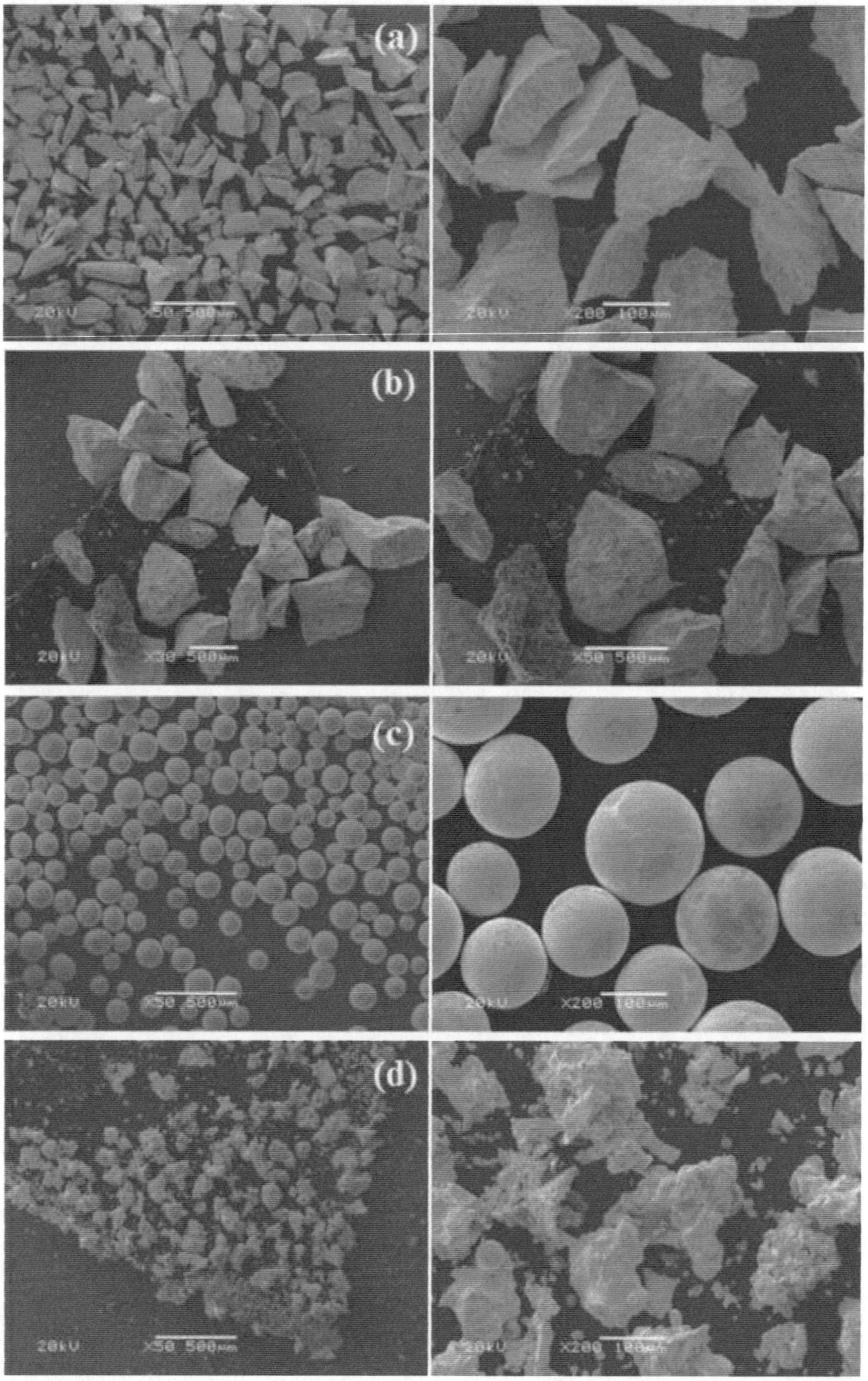

Figure 1. The microstructure of tungsten carbide powders: **(a)** – Relit LKV "Z"-2; **(b)** – Relit LKV "Z"-6; **(c)** – Relit LKV "c"; **(d)** – VSNGN-88.

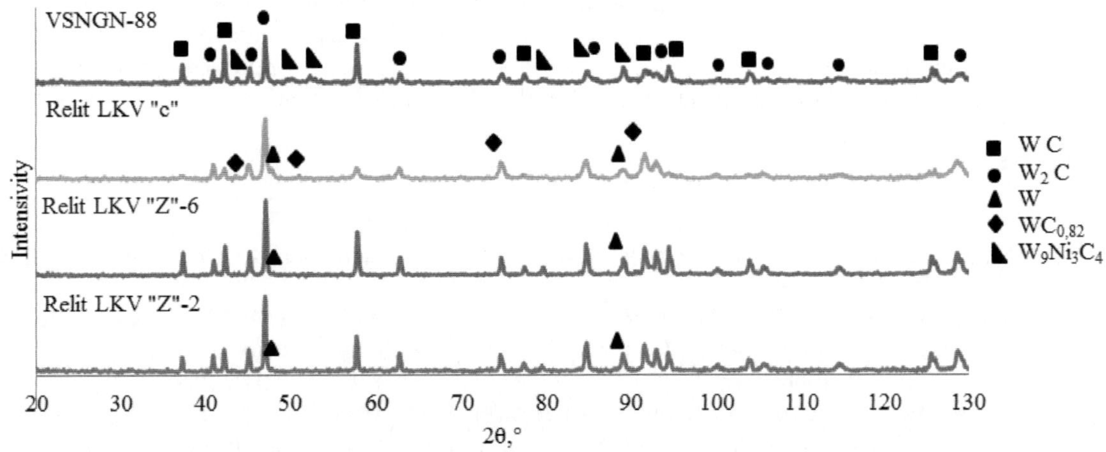

Figure 2. X-ray diffraction pattern of powders of tungsten carbides.

Table 4. The quantitative ratio of phases in the samples.

Grade	Mass fraction of phases, %				
	WC	W_2C	W	$WC_{0.82}$	$W_9Ni_3C_4$
Relit LKV "Z"-2	31.5	64.8	3.8	–	–
Relit LKV "Z"-6	35.5	59.4	5.1	–	–
Relit LKV "c"	20.9	66.7	6.7	5.7	–
VSNGN-88	36.1	37.4	–	–	26.5

Powder VSNGN-88 has a highly-developed surface of the particles compared to the other, and thus is able better than others to stay in the matrix. In this powder, the $W_9Ni_3C_4$ phase was detected, as well as a decrease in the content of tungsten semicarbide in half, which in our opinion should lead to a reduction in its performance characteristics compared to other powders. However, from literature it is known that the cladding of WC particles with nickel contributes to an increase in operational characteristics [8]. Therefore, this issue remains controversial and requires comparison of coatings containing this powder.

The powder Relit LKV "Z"-2 and Relit LKV "Z"-6 have the same phase composition, but the content of W_2C of Relit LKV "Z"-2 (64.8 wt. %) is higher than the content in relit LCV "Z"-6 (59.4 wt. %). Particle size distribution of these powders is significantly different. Presumably, the larger grains of the powder of Relit LKV "Z"-6 the result of the work of the drilling tool will be longer in the matrix.

3. Conclusions

As a result of studies have explored the morphology of carbide particles which can be used on the bit body. The most promising for use in wear-resistant coatings on the PDC drill body are powder of the grade Relit LKV "c" (due to their spherical shape), LKV "Z"-6 (due to larger particle size) and VSNGN-88 (due to the complex chemical composition). Such powders can be recommended for use as wear-resistant particles in the surface on drill bit bodies for carrying out model comparative abrasive tests.

Acknowledgment

This work was carried out within the framework of the federal target project of the program "Research and development in priority areas of development of the scientific and technological complex of

Russia for 2014-2020" on the topic "Development of technology for manufacturing high-performance horizontal and tilt drilling drill bits for the oil and gas industry" agreement No. 14.581.21.0012 (unique identifier of the agreement RFMEFI58115X0012) with financial support of applied scientific research by the Ministry of Education and Science of the Russian Federation.

References

[1] Sukhova E V, Syrovatko V A and Syrovatko Yu V 2011 *Science and transport progress. Bulletin of Dnipropetrovsk national university of railway transport* **37** 238–43
[2] Polushin N I, Ovchinnikova M S and Maslov A L 2014 *Advanced Materials Research* **1040** 199–201
[3] Bondarenko V P, Isonkin A M, Bogdanov R K and Matveichuk A A 2010 *Rock-destroying and metal-processing tool – technique and technology of its production* **13** 168–73
[4] Polushin N I, Ovchinnikova M S, Maslov and Kuchina I Yu 2015 *Proceedings of the Universities. Chemistry and Chemical Technology* **58**(7) 55–7
[5] Polushin N I, Bogatirev A V, Laptev A I and Sorokin M N 2016 *Proceedings of the Universities. Powder metallyrgy and functional coatings* **1** 60–6
[6] Ryabtsev I A, Panfilov A I, Babinets A A, Ryabtsev I I, Gordan G N and Babijchuk I L 2015 *Automatic welding* **5–6** 84–8
[7] Kiffer R and Benezovsky F 1971 *Hard alloys* 392
[8] Samsonov G V, Vistryanyuk V K and Chaplygin F I 1974 *Tungsten carbides* 236